S0-CLP-794

ADVANCES IN LIPID RESEARCH

Volume 25

Sphingolipids
Part A: Functions and Breakdown Products

Advances in Lipid Research

Volume 25

Sphingolipids
Part A: Functions and Breakdown Products

Edited by

ROBERT M. BELL
Department of Biochemistry
Duke University Medical Center
Durham, North Carolina

YUSUF A. HANNUN
Departments of Medicine and Cell Biology
Duke University Medical Center
Durham, North Carolina

ALFRED H. MERRILL, JR.
Department of Biochemistry
Emory University School of Medicine
Atlanta, Georgia

SERIES EDITORS

RICHARD J. HAVEL
Cardiovascular Research Institute
University of California, San Francisco
San Francisco, California

DONALD M. SMALL
Department of Biophysics
Boston University
Boston, Massachusetts

ACADEMIC PRESS, INC.
A Division of Harcourt Brace & Company

San Diego New York Boston London Sydney Tokyo Toronto

This book is printed on acid-free paper. ∞

Academic Press, Inc.
1250 Sixth Avenue, San Diego, California 92101-4311

United Kingdom Edition published by
Academic Press Limited
24–28 Oval Road, London NW1 7DX

International Standard Serial Number: 0065-2849

International Standard Book Number: 0-12-024925-1

PRINTED IN THE UNITED STATES OF AMERICA
93 94 95 96 97 98 BB 9 8 7 6 5 4 3 2 1

CONTENTS

The Novel Second Messenger Ceramide: Identification, Mechanism of Action, and Cellular Activity

Yusuf A. Hannun, Lina M. Obeid, and Robert A. Wolff

Ceramide: A Novel Second Messenger

Shalini Mathias and Richard Kolesnick

Ceramide-Activated Protein Phosphatase: Partial Purification and Relationship to Protein Phosphatase 2A

Rick T. Dobrowsky and Yusuf A. Hannun

The Role of Sphingosine in Cell Growth Regulation and Transmembrane Signaling

Sarah Spiegel, Ana Olivera, and Robert O. Carlson

Sphingolipid Regulation of the Epidermal Growth Factor Receptor

Roger J. Davis

Gangliosides and Glycosphingolipids as Modulators of Cell Growth, Adhesion, and Transmembrane Signaling

Sen-itiroh Hakomori and Yasuyuki Igarashi

PART II
Biological and Pharmacological Functions

Gangliosides as Receptors for Bacterial Enterotoxins

Peter H. Fishman, Tadeusz Pacuszka, and Palmer A. Orlandi

Verotoxins and Their Glycolipid Receptors

Clifford A. Lingwood

Glycosphingolipid Tumor Antigens

Pam Fredman

Gangliosides and Modulation of the Function of Neural Cells

Guido Tettamanti and Laura Riboni

Protection by Gangliosides against Glutamate Excitotoxicity

H. Manev, A. Guidotti, and E. Costa

The Role of Glycosphingolipids in Hypoxic Cell Injury

Ady Kendler and Glyn Dawson

Bioactive Gangliosides: Differentiation Inducers for Hematopoietic Cells and Their Mechanism(s) of Actions

Masaki Saito

CONTRIBUTORS

Numbers in parentheses indicate the pages on which the authors' contributions begin.

Robert M. Bell (1, 27), Department of Biochemistry and Section of Cell Growth, Regulation, and Oncogenesis, Duke University Medical Center, Durham, North Carolina 27710

Robert O. Carlson (105), Department of Biochemistry, Georgetown University Medical Center, Washington, D.C. 20007

E. Costa (269), Fidia-Georgetown Institute for the Neurosciences, Georgetown University Medical School, Washington, D.C. 20007

Roger J. Davis (131), Howard Hughes Medical Institute, Program in Molecular Medicine, Department of Biochemistry and Molecular Biology, University of Massachusetts Medical School, Worcester, Massachusetts 01605

Glyn Dawson (289), The Joseph P. Kennedy Jr. Mental Retardation Research Center, Departments of Pediatrics, Biochemistry, and Molecular Biology, and the Committee on Neurobiology, University of Chicago, Chicago, Illinois 60637

Rick T. Dobrowsky (91), Department of Medicine, Duke University Medical Center, Durham, North Carolina 27710

Peter H. Fishman (165), Membrane Biochemistry Section, Laboratory of Molecular and Cellular Neurobiology, National Institute of Neurological Disorders and Stroke, National Institutes of Health, Bethesda, Maryland 20892

Pam Fredman (213), Department of Psychiatry and Neurochemistry, University of Göteborg, Mölndals sjukhus, S-43180 Mölndal, Sweden

A. Guidotti (269), Fidia-Georgetown Institute for the Neurosciences, Georgetown University Medical School, Washington, D.C. 20007

Sen-itiroh Hakomori (147), The Biomembrane Institute, Seattle, Washington 98119, and Department of Pathobiology, University of Washington, Seattle, Washington 98195

Yusuf A. Hannun (1, 27, 43, 91), Departments of Medicine and Cell Biology, Duke University Medical Center, Durham, North Carolina 27710

Yasuyuki Igarashi (147), The Biomembrane Institute, Seattle, Washington 98119, and Department of Pathobiology, University of Washington, Seattle, Washington 98195

Ady Kendler (289), The Joseph P. Kennedy Jr. Mental Retardation Research Center, Departments of Pediatrics, Biochemistry, and Molecular Biology, and the Committee on Neurobiology, University of Chicago, Chicago, Illinois 60637

Richard Kolesnick (65), Laboratory of Signal Transduction, Memorial Sloan-Kettering Cancer Center, New York, New York 10021

Clifford A. Lingwood (189), Department of Microbiology, The Hospital for Sick Children, Toronto, Ontario, Canada M5G 1X8

H. Manev (269), Fidia-Georgetown Institute for the Neurosciences, Georgetown University Medical School, Washington, D.C. 20007

Shalini Mathias (65), Laboratory of Signal Transduction, Memorial Sloan-Kettering Cancer Center, New York, New York 10021

Alfred H. Merrill, Jr. (1), Department of Biochemistry, Emory University School of Medicine, Atlanta, Georgia 30322

Lina M. Obeid (43), Department of Medicine, Duke University Medical Center, Durham, North Carolina 27710

Ana Olivera (105), Department of Biochemistry, Georgetown University Medical Center, Washington, D.C. 20007

Palmer A. Orlandi (165), Membrane Biochemistry Section, Laboratory of Molecular and Cellular Neurobiology, National Institute for Neurological Disorders and Stroke, National Institutes of Health, Bethesda, Maryland 20892

Tadeusz Pacuszka (165), Membrane Biochemistry Section, Laboratory of Molecular and Cellular Neurobiology, National Institute of Neurological Disorders and Stroke, National Institutes of Health, Bethesda, Maryland 20892

Laura Riboni (235), Department of Medical Chemistry and Biochemistry, The Medical School, University of Milan, 20133 Milan, Italy

Masaki Saito (303), Division of Hemopoiesis, Institute of Hematology, Jichi Medical School, Tochigi-ken 329-04, Japan

Sarah Spiegel (105), Department of Biochemistry, Georgetown University Medical Center, Washington, D.C. 20007

Guido Tettamanti (235), Department of Medical Chemistry and Biochemistry, The Medical School, University of Milan, 20133 Milan, Italy

Robert A. Wolff (43), Department of Medicine, Duke University Medical Center, Durham, North Carolina 27710

PREFACE

The sphingolipids were aptly named in 1884 by their discoverer, J. L. W. Thudichum, after the riddle of the ancient Greek Sphinx which guarded the city of Thebes. The structural complexity of the sphingolipids now exceeds 400 distinct molecules; this portends important functions. The first of these functions is the role of glycosphingolipids on the cell surface as receptors and as modulators of growth factor receptors. In 1986, we reported on the possible role of sphingosine, a breakdown product of complex sphingolipids, on cellular regulation and its inhibitory effects on protein kinase C. These findings marked the beginning of a pathway leading to the discovery of a variety of intracellular sphingolipid metabolites that function as lipid second messengers/lipid mediators. The first bona fide sphingolipid cycle, "the sphingomyelin cycle," was discovered in 1989; ceramide appears to be a second messenger. Since then, a variety of sphingosine derivatives, including *N,N*-dimethylsphingosine, sphingosine phosphate, and sphingosine phosphorylcholine, have been reported to be bioactive.

Thus, the answer to the sphingolipid riddle posed long ago by Thudichum appears, in part, to reflect the fact that complex sphingolipids serve as reservoirs for the production of bioactive molecules that regulate cellular responses, which encompasses the regulation of cell growth and differentiation, mitogenesis, calcium mobilization, regulation of protein kinases and phosphatases, and apoptosis. These studies have led to reinterpretations of the significance of findings 20 years old or older on the metabolism of the sphingolipids and to renewed interest and understanding of the biosynthesis, catabolism, and role of breakdown products in cellular regulation.

These findings offer promise for new molecular insights into tumor growth and metastasis (glycosphingolipids become dominant tumor antigens), neural functions, diabetes, obesity, hypoxic injury, atherosclerosis, and inflammatory disorders including psoriasis. Defects in these metabolic pathways, as well as their disruption by toxic agents in the environment, should emerge as causes of diseases. As recent data define novel pathways and pinpoint targets for potential therapeutic intervention, these findings place the sphingolipids and their roles in cells in parallel with the more advanced studies on the glycerolipids. Still, much remains to be done to define the extent of the set of intracellular bioactive lipids, second messengers, and lipid mediators derived from the sphingolipids. Our intuition suggests that an extensive number of sphingolipid metabolites may emerge. The field may be poised in a manner similar to the early days of the eicosanoids, as novel metabolic pathways that generate and attenuate these messengers await discovery and clarification. Most functions of the cellular sphingolipids remain to be elucidated. The role of sphingolipids as dietary components represents another frontier for future research.

We are grateful to Dr. Donald Small for suggesting the subject matter for Volumes 25 and 26 of *Advances in Lipid Research* and to the authors of the 32 chapters that constitute these two volumes for the many new ideas contained in their contributions. The authors appreciate the encouragement of their colleagues, students, and postdocs in planning and editing these volumes. We anticipate further exciting discoveries that will advance and clarify this new field of research.

ROBERT M. BELL
YUSUF A. HANNUN
ALFRED H. MERRILL, JR.

ADVANCES IN LIPID RESEARCH, VOL. 25

Introduction: Sphingolipids and Their Metabolites in Cell Regulation

ALFRED H. MERRILL, Jr.,* YUSUF A. HANNUN,† AND ROBERT M. BELL‡

*Department of Biochemistry
Emory University School of Medicine
Atlanta, Georgia 30322

†Departments of Medicine and Cell Biology
Duke University Medical Center
Durham, North Carolina 27710

‡Department of Biochemistry
and Section of Cell Growth, Regulation, and Oncogenesis
Duke University Medical Center
Durham, North Carolina 27710

I. Introduction

Sphingolipids constitute one of the most chemically, and functionally, diverse classes of biomolecules. Since their first description by Thudichum in 1884, the structures of hundreds of sphingolipids have been elucidated and the basic features of their biosynthesis, transport, and turnover have been established; the most basic features are summarized in Fig. 1 (for background information about sphingolipids, consult Sweeley, 1991; Kanfer and Hakomori, 1983; and relevant articles in this volume and Vol. 26 of this series). Sphingolipids help define the structural properties of membranes, lipoproteins, and the water barrier of skin (Barenholz and Thompson, 1980; and articles by Bouhours *et al.* and Chatterjee in Vol. 26); participate in cell–substratum interactions and cell–cell communication (Hakomori and Igarashi, this volume), including recognition of cells by some microorganisms, viruses, and antibodies (Fishman *et al.* and Lingwood, this volume); interact with receptors to affect cellular responses to growth factors and other agonists (Davis and Hakomori and Igarashi, this volume); and influence intracellular signal transduction systems (see articles by Hannun and Bell, Dobrowsky and Hannun, Hakomori and Igarashi, Hannun *et al.*, Mathias and Kolesnick, and Spiegel, *et al.* in this volume). Sphingolipids also appear to serve as membrane anchors for some proteins (Stadler *et al.*, 1989; Lederkremer *et al.*, 1990; Takeda, Vol. 26). The importance of these compounds is revealed by their structural sophistication, the complex changes that they undergo with development (Bouhours *et al.* and Fredman, Vol. 26), and their widespread utilization by eukaryotic organisms (Dennis and Wiegandt and Lester and Dickson, Vol. 26).

Nonetheless, most of the biological functions of sphingolipids are still not known, and very little is known about how these processes are regulated. Although sphingolipids remain enigmatic, recent findings are beginning to explain how they control various cell functions and to reveal why they undergo changes with cell growth, differentiation, and disease. We have emphasized these regulatory aspects of sphingolipid structure, metabolism, and function in selecting topics and authors for articles in this volume of *Advances in Lipid Research.* This introductory article presents an overview from this perspective and, where possible, directs the reader to articles that provide more detailed information.

II. Sphingolipids as Modulators of Cell Behavior

A. Complex Sphingolipids

1. Interactions between Sphingolipids and Cell Surface Proteins

Many of the known functions of complex sphingolipids involve interactions that occur on the cell surface, with much (but not all) of the sphingolipids being thought to reside in the external leaflet of the plasma membrane (Miller-Podraza *et al.*, 1982). These types of interactions have been described in this volume with respect to regulation of growth factor receptors (Davis and Hakomori and Igarashi, this volume); cell adhesion and cell–cell communication (Hakomori and Igarashi and Tettamanti and Riboni, this volume); recognition of cells by microbial toxins (Fishman *et al.* and Lingwood, this volume); and expression of glycosphingolipid tumor antigens (Fredman, this volume).

One of the likely mechanisms for the regulation of cell behavior by sphingolipids is for them to interact directly or indirectly with cell surface receptors (Zeller and Marchase, 1992; Davis, Hakomori and Igarashi, and Tettamanti and Riboni, this volume). It has been relatively easy to show that some sphingolipids (particularly gangliosides) can affect cell growth because they form water-soluble

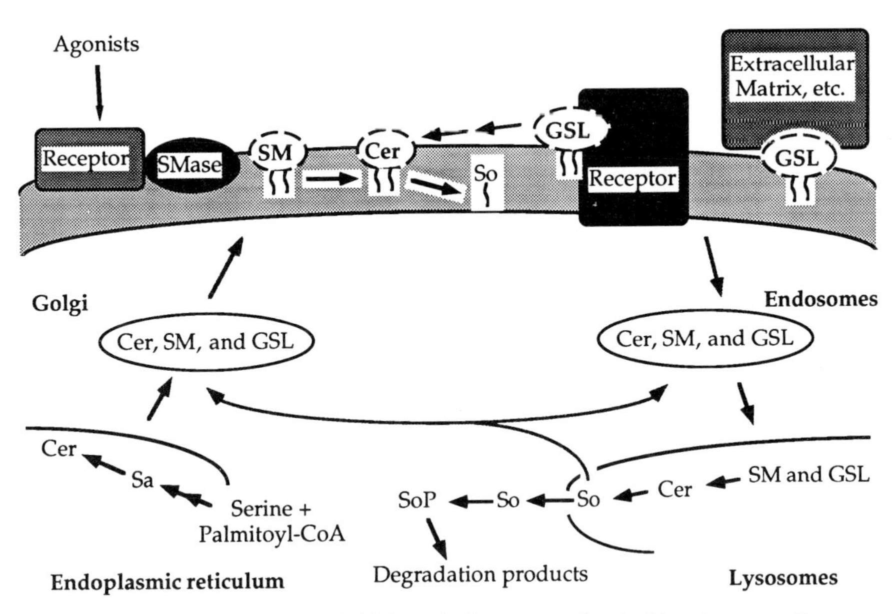

FIG. 1. An overview of sphingolipid biosynthesis, transport, function(s), and turnover. *De novo* sphingolipid biosynthesis begins in the endoplasmic reticulum with the condensation of serine and palmitoyl-CoA to form 3-ketosphinganine, sphinganine (Sa), *N*-acylsphinganines (dihydroceramides) and, eventually, ceramides (Cer), which are incorporated into glycosphingolipids (GSL) in the endoplasmic reticulum and Golgi, and sphingomyelin (SM), which is thought to be synthesized in the Golgi (and plasma membranes). The sphingolipids are transported mostly to the plasma membranes, but are also localized in secretory vesicles, endosomes, lysosomes, and some other intracellular sites (including the nuclear membrane) that are not shown in this diagram. In the plasma membrane, sphingolipids contribute to membrane structure and serve as cell surface receptors for proteins of the extracellular matrix (e.g., for cell–substratum and cell–cell interactions), and act as modulators of receptors such as the epidermal growth factor receptor. Surface sphingolipids are internalized and recycled, remodeled, or turned over in diverse sites. Sphingolipid hydrolases are not only present in lysosomes, but also, in plasma membranes [e.g., sphingomyelinase (SMase), ceramidase, and sialidase] and probably other intracellular compartments. Recent studies have found that various agonists activate sphingomyelinase (although not necessarily the plasma membrane SMase, as shown) to produce ceramide. After hydrolysis of complex sphingolipids to sphingosine (So), the long-chain bases leave the lysosomes for reutilization or phosphorylation to sphingosine 1-phosphate (So-P), which is cleaved to trans-2-dexadecenal and ethanolamine phosphate.

micelles and, unlike most other lipids, can be delivered to cells. Gangliosides G_{M3} or G_{M1} inhibit growth through extension of the G_1 phase of the cell cycle (Laine and Hakomori, 1973; Keenan *et al.*, 1975) and make cells refractory to stimulation by fibroblast growth factor (Bremer and Hakomori, 1982), platelet-derived growth factor (Bremer *et al.*, 1984), and epidermal growth factor (EGF) (Bremer *et al.*, 1986). Complementary to these findings, addition of the β subunit of cholera toxin to bind ganglioside G_{M1} is mitogenic for rat thymocytes (Spiegel *et al.*, 1985) and quiescent 3T3 cells, and potentiates the response of the latter of EGF, PDGF, and insulin (Spiegel and Fishman, 1987). Similar effects have been seen with antibodies to gangliosides. In contrast, the β subunit is inhibitory for rapidly growing 3T3 cells and *ras*-transformed 3T3 cells; therefore, gangliosides have been described as bimodal regulators of cell growth (Spiegel and Fishman, 1987).

Gangliosides G_{M3} and G_{M1} inhibit PDGF binding and PDGF-induced protein phosphorylation on tyrosine (Bremer *et al.*, 1984); in other studies, G_{M3} has been found to inhibit EGF-induced phosphorylation without inhibition of EGF binding (Bremer *et al.*, 1986). Therefore, inhibition of growth by these gangliosides appears to involve direct interactions with the PDGF and EGF receptors. This is also suggested by the finding by Hanai *et al.* (1988a,b) that G_{M3} copurifies with the EGF receptor from A5S cells, which are only weakly responsive to EGF. G_{M3} did not copurify with the EGF receptor of a subclone (A5I) that retains responsiveness to EGF, although more polar gangliosides, including lyso-G_{M3}, were found in small amounts in the cells. Song *et al.* (1991) confirmed that G_{M3} inhibited and de-*N*-acetyl-G_{M3} stimulated EGF receptor autophosphorylation in the presence of Triton X-100, but only G_{M3} inhibited EGF-stimulated receptor autophosphorylation in membranes. This group has obtained other evidence for a direct interaction between the receptor and G_{M3} using fluorescence spectroscopy: the mobility of a fluorescently labeled G_{M3} is restricted in membranes that contain the EGF receptor.

Using a mutant CHO cell line that has a conditional defect in ganglioside biosynthesis and stable cell lines expressing the EGF receptor, Davis and coworkers (Davis, this volume) also found that modulation of the physiological levels of gangliosides (primarily G_{M3}) modulated signal transduction by the EGF receptor. Little else is known about how gangliosides actually influence the receptor (and cell growth) *in vivo,* although it has been known for some time (Chatterjee *et al.*, 1973) that there is greater incorporation of galactose into gangliosides and neutral glycolipids during late M and/or early G_1 phases of the cell cycle of KB cells. Also, Usuki *et al.* (1988a,b) have associated the turnover of G_{M3} with cell growth with removal of the sialic acid by a G_{M3} sialidase activity found in the culture medium of human fibroblasts. To assess the importance of extracellular sialidase(s) to cell growth, fibroblasts were incubated with a sialidase inhibitor (2-deoxy-2,3-dehydro-*N*-acetylneuraminic acid), which reversibly inhibited growth and the turnover of labeled G_{M3} (Usuki *et al.*, 1988b). Thus, it appears that G_{M3} turnover to lactosylceramide is more active for growing cells—to release them from the growth inhibitory effects of the intact ganglioside—but that the sialidase activity is low at confluence (and G_{M3} turnover is reduced) to contribute to the downregulation of growth factor receptor(s) (Sweeley, Vol. 26).

Why might cells utilize gangliosides (or other sphingolipids) to modulate the behavior of receptors? Possibly because these molecules can be made and turned over rapidly to enable cells to "fine tune" their responsiveness to external stimuli. Another advantage might be that some glycolipids tend to aggregate, producing local concentrations in the molar range so that receptors need not bind them with high affinity (Thompson and Tillack, 1985).

2. *Effects of Complex Sphingolipids on Cellular Kinases*

In addition to inhibition of receptor kinases, gangliosides have been found to have complex effects on protein phosphorylation in membranes from brain (Goldenring *et al.,* 1985; Nakaoka *et al.,* 1992), including stimulation of the Ca^{2+}/calmodulin-dependent kinase system (Goldenring *et al.,* 1985; Tsuji *et al.,* 1985; Cimino *et al.,* 1987), and in muscle (Chan, 1989). A ganglioside-stimulated protein kinase, purified from guinea pig brain (Chan, 1987), phosphorylates exogenous polypeptides such as Leu-Arg-Arg-Ala-Ser-Leu-Gly and undergoes autophosphorylation. The kinase does not require Ca^{2+} for activity and appears to be distinct from cAMP-dependent protein kinase, protein kinase C, and Ca^{2+}/calmodulin-dependent kinase(s). Gangliosides have been reported to inhibit protein kinase C (Kim *et al.,* 1986; Kreutter *et al.,* 1987) under visicular assay conditions, but not in mixed micelles (Hannun and Bell, 1987).

Tsuji *et al.* (1988a,b) have noted that addition of nanomolar concentrations of ganglioside G_{Q1b} to a human neuroblastoma cell line (GOTO) stimulated the phosphorylation of at least three cell surface-associated proteins by ecto-type kinase(s) (i.e., extracellular protein kinases). Since G_{Q1b} was effective at low concentrations, exhibited a ganglioside specificity resembling that for neuritogenesis, and activated kinase(s) on the outer surface of the cell (where most cellular gangliosides are thought to reside), Tsuji and co-workers suggest that this may represent the system that is associated with the neuritogenic effect of this ganglioside. Using rat brain membranes, Nakaoka *et al.* (1992) found that G_{Q1b} induces both phosphorylation and release of phosphate from a 72-kDa phosphoprotein, indicating that there may be a bimodal regulation of this protein.

3. *Influence on the Behavior of Other Lipids*

In addition to affecting the behavior of proteins, sphingolipids also appear to influence the metabolism of other lipids, especially phosphatidylcholine and cholesterol. There is often an inverse correlation between the amounts of sphingomyelin and phosphatidylcholine in tissues (Barenholz and Thompson, 1980; Barenholz and Gatt, 1982), which is most likely due to the pathway for sphingomyelin synthesis via transfer of the phosphocholine group from phosphatidylcholine to ceramide, yielding diacylglycerol as a by-product (Marggraf and Kanfer, 1984, 1987; Rosenwald and Pagano, Vol. 26). This implies that the relative amounts of phosphatidylcholine and sphingomyelin will be determined by the relationship between these choline-containing lipids and the amounts of available ceramide and diacylglycerol.

A considerable amount of evidence has established that sphingomyelin affects the behavior of cholesterol (and vice versa). The levels of sphingomyelin are closely correlated with the amounts of cholesterol in different membranes (Patton, 1970); sphingomyelin influences cholesterol movement among membranes (Lange *et al.*, 1979; Wattenberg and Silbert, 1983; Fugler *et al.*, 1985; Bittman, 1988; Stein *et al.*, 1988), including the distribution of cholesterol among membranes (Wattenberg and Silbert, 1983; Yeagle and Young, 1986); and sphingomyelin alters the interaction between cholesterol and proteins (Stevens *et al.*, 1986) and cholesterol metabolism (Slotte *et al.*, 1990), including the activity of HMG-CoA reductase (Gupta and Rudney, 1991). Interactions between cholesterol and sphingomyelin may be one of the major factors in regulating cholesterol esterification because it does not appear that the amount of acyl-CoA:cholesterol acyltransferase is regulatory (Cadigan and Chang, 1988). The converse is also true, that is, cholesterol influences sphingomyelin metabolism (Nikolova-Karakashian *et al.*, 1992). Sphingomyelin also affects low-density lipoprotein (LDL) binding and utilization (Gatt and Bierman, 1980; Xu and Tabas, 1991). These facets of sphingomyelin metabolism have been discussed by Chatterjee (Vol. 26).

It has been proposed that sphingolipids are involved in the intracellular trafficking of other lipids and proteins (Simons and Wandinger-Ness, 1990; Rosenwald and Pagano, Vol. 26).

B. Ceramides and Sphingosines as "Lipid Second Messengers"

The discovery that sphingosine (Hannun *et al.*, 1986) is a potent inhibitor of protein kinase C introduced the paradigm that hydrolysis products of sphingolipids may serve as "lipid second messengers," in analogy to the bioactive products that are formed from the turnover of glycerolipids. Subsequent work has expanded considerably the number of sphingolipids that elicit cell responses and can serve as candidate second messengers (ceramides, sphingosine 1-phosphate, lysosphingolipids, *N*-methylsphingosines, among others) and the number and diversity of the systems that are affected (Hannun and Bell, Dobrowsky and Hannun, Hakomori and Igarashi, Hannun *et al.*, Mathias and Kolesnick, Spiegel *et al.*, and Tettamanti and Riboni, this volume).

1. *Ceramide-Activated Protein Kinases and Phosphatases and the Sphingomyelin/Ceramide Cycle*

Recent investigations by Hannun and co-workers have discovered a sphingomyelin/ceramide cycle that meets the criteria of a lipid second messenger pathway (see articles by Hannun and Bell and Hannun *et al.* in this volume). The initial observations (Okazaki *et al.*, 1989, 1990) were that a significant portion of the cellular sphingomyelin is hydrolyzed to ceramide and phosphorylcholine when HL-60 cells are treated with 1α,25-dihydroxyvitamin D_3 to induce differentiation. This response was maximal after approximately 2 hours, and by 4 hours the amounts of sphingomyelin and ceramide had returned to basal levels, suggesting the existence of a hydrolysis/resynthesis "cycle." Vitamin D_3 increased the activity of a neutral sphingomyelinase, and the response could be mimicked by treating cells with a bacterial sphingomyelinase. Ceramide formation appeared to mediate at least some aspects of vitamin D_3-induced differentiation, because addition of a cell-permeable ceramide (*N*-acetylsphingosine) resulted in similar changes in cell phenotype (Okazaki *et al.*, 1989). This sphingomyelin cycle appears to be involved in the action of two other factors that induce differentiation of HL-60 cells: tumor necrosis factor (TNF) and γ-interferon (Kim *et al.*, 1991). Other agents that induce differentiation (retinoic acid, phorbol myristate acetate, and dibutyryl cyclic AMP), but to granulocytes rather than monocytes, did not induce sphingomyelin turnover to ceramide. Studies with GH4C1 cells has found that retinoic acid increased ceramide levels, but probably not by inducing sphingomyelin turnover (Kalen *et al.*, 1992).

Ceramide has been found to activate a serine/threonine protein phosphatase (Dobrowsky and Hannun, this volume). The characteristics of the ceramide-activated protein phosphatase (CAPP) are similar to the subgroup 2A protein phosphatases (i.e., it is cation independent and is sensitive to inhibition by okadaic acid). CAPP is activated by a variety of ceramides with different hydrophobic moieties; however, it is more stereospecific (Bielawska *et al.*, 1992).

Kolesnick and co-workers have shown that ceramide also activates protein kinase(s) (see article by Mathias and Kolesnick in this volume). These studies (Goldkorn *et al.*, 1991) arose from investigation of the mechanism of sphingosine activation of the epidermal growth factor receptor, which involves phosphorylation on threonine 669 (see article by Davis). Goldkorn *et al.* (1991) found that ceramide was able to induce this phosphorylation, and that at the concentrations of sphingosine that were active, much of the sphingosine had been metabolized to ceramide by the cells. Subsequent studies (Mathias *et al.*, 1991) utilizing a polypeptide substrate (representing amino acids 663–681 of the EGF receptor), found a Mg^{2+}-dependent, ceramide-activated kinase activity in the membrane fraction from A431 cells. Activation of this kinase activity has been demonstrated in a cell-free system (Dressler *et al.*, 1992) in which postnuclear supernatants from HL-60 cells were treated with TNF to induce sphingomyelin turnover to ceramide, and could be mimicked by adding exogenous sphingomyelinase to the preparation instead of TNF.

In addition to their reuse to synthesize complex sphingolipids (see articles by Jeckel and Wieland, Rosenwald and Pagano, and Sandhoff and van Echten in

Vol. 26), cellular ceramides can undergo hydrolysis to sphingosine (see discussion below and the article by Hassler and Bell) or be phosphorylated by a Ca^{2+} dependent kinase (Bajjalieh *et al.,* 1989; Dressler and Kolesnick, 1990). Little is known about this latter, more recently discovered product. One can speculate that ceramide 1-phosphate is bioactive because Truett and King (Vol. 26) have described that some bacteria and arthropods (including the brown recluse spider) produce a sphingomyelinase D that produces a severe and prolonged inflammatory response when injected into rabbits.

The utilization of sphingomyelin turnover, particularly the involvement of a neutral sphingomyelinase, is intriguing because a neutral sphingomyelinase activity has been known to exist for some time (Schneider and Kennedy, 1967; Gatt, 1976; Rao and Spence, 1976; see articles by Chatterjee and Spence in Vol. 26) but its function has been unknown. *In vitro,* it has been shown to be activated by volatile anesthetics (Pellkofer and Sandoff, 1980; Mooibroeck *et al.,* 1985) with hydrolysis of greater than 80% of the plasma membrane sphingomyelin of neuroblastoma cells in culture. Among the other interesting observations about this enzyme are that the levels in brain (Spence and Burgess, 1976) and liver (Petrova *et al.,* 1988) and the activity in the colon are affected by neoplastic transformation (Dudeja *et al.,* 1986). This activity may be distinct from the sphingomyelinase that is activated by TNF because Hannun and co-workers believe that this activity may be intracellular (Hannun *et al.,* this volume). There are apparently many pathways for sphingomyelin turnover when one includes reversal of sphingomyelinase, (several) neutral and acidic sphingomyelinases, and (in some systems) a sphingomyelinase D. While the acidic sphingomyelinase probably functions largely in the lysosomal turnover of sphingolipid, Kolesnick (1987) has observed that the activity of an acidic sphingomyelinase can be increased by diacylglycerol but not phorbol ester, and this has been suggested to participate in the activation of NF-κB by TNF (Schutze *et al.,* 1992).

2. *Long-Chain (Sphingoid) Bases and Sphingosine Phosphate as Bioactive Products of Sphingolipid Biosynthesis and Turnover*

Since the initial discoveries that sphingosine and other long-chain bases inhibit protein kinase C *in vitro* and cellular responses to protein kinase C activators in platelets (Hannun *et al.,* 1986), neutrophils (Wilson *et al.,* 1986), and HL-60 cells (Merrill *et al.,* 1986), over 100 different cellular systems have been found to be affected by these compounds. Sphingosine is one of the most potent inhibitors of protein kinase C that is found in mammalian cells, and acts on an equimolar basis with 1,2-dioleoylglycerol (Hannun *et al.,* 1986). It is competitive with diacylglycerol, phorbol dibutyrate (PDB), and Ca^{2+}, and blocks protein kinase C activation by unsaturated fatty acids and other lipids (Wilson *et al.,* 1986; Oishi *et al.,* 1988). The inhibition depends on the mole percentage of the sphingosine with respect to the other lipids or detergents, as typically occurs with molecules that partition into membranes; however, the exact mechanism of protein kinase C inhibition is not known. It is possible that positively charged long-chain bases may localize in the same region of the membrane as acidic lipids (e.g., phosphatidylserine), which are required for maximal activity, thereby blocking binding and/or activity (Hannun *et al.,* 1986; Bazzi and Nelsestuen, 1987; Rando, 1988; Merrill *et al.,* 1989). Sphingosine appears to interact with the membrane-binding domain of protein kinase C because it is not inhibitory for catalytic fragment when lipids are present (Nakadate *et al.,* 1988). Higher concentrations of sphingosine also inhibit activation by oleic acid, which has been shown to occur with a non-membrane-associated form of protein kinase C (El Touny *et al.,* 1990).

The inhibition is roughly the same with all of the major long-chain bases that are found naturally (Hannun *et al.,* 1986; Merrill *et al.,* 1989), including some lysosphingolipids (Hannun and Bell, 1987). Recent studies of other structure/function relationships (Merrill *et al.,* 1989) revealed that the most potent inhibitors have an alkyl chain length of 18 carbon atoms, which is the prevalent chain length *in vivo* (Karlsson, 1970a,b). Subtle differences among the four stereoisomers of sphingosine, *N*-methyl derivatives, and simpler alkylamines (e.g., stearylamine) have been seen; however, the inhibition of protein kinase C is not limited to the D-*erythro*-sphingosine, the major naturally occurring stereoisomer. This may indicate that there is not a specific binding site for long-chain bases on protein kinase C; however, it is premature to draw this conclusion because all four stereoisomers provide the same headgroup conformation and vary only in the position of the alkyl chain (Merrill *et al.,* 1989). It has been concluded that the positive charge on sphingosine is involved in inhibition of protein kinase C (Hannun *et al.,* 1986; Hannun and Bell, 1989; Merrill *et al.,* 1989; Bottega *et al.,* 1989) and it should be borne in mind that sphingosine has a lower pK_a than simple alkylamines; thus, sphingosine is probably present in both neutral and ionized forms at physiological pH. This may additionally explain the ability of sphingosine to be taken up by cells and affect intracellular targets even though, as a charged species, it would have been predicted to have difficulty crossing the plasma membrane.

Besides protein kinase C, many other cell regulatory pathways have been found to be affected by long-chain bases. In fact, these are too numerous to describe fully in this introduction; therefore, systems that are affected by low concentrations (i.e., low micromolar), or ones with particular relevance to signal transduction, are mainly discussed. Sphingosine and other alkylamines are potent inhibitors of phosphatidic acid phosphohydrolase (Aridor-Piterman *et al.,* 1992; Lavie *et al.,* 1990; Mullmann *et al.,* 1991; Jamal *et al.,* 1991), which is responsible for the formation of diacylglycerols (and alkylacylglycerols) as intermediates of glycerolipid biosynthesis and signal transduction pathways utilizing phospholipase D. Studies with human neutrophils have found that sphingosine is as much as 10-fold more potent at inhibiting phosphatidic acid phosphohydrolase (and thereby blocking the neutrophil respiratory burst) than as an inhibitor of protein kinase C (Perry *et al.*, 1992).

Sphingosine activates a phosphatidylethanolamine-specific phospholipase D (Kiss and Crilly, 1991) and sphingosine has been reported to activate, and sphingomyelin to inhibit, phospholipase C δ (Pawelczyk and Lowenstein, 1992). The sphingosine concentrations that are required for activation are high, whereas cellular levels of sphingomyelin may be adequate to allow an effect on this activity *in vivo*. At high concentrations, sphingosine also stimulates a key regulatory enzyme of phosphatidylcholine biosynthesis, CTP-choline phosphate cytitylyltransferase (Sohal and Cornell, 1990).

Sphingosine inhibits the Na,K-ATPase with approximately the same dose response as for protein kinase C inhibition (Oishi *et al.*, 1990). Long-chain bases have also been reported to affect several other ion transport systems that are thought to be regulated by protein kinase C (Connor *et al.*, 1988; Soliven *et al.*, 1988; Conn *et al.*, 1989; Gillies *et al.*, 1989; Ling and Eaton, 1989). Sphingosine and sphingosylphosphorylcholine induce calcium release from intracellular stores in smooth muscle cells permeabilized with saponin (Ghosh *et al.*, 1990) and sphingosine and sphingosine 1-phosphate stimulated increases in intracellular calcium in Swiss 3T3 cells (Zhang *et al.*, 1991). These observations have led to the hypothesis that a sphingosine metabolite (possibly sphingosine 1-phosphate) might act as a regulator of calcium mobilization (see article by Spiegel *et al.* in this volume). In contrast, sphingosine is a potent inhibitor of sarcoplasmic reticulum calcium release in response to caffeine, doxorubicin, and other agents (Sabbadini *et al.*, 1992).

Sphingosine inhibits coagulation initiated by lipopolysaccharide-stimulated human monocytes by inhibiting factor VII binding (Conkling *et al.*, 1989). It inhibits thyrotropin-releasing hormone binding to GH_3 cells (Winicov and Gershengorn, 1988), but the IC_{50} is fairly high (i.e., 63 μM compared to a typical IC_{50} of 1 to 3 μM for protein kinase C in a comparable number of cells) and exhibits an atypical difference in potency for sphinganine versus sphingosine. Sphingosine inhibits a calmodulin-dependent kinase (Jefferson and Schulman, 1988); however, the cellular effects were only reported with very high concentrations of sphingosine (suggesting that they, too, may be less potent than for protein kinase C), and sphingosine has been reported to inhibit c-*src* and v-*src* kinases, but at concentrations of 330 and 660 μM (Igarashi *et al.*, 1989). At high concentrations or in the absence of membranes, long-chain bases probably have nonspecific effects due simply to their charge and/or amphipathic nature. Furthermore, sphingosine is metabolized to ceramide and sphingosine 1-phosphate, which are themselves bioactive. It is likely, for example, that the inhibition of fibronectin release by human lung fibroblasts by sphingosine (Scheidl *et al.*, 1992) is due to ceramide formation because it occurs under conditions where protein kinase C inhibition is not evident, it is not mimicked by calphostin C (another protein kinase C inhibitor), and fibronectin release was inhibited by *N*-acetyl-sphingosine.

Sphingosine has also been shown to *activate* some protein kinases. Davis and co-workers (Faucher *et al.,* 1988; Davis *et al.,* 1988; Northwood and Davis, 1988; see article by Davis, this volume) demonstrated that sphingosine increases the phosphorylation at threonine 669 (the MAP kinase site) of the EGF receptor through a pathway that appears to be independent of protein kinase C (it also blocks the protein kinase C-dependent phosphorylation of threonine 654). As noted earlier, the stimulation of protein phosphorylation on threonine 669 by low concentrations of sphingosine may be due to the formation of ceramide and activation of a ceramide-activated kinase. Nonetheless, higher levels of sphingosine and N-methylated long-chain bases also directly induce phosphorylation on threonine 669, which indicates that the kinase(s) recognizes the sphingosyl moiety even when the amino group is modified (see article by Mathias and Kolesnick, this volume). Wedegaertner and Gill (1989) have shown that sphingosine increases the activity of the cytoplasmic tyrosine kinase domain of the EGF receptor to equal or greater than that of the ligand-activated holo EGF receptor, and have suggested that sphingosine may mimic the effect of EGF to induce a conformation that is optimal for tyrosine kinase activity.

A common hindrance in studies using sphingosine is that long-chain bases can be growth inhibitory and cytotoxic. For CHO cells, inhibition of protein kinase C appears to be involved based on the concentration dependency, structural specificity, protein phosphorylation patterns (especially when wild-type cells were compared to a cell line selected for partial resistance to the toxicity of long-chain bases), and the apparent *lack* of perturbation of cellular acidic compartments—the most likely site of an artifactual effect of these types of compounds (Merrill *et al.,* 1989; Stevens *et al.,* 1990a). It has been known for some time that even simpler long-chain bases such as stearylamine (Campbell, 1983), which have been used in preparing stable liposomes for drug delivery, can be cytotoxic. While it is difficult to define the specific molecular events of cell death, protein kinase C is thought to influence growth, ion transporters, and other systems that could cause this phenotype. Virus replication, as another type of "growth," is also affected by sphingosine. Nutter *et al.* (1987) have reported that sphingosine inhibits the induction of DNA polymerase and DNase activities in Epstein–Barr virus-infected cells treated with phorbol esters and *n*-butyrate. Interestingly, Van Veldhoven *et al.* (1992) have found that ceramide, but not sphingosine, is increased in HIV-infected cells.

Considering the large number of cellular systems that can be affected by exogenously added sphingosine, one would predict that this compound is utilized as a regulator of cell functions. In contrast to ceramide, however, no studies have conclusively linked agonist-induced changes in the cellular amounts of sphingosine with a biological response. All cells examined to date contain free sphingosine, although the amounts are small (on the order of 1–10 nmol per gram of tissue, wet weight; or 10 to 100 pmol/10^6 cells) (for examples, see Merrill *et al.*, 1986, 1988; Wilson *et al.*, 1988; Kobayashi *et al.*, 1988; Wertz and Downing, 1989; Van Veldhoven *et al.*, 1989). Studies with both intact cells (Wilson *et al.*, 1988) and isolated plasma membranes (Slife *et al.*, 1989) have found that endogenous cellular sphingolipids are hydrolyzed to free sphingosine. In the case of rat liver plasma membranes, the source of the free sphingosine appears to be hydrolysis of sphingomyelin to ceramide by a Mg^{2+}-dependent sphingomyelinase followed by further hydrolysis of a portion of the ceramide (Slife *et al.*, 1989). It is unclear what factors determine the extent to which ceramides are cleaved to sphingosine.

The cellular concentrations of sphingosine are usually within one order of magnitude of the amounts that affect these systems, and these probably underestimate the cellular concentrations because most of the free sphingosine is localized in the plasma membrane, whereas exogenous sphingosine may be less localized (Slife *et al.*, 1989). Nonetheless, it is not known whether or not these levels of sphingosine affect any of the systems that have been shown to be sensitive to activation or inhibition by exogenously added long-chain bases.

III. Perspectives on the Regulation of Cell Function by Sphingolipids

A. Models for the Biological Functions of Sphingolipids

There are many ways in which sphingolipid synthesis and turnover could help regulate cell functions, and evidence is accumulating in favor of these hypotheses. Some possibilities have been depicted in Figs. 1 and 2, using sphingomyelin and an unspecified glycosphingolipid. An attractive feature of sphingolipids is that unique compounds might be associated with specific receptors and their targets, and they could also be turned over to produce products that act on a large number of additional sites. Thus, one can envisage that sphingolipids serve as cell regulators in diverse ways in multiple intracellular sites:

1. *Sphingolipids could regulate cell behavior at the surface of the cell,* where they bind to extracellular matrix proteins and receptors (as well as bacterial toxins, antibodies, etc.). The effects of the sphingolipids might be to form a classical receptor–ligand interaction (ganglioside G_{M3} appears to be a good prototype for this type of action in its modulation of the EGF receptor), or to define regions of the membrane with surface characteristics that aid in receptor binding and responses, internalization, recycling, etc., or in anchoring proteins to the cell

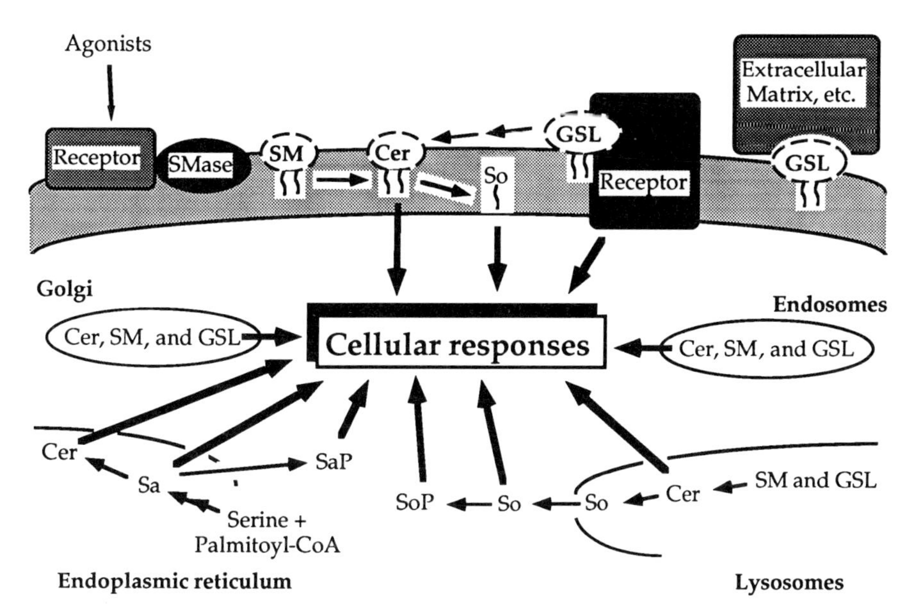

FIG. 2. Illustration of some of the pathways through which bioactive sphingolipids are formed in cells. The pathways of Fig. 1 have been redrawn to focus on metabolic steps that yield the bioactive compounds sphingosine (and sphinganine), sphingosine(anine) 1-phosphate, (dihydro)ceramides, and various complex sphingolipids. Not only are there multiple sphingolipids that can affect cell behavior, there are a large number of targets for these compounds. For examples, sphingosine and sphinganine inhibit protein kinase C and phosphatidic acid phosphohydrolase; sphingosine 1-phosphate stimulates increases in intracellular calcium; ceramides activate protein kinase(s) and protein phosphatase(s); gangliosides activate protein kinases and modulate the activity of a number of growth factor receptors; and, various sphingolipids interact with matrix proteins to mediate cell–cell and cell–substratum interactions. For a more complete discussion of these and other targets, consult the text.

surface. There may be instances in which the *loss* of an interaction is important. For example, cells might be able to interrupt their binding to the extracellular matrix or to a neighboring cell by stopping synthesis of (and internalizing or turning over) the glycolipids that are involved in the interaction. As has become evident over recent years, the products of sphingolipid turnover are also bioactive; therefore, they might serve as second messengers to communicate these surface changes to the inside of the cell.

2. *Sphingolipids could regulate cell behavior by changes that occur in the plasma membrane and/or endosomes,* where complex sphingolipids and products of their turnover could stimulate or inhibit other domains of the receptor kinases or the response of G proteins, other protein kinases and protein phosphatases, ion transporters, etc. Some sphingolipid hydrolysis products (ceramide

and sphingosine) can move across biological membranes; therefore, turnover of sphingolipids on the extracellular leaflet could generate modulators of proteins on the internal leaflet. Ceramide and sphingosine clearly affect more than one target [e.g., sphingosine inhibits protein kinase C and phosphatidic acid phosphohydrolase, among others, but activates the EGF receptor kinase and other protein kinase(s), and ceramides activate protein phosphatase(s)]; therefore, formation of these compounds would appear to result in a pleiotropic response. The hydrolysis of complex sphingolipids with one function (e.g., inhibition of a growth factor receptor) to bioactive products such as free sphingosine (with the opposite function) could provide a powerful switching mechanism.

3. *Sphingolipids could regulate cell behavior at intracellular sites that are sensitive to products of sphingolipid turnover.* Some sphingolipids (e.g., lysosphingolipids and free long-chain bases) are able to move rapidly among membranes and might, therefore, affect targets at sites distant from their site of formation. For example, sphingosine that is released during the lysosomal catabolism of sphingolipids might subsequently interact with targets in other locations in the cell. Or sphinganine, which is not thought to accumulate normally, might increase under aberrant conditions and affect sensitive targets. Sphingolipids can also be converted to other bioactive species, such as sphingosine 1-phosphate, *N*-methylated sphingosines, ceramide 1-phosphate, and lysosphingolipids, that can affect multiple intracellular targets.

B. Implications for Diseases Due to Disorders of Sphingolipid Metabolism

There are multiple steps at which disruption of normal sphingolipid synthesis and turnover might result in aberrant cell behavior.

1. Synthesis of the Ceramide Backbone

The initial precursors of *de novo* sphingolipid biosynthesis are palmitoyl-CoA and serine, which are combined to form 3-ketosphinganine by serine palmitoyltransferase, a pyridoxal 5′-phosphate-dependent enzyme. This enzyme has been shown to be critical for cell survival by recent work with yeast cells with a defective enzyme (selected by auxotrophy for exogenously provided long-chain bases) (Lester and Dickson, Vol. 26) and by studies with a temperature-sensitive enzyme in CHO cells (Hanada *et al.*, 1990). The 3-ketosphinganine produced by serine palmitoyltransferase is rapidly reduced to sphinganine by an NADPH-dependent reductase and converted to (dihydro)ceramides; the 4-*trans* double bond of ceramides appears to be introduced after fatty acids have been added to sphinganine (Ong and Brady, 1973; Wang and Merrill, 1986). Recent studies have discovered a family of diseases that appear to be caused by inhibition of ceramide formation at the level of the acylation of the long-chain bases. The inhibitors, called fumonisins, are produced by *Fusarium moniliforme* and related molds (Merrill *et al.*, Vol. 26). Consumption of fumonisins or feed contaminated with *F. moniliforme* is known to result in equine leukoencephalomalacia, porcine pulmonary syndrome, hepatotoxicity and hepatocarcinogenicity in rats, and other diseases of agricultural and laboratory animals, and has been correlated with human esophageal cancer. The fumonisins inhibit ceramide synthase and thereby block complex sphingolipid biosynthesis *de novo* (Wang *et al.*, 1991). Perhaps as importantly (or more), there is an accumulation of sphinganine (and at later timepoints, sphingosine) to levels that may affect some of the targets that have been identified by studies with exogenously added long-chain bases. Defects in ceramide synthesis have also been implicated recently in disease by the finding that Trembler mice have very low levels of ceramide synthase (Boiron-Sargueil *et al.*, 1992), and perhaps by the observation of unusually high amounts of sphinganine (which usually only appears in significant amounts as an intermediate of sphingolipid synthesis) in fibroblasts from patients with Niemann Pick's disease type C (Goldin *et al.*, 1992). Kendler and Dawson (1992) (also see their article in this volume) have also found that hypoxic injury to rat oligodendrocytes causes ceramide to accumulate in the endoplasmic reticulum.

2. *Addition of Sphingolipid Headgroups*

Complex sphingolipid biosynthesis by mammals involves the transfer of the phosphorylcholine headgroup from phosphatidylcholine to ceramide to form sphingomyelin (Mathias and Kolesnick, this volume; Rosenwald and Pagano, Vol. 26) and the stepwise addition of carbohydrate groups (as well as sulfate, for the sulfatides) to form glycosphingolipids (Sandhoff and van Echten, Rosenwald and Pagano, Young, and Jeckel and Wieland, Vol. 26). The regulation of complex sphingolipid synthesis appears to reside in the rate of ceramide formation and its partitioning into different products by the relative activities, and subcellular localization, of sphingomyelin synthase and the glycosyltransferases. Little is known about the extent to which disruption of these pathways by genetic mutation or environmental factors leads to disease. Recent studies with an inhibitor of the initial glycosylation of ceramide (termed PDMP) have uncovered interesting changes in cell behavior that may lead to a better understanding of this aspect (Radin *et al.*, Vol. 26).

3. *Sphingolipid Turnover*

The catabolism of cellular sphingolipids has been studied extensively because of the many well-known genetic diseases that arise from defects in sphingolipid hydrolysis (Sweeley, 1991). Sphingolipid turnover involves removal of the sugars by exoglycosidases (and some recently discovered endoglycoceramidases) and the phosphorylcholine of sphingomyelin by phospholipases. These processes can be complex (for example, accessory proteins are sometimes required) and involve both acidic (lysosomal) sphingomyelinase as well as enzymes with neutral to alkaline pH optima (Chatterjee, Spence, and Truett and King, Vol. 26). As described above, sphingomyelinases are thought to be part of signal transduction pathways for TNF and probably other agonists (Hannun *et al.* and Mathias and Kolesnick, this volume).

After removal of the sphingolipid headgroup, some recycling of sphingolipids in whole or part has been seen in many studies (Sonderfeld *et al.,* 1985; Trinchera *et al.,* 1988; Kok *et al.,* 1989). Ceramides can also undergo phosphorylation by a ceramide kinase (Bajjalieh *et al.,* 1989; Dressler and Kolesnick, 1990), the purpose of which is unknown. Cleavage of the amide-linked fatty acid can occur by (at least) three ceramidases with acidic, neutral, and alkaline pH optima (Spence *et al.,* 1986; Hassler and Bell, Vol. 26). The sphingosine that is released by hydrolysis of ceramide can be converted back to ceramides or degraded, which involves phosphorylation at position 1 by sphingosine kinase (Buehrer and Bell, Vol. 26), followed by lytic cleavage to ethanolamine phosphate and *trans*-2-hexadecanal (Van Veldhoven and Mannaerts, Vol. 26). Sphingosine appears to leave the lysosomes to interact with the kinase and lyase, which are not thought to be lysosomal enzymes. Igarashi and Hakomori (1989) have also reported methylation of sphingosine, but it is unclear if this is a pathway specific for sphingosine or one of the general detoxification reactions of cells.

As far as the authors are aware, diseases due to defects in these steps of nonlysosomal sphingolipid turnover have not been found. However, one suspects that they may surface when more is learned about these processes.

C. Implications for Endogenous and Exogenous Sphingolipids in Disease Prevention and Treatment

There is an extensive literature on the effects of gangliosides on neuronal cell function (Tettamanti and Riboni, this volume) and their role in hypoxic injury (Kendler and Dawson, this volume) and other forms of neural injury (Manev *et al.,* this volume). Gangliosides and related compounds have been tested as therapeutic agents in animals (Manev *et al.,* this volume), and a recent report has proposed ganglioside G_{M1} as a potential therapeutic agent in Parkinson's disease (Schncider *et al.,* 1992).

The suggested relationship between sphingomyelin and cholesterol (reviewed earlier in this article) indicates that changes in the metabolism of sphingolipids might alter atherogenesis. Williams *et al.* (1987) found that the activity of serine palmitoyltransferase increases in aorta during atherogenesis; therefore, inhibitors of this enzyme, such as cycloserine and β-haloalanines (Medlock and

Merrill, 1988) may influence the progression of this disease. Imaizumi *et al.* (1992) have reported that feeding rats diets containing sphingolipids increases the amount of sphingomyelin in serum, while decreasing serum triglycerides and liver (but not serum) cholesterol esters.

As an inhibitor of protein kinase C, sphingosine and other long-chain bases have potential as antitumor agents (Hannun *et al.,* 1986). Sphingosine blocks the induction of ornithine decarboxylase by phorbol esters in mouse skin (Gupta *et al.,* 1988; Enkvetchakul *et al.,* 1989), one biochemical marker of tumor promotion; however, it did not reduce the number of tumors in a longer term study (Enkvetchakul *et al.,* 1992). In a cell culture model of transformation (mouse C3H10T1/2 cells) (Borek *et al.,* 1991), sphingosine and sphinganine reduced cell transformation in response to γ-irradiation and PMA. In a recent carcinogenesis study with mice treated with *N,N*-dimethylhydrazine (DMH) to induce colon tumors, Dillehay *et al.* (1993) found that including milk sphingomyelin in the diet reduced the number of aberrant colonic crypts in short-term studies and decreased the tumor incidence by approximately one-third. Another study of mammary carcinogenesis induced by 7,12-dimethylbenz[*a*]anthracene (Parenteau *et al.,* 1992) found a reduced number of tumors when the diets were supplemented with stearylamine, which is not a sphingoid base but inhibits protein kinase C much like sphingosine (Merrill *et al.,* 1989). Because sphingolipids are natural constituents of foods, it is possible that the ingestion and digestion of sphingolipids may be a factor in diet and cancer. Spiegel and co-workers (article in this volume) have shown that long-chain bases are mitogenic at low concentrations for Swiss 3T3 cells (Zhang *et al.,* 1990), and there is evidence for both growth stimulation and inhibition with CHO cells (Stevens *et al.,* 1989). When applied to mouse skin, sphingosine increased the number of papillomas, which may indicate that it can be growth stimulatory *in vivo* (Enkvetchakul *et al.,* 1992). Although the latter study found no increase in the number of carcinomas in the sphingosine-treated group, more studies are needed to elucidate the extent to which these compounds may be mitogenic *in vivo.*

Long-chain bases have also been found to induce differentiation and, when used in combination with other agents that induce differentiation, to result in cells with a more fully differentiated phenotype (Stevens *et al.,* 1989; 1990b; Yung *et al.,* 1992). Sphingosines, N-methylated sphingosines, and sphingosine 1-phosphate have been found to reduce tumor growth and metastasis (Sadahira *et al.,* 1992; Hakomori and Igarashi, this volume).

Based on the diverse signal transduction pathways affected by sphingolipids, one can envision that these compounds—or, more likely, new compounds based on the major pharmacophors of the bioactive sphingolipids—will prove to be useful in the prevention and treatment of a wide range of diseases, including inflammation and psoriasis.

IV. Comments on the Nomenclature of Sphingolipids

A. Long-Chain (Sphingoid) Bases

Sphingolipids are named for the "sphingosin" backbone first described by Thudichum in 1884, which is now known to encompass a large group of structurally similar long-chain (sphingoid) bases. Sphingosine, the prevalent long-chain base of many mammalian sphingolipids, has 18 carbon atoms and the stereochemistry shown in Fig. 3. The term sphingosine is sometimes used to designate any combination of long-chain (sphingoid) bases; however, this is uncommon and confusing. It would appear to be preferable to describe the entire class by a separate name, such as "sphingoid bases" in analogy to use of the term "eicosanoids."

Also encountered are homologs of other chain lengths, species without the 4,5-*trans* double bond (or with additional double bonds at other positions along the alkyl chain), compounds with a hydroxyl at position 4, and various branched-chain species (examples of these compounds are shown in Fig. 4). These compounds are known to undergo a considerable amount of rearrangement and decomposition when handled under acidic conditions, yielding a large number of additional long-chain bases, some of which have been described by Karlsson (1970a,b). Most of the long-chain bases that are commercially available contain at least small amounts of these naturally occurring species or, if prepared synthetically, mixed stereoisomers.

The nomenclature of these long-chain (sphingoid) bases is awkward and should be reexamined. The recommended names are *trans*-4-sphingenine (for sphingosine), sphinganine (for dihydrosphingosine), and 4-hydroxysphinganines (for "phytosphingosines"). In our experience, the subtle differences in the spelling and pronunciation of sphingenine and sphinganine lead to confusion. This is avoided if the name sphingosine is retained, and the saturated compounds are referred to as sphinganines or dihydrosphingosines. The name 4-hydroxy-

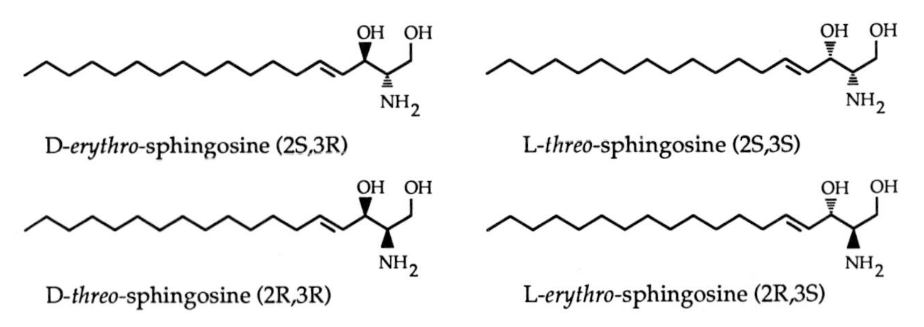

FIG. 3. Conformations of the four stereoisomers of sphingosine.

Examples of long-chain (sphingoid) bases known to be present in animal tissues

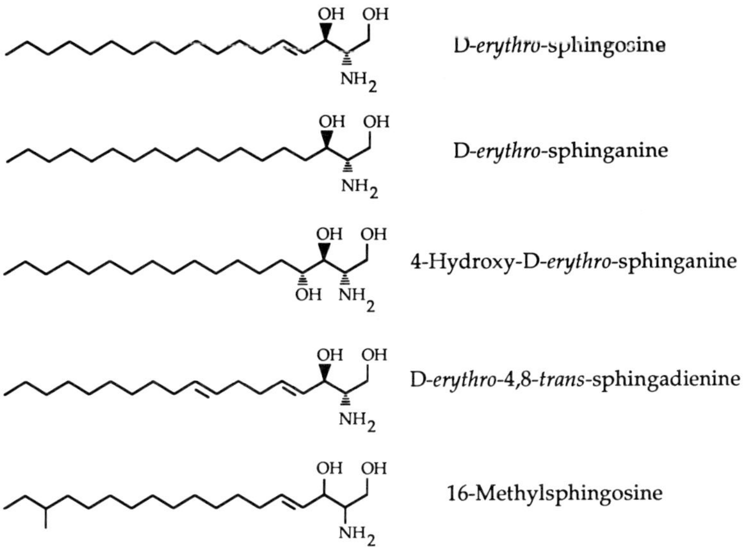

D-*erythro*-sphingosine

D-*erythro*-sphinganine

4-Hydroxy-D-*erythro*-sphinganine

D-*erythro*-4,8-*trans*-sphingadienine

16-Methylsphingosine

Related compounds found in other organisms

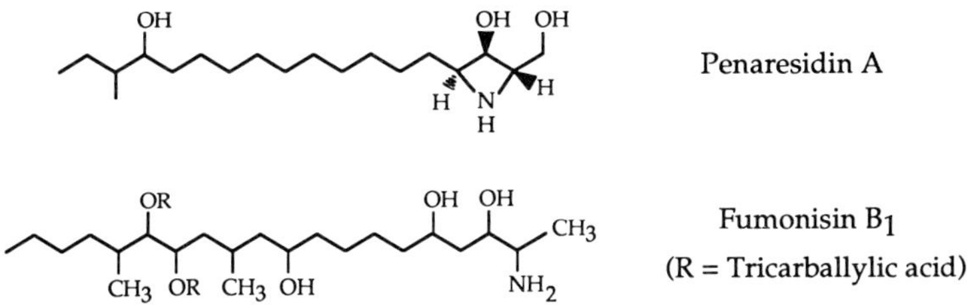

Penaresidin A

Fumonisin B_1
(R = Tricarballylic acid)

FIG. 4. Examples of structural variation in long-chain (sphingoid) bases and related compounds. Shown are the structures of the three major long-chain bases of mammalian tissues (sphingosine, sphinganine, and hydroxysphinganine), which are also found in various alkyl chain lengths. Animals have additional long-chain bases with hydroxyl groups, double bonds, and branched chains as shown by two examples (sphingadienine and 16-methylsphingosine). New classes of compounds have recently been discovered in sponges (the penaresidin) and molds (the fumonisins).

sphinganine is preferable over phytosphingosine because this type of long-chain base is not a "sphingosine" (i.e., it does not have a 4,5-*trans* double bond) and it is produced by mammals (Crossman and Hirschberg, 1977, 1984). Unless otherwise stated, the alkyl chain length of sphingosine, sphinganine, and 4-hydroxysphinganine is implied to be 18 carbons because this is the most prevalent homolog; other alkyl-chain lengths can be designated by preceding the name with the number of carbon atoms and double bonds, as in C20 : 1 for the 20-carbon sphingosine.

A number of novel sphingosine derivatives have been isolated from the sponge *Penares* sp. (Kobayashi *et al.,* 1991) (Fig. 4). These azetidine alkaloids are potent actomyosin ATPase activators.

B. Ceramides

Short-chain ceramides are coming into increased use because they are more easily delivered to cells than are the longer-chain compounds; therefore, there should be a convenient and clear nomenclature for such analogs. Where the backbones of these ceramides are sphingosines or sphinganines, these could be referred to as ceramides or dihydroceramides, respectively. Thus, the length of the amide-linked fatty acid can be indicated by its carbon number, such as C_2-ceramide for *N*-acetylsphingosine. This creates some confusion about how to designate the type of long-chain base when the chain lengths of both are altered. Until an unambiguous nomenclature is adopted, authors should define their usage in the abstracts and text, and try to avoid confusion by not using similar abbreviations for different compounds.

C. Complex Sphingolipids

Few studies to date have been able to elucidate how variation of the type of long-chain base and/or amide-linked fatty acid affects the biological functions of complex sphingolipids; however, it is clear that the physical properties, metabolism, and intracellular trafficking of sphingolipids depends on these features (Barenholz and Thompson, 1980; Valsecchi *et al.,* 1992; Bassi and Sonnino, 1992; Levade *et al.,* 1991; Tettamanti and Riboni and Rosenwald and Pagano, this volume). As described elsewhere in this volume by Hannun *et al.,* evidence is surfacing that indicates that ceramides, but not dihydroceramides, can modulate cell growth. Considering that over 60 different long-chain bases and dozens of fatty acids have been found in the ceramide backbones of at least 300 complex sphingolipids, there appears to be a considerable amount of structural information that can be used for cell regulation.

Acknowledgments

The authors thank their students, postdoctoral fellows, technicians, and collaborators, as well as the authors of the other articles in these volumes of *Advances in Lipid Research*, for their contributions to the findings and ideas summarized in this introductory review.

References

Aridor-Piterman, O., Lavie, Y., and Liscovitch, M. (1992). *Eur. J. Biochem.* **204,** 561–568.

Bajjalieh, S. H., Martin, T. F. J., and Floor, F. (1989). *J. Biol. Chem.* **264,** 14354–14360.

Barenholz, Y., and Gatt, S. (1982). *In* "Phospholipids" (J. N. Hawthorne and G. B. Ansell, eds.). Chapter 4. Elsevier, Amsterdam.

Barenholz, Y., and Thompson, T. E. (1980). *Biochim. Biophys. Acta* **604,** 129–158.

Bassi, R., and Sonnino, S. (1992). *Chem. Phys. Lipids* **62,** 1–9.

Bazzi, M. D., and Nelsestuen, G. L. (1987). *Biochem. Biophys. Res. Commun.* **146,** 203–207.

Bielawska, A., Linardič, C. M., and Hannun, Y. A. (1992). *J. Biol. Chem.* **267,** 18493–18497.

Bittman, R. (1988). "Biology of Cholesterol," Chapter 8. CRC Press, Boca Raton, FL.

Boiron-Sargueil, F., Heape, A., and Cassagne, C. (1992). *J. Neurochem.* **59,** 652–656.

Borek, C., Ong, A., Stevens, V. L., Wang, E., and Merrill, A. H., Jr. (1991). *Proc. Natl. Acad. Sci. U.S.A.* **88,** 1953–1957.

Bottega, R., Epand, R. M., and Ball, E. H. (1989). *Biochem. Biophys. Res. Commun.* **164,** 102–107.

Bremer, E. G., and Hakomori, S. I. (1982). *Biochem. Biophys. Res. Commun.* **106,** 711–718.

Bremer, E. G., Hakomori, S. -I., Bowen-Pope, D. F., Raines, E., and Ross, R. (1984). *J. Biol. Chem.* **259,** 6818–6825.

Bremer, E. G., Schlessinger, J., and Hakomori, S.-I. (1986). *J. Biol. Chem.* **261,** 2434–2440.

Cadigan, K. M., and Chang, T. Y. (1988). *J. Lipid Res.* **29,** 1683–1692.

Campbell, P. I. (1983). *Cytobios* **37,** 21–26.

Chan, K.-F. J. (1987). *J. Biol. Chem.* **262,** 5248–5255.

Chan, K.-F. J. (1989). *J. Biol. Chem.* **264,** 18632–18637.

Chatterjee, S., Sweeley, C. C., and Velicer, L. F. (1973). *Biochem. Biophys. Res. Commun.* **54,** 585–595.

Cimino, M., Benfenati, F., Farabegoli, C., Cattabeni, F., Fuxe, K., Agnati, L. F., and Toffano, G. (1987). *Acta Physiol. Scand.* **130,** 317–325.

Conkling, P. R., Patton, K. L., Hannun, Y. A., Greenberg, C. S., and Weinberg, J. B. (1989). *J. Biol. Chem.* **264,** 18440–18444.

Conn, P. J., Strong, J. A., and Kaczmarek, L. K. (1989). *J. Neurosci.* **9,** 480–487.

Connor, J. A., Wadman, W. J., Hockberger, P. E., and Wong, R. K. S. (1988). *Science* **240,** 649–653.

Crossman, M. W., and Hirschberg, C. B. (1977). *J. Biol. Chem.* **252,** 5815–5819.

Crossman, M. W., and Hirschberg, C. B. (1984). *Biochim. Biophys. Acta* **795,** 411–416.

Davis, R. J., Gironës, N., and Faucher, M. (1988), *J. Biol. Chem.* **263,** 5373–5379.

Dillehay, D. L., Crall, K. J., Webb, S. K., Schmelz, E., and Merrill, A. H., Jr. (1993). *FASEB J.* **7,** A69.

Dressler, K. A., and Kolesnick, R. N. (1990). *J. Biol. Chem.* **265,** 14917–14921.

Dressler, K. A., Mathias, S., and Kolesnick, R. N. (1992). *Science* **255,** 1715–1718.

Dudeja, P. K., Dahiya, R., and Brasitus, T. A. (1986). *Biochim. Biophys. Acta* **863,** 309–312.

El Touny, S., Khan, W., and Hannun, Y. A. (1990). *J. Biol. Chem.* **265,** 16437–16443.

Enkvetchakul, B., Merrill, A. H., Jr., and Birt, D. F. (1989). *Carcinogenesis (London)* **10,** 379–381.

Enkvetchakul, B., Barnett, T., Liotta, D. C., Geisler, V., Menaldino, D., Merrill, A. H., Jr., and Birt, D. F. (1992). *Cancer Lett.* **62,** 35–42.

Faucher, M., Girones, N., Hannun, Y. A., Bell, R. M., and Davis, R. J. (1988). *J. Biol. Chem.* **263,** 5319–5327.

Fugler, L., Clejan, S., and Bittman, R. (1985). *J. Biol. Chem.* **260,** 4098–4102.

Gatt, S. (1976). *Biochem. Biophys. Res. Commun.* **68,** 235–241.

Gatt, S., and Bierman, E. L. (1980). *J. Biol. Chem.* **255,** 3371–3376.

Ghosh, T. K., Bian, J., and Gill, D. L. (1990). *Science* **248,** 1653–1656.

Gillies, R. J., Martinez, R., Sneider, J. M., and Hoyer, P. B. (1989). *J. Cell. Physiol.* **139,** 125–130.
Goldenring, J. R., Otis, L. C., Yu, R. K., and DeLorenzo, R. J. (1985). *J. Neurochem.* **44,** 1129–1134.
Goldin, E., Roff, C. F., Miller, S. P. F., Rodriguez-Lafrasse, C., Vanier, M. T., Brady, R. O., and Pentchev, P. G. (1992). *Biochim. Biophys. Acta* **1127,** 303–311.
Goldkorn, T., Dressler, K. A., Muindi, J., Radin, N. S., Mendelsohn, J., Menaldino, D., Liotta, D., and Kolesnick, R. N. (1991). *J. Biol. Chem.* **266,** 16092–16097.
Gupta, A. K., and Rudney, H. (1991). *J. Lipid Res.* **32,** 125–136.
Gupta, A. K., Fischer, G. J., Elder, J. T., Nickoloff, B. J., and Voorhees, J. J. (1988). *J. Invest. Dermatol.* **91,** 486–491.
Hanada, K., Nishijima, M., and Akamatsu, Y. (1990). *J. Biol. Chem.* **265,** 22137–22142.
Hanai, N., Dohi, T., Nores, G. A., and Hakomori, S.-I. (1988a). *J. Biol. Chem.* **263,** 6296–6301.
Hanai, N., Nores, G. A., MacLeod, C., Torres-Mendez, C.-R., and Hakomori, S.-I. (1988b). *J. Biol. Chem.* **263,** 10915–10921.
Hannun, Y. A., and Bell, R. M. (1987). *Science* **235,** 670–674.
Hannun, Y. A., and Bell, R. M. (1989). *Science* **243,** 500–507.
Hannun, Y. A., Loomis, C. R., Merrill, A., and Bell, R. M. (1986). *J. Biol. Chem.* **261,** 12604–12609.
Hannun, Y. A., Greenberg, C. S., and Bell, R. M. (1987). *J. Biol. Chem.* **262,** 13620–13626.
Igarashi, Y., and Hakomori, S.-I. (1989). *Biochem. Biophys. Res. Commun.* **164,** 1411–1414.
Igarashi, Y., Hakomori, S.-I., Toyokuni, T., Dean, B., Fujita, S., Sugimoto, M., Ogawa, T., El-Ghendy, K., and Racker, E. (1989). *Biochemistry* **28,** 6796–6800.
Imaizumi, K., Tominaga, A., Sato, M., and Sugano, M. (1992). *Nutr. Res.* **12,** 543–548.
Jamal, H., Martin, A., Gomez-Munoz, A., and Brindley, D. N. (1991). *J. Biol. Chem.* **266,** 2988–2996.
Jefferson, A. B., and Schulman, H. (1988). *J. Biol. Chem.* **263,** 15,241–15,244.
Kalen, A., Borchardt, R. A., and Bell, R. M. (1992). *Biochim. Biophys. Acta* **1125,** 90–96.
Kanfer, J. N., and Hakomori, S., eds. (1983). "Sphingolipid Biochemistry." Plenum, New York.
Karlsson, K.-A. (1970a). *Chem. Phys. Lipids* **5,** 6–43.
Karlsson, K.-A. (1970b). *Lipids* **5,** 878–891.
Keenan, T. W., Schmid, E., Franke, W. W., and Wiegandt, H. (1975). *Exp. Cell Res.* **92,** 259–270.
Kendler, A., and Dawson, G. (1992). *J. Neurosci. Res.* **31,** 205–211.
Kim, J. Y. H., Goldenring, J. R., DeLorenzo, R. J., and Yu, R. K. (1986). *J. Neurosci. Res.* **15,** 159–166.
Kim, M.-Y., Linardič, C., Obeid, L., and Hannun, Y. A. (1991). *J. Biol. Chem.* **266,** 484–489.
Kiss, Z., and Crilly, K. S. (1991). *Eur. J. Biochem.* **197,** 785–790.
Kobayashi, J., Cheng, J.-F., Ishibashi, M., Walchli, M. R., Yamamura, S., and Ohizumi, Y. (1991). *J. Chem. Soc., Perkin Trans. 1,* 1135–1137.
Kobayashi, T., Mitsuo, K., and Goto, I. (1988). *Eur. J. Biochem.* **171,** 747–752.
Kok, J. W., Eskelinen, S., Hoekstra, K., and Hoekstra, D. (1989). *Proc. Natl. Acad. Sci. U.S.A.* **86,** 9896–9900.
Kolesnick, R. N. (1987). *J. Biol. Chem.* **262,** 16759–16762.
Kreutter, D., Kim, J. Y. H., Goldenring, J. R., Rasmussen, H., Ukomandu, C., DeLorenzo, R. J., and Yu, R. K. (1987). *J. Biol. Chem.* **262,** 1633–55912.
Laine, R. A., and Hakomori, S.-I. (1973). *Biochem. Biophys. Res. Commun.* **54,** 1039–1045.
Lange, Y., D'Alessandro, J. S. D., and Small, D. M. (1979). *Biochim. Biophys. Acta* **556,** 388–398.
Lavie, Y., Piterman, O., and Liscovitch, M. (1990). *FEBS Lett.* **277,** 7–10.
Lederkremer, R. M., Lima, C., Ramirez, M. I., and Casal, O. L. (1990). *Eur. J. Biochem.* **192,** 337–345.
Levade, T., Gatt, S., and Salvayre, R. (1991). *Biochem. J.* **275,** 211–217.
Ling, B. N., and Eaton, D. C. (1989). *Am. J. Physiol.* **256,** F1094–F1103.
Marggraf, W. D., and Kanfer, J. N. (1984). *Biochim. Biophys. Acta* **793,** 346–353.
Marggraf, W. D., and Kanfer, J. N. (1987). *Biochim. Biophys. Acta* **897,** 57–68.

Mathias, S., Dressler, K. A., and Kolesnick, R. N. (1991). *Proc. Natl. Acad. Sci. U.S.A.* **88,** 10009–10013.
Medlock, K. A., and Merrill, A. H., Jr. (1988). *Biochemistry* **27,** 7079–7084.
Merrill, A. H., Jr., Sereni, A. M., Stevens, V. L., Hannun, Y. A., Bell, R. M., and Kinkade, J. M., Jr. (1986). *J. Biol. Chem.* **261,** 12610–12615.
Merrill, A. H., Jr., Wang, E., Mullins, R. E., Jamison, W. C. L., Nimkar, S., and Liotta, D. C. (1988). *Anal. Biochem.* **171,** 373–381.
Merrill, A. H., Jr., Nimkar, S., Menaldino, D., Hannun, Y. A., Loomis, C., Bell, R. M., Tyagi, S. R., Lambeth, J. D., Stevens, V. L., Hunter, R., and Liotta, D. C. (1989). *Biochemistry* **28,** 3138–3145.
Miller-Podraza, H., Bradley, R. M., and Fishman, P. H. (1982). *Biochemistry* **21,** 3260–3265.
Mooibroeck, M. J., Cook, H. W., Clarke, J. T. R., and Spence, M. W. (1985). *J. Neurochem.* 44, 1551–1558.
Mullmann, T. J., Siegel, M. I., Egan, R. W., and Billah, M. M. (1991). *J. Biol. Chem.* **266,** 2013–2016.
Nakadate, T., Jeng, A. Y., and Blumberg, P. M. (1988). *Biochem. Pharmacol.* **37,** 1541–1545.
Nakaoka, T., Tsuji, S., and Nagai, Y. (1992). *J. Neurosci. Res.* **31,** 724–730.
Nikolova-Karakashian, M. N., Petkova, H., and Koumanov, K. S. (1992). *Biochimie* **74,** 153–159.
Northwood, I. C., and Davis, R. J. (1988). *J. Biol. Chem.* **263,** 7450–7453.
Nutter, L. M., Grill, S. P., Li, J. S., Tan, R. S., and Cheng, Y. C. (1987). *Cancer Res.* **47,** 4407–4412.
Oishi, K., Raynor, R. L., Charp, P. A., and Kuo, J. F. (1988). *J. Biol. Chem.* **263,** 6865–6871.
Oishi, K., Zheng, B., and Kuo, J. F. (1990). *J. Biol. Chem.* **265,** 70–75.
Okazaki, T., Bell, R. M., and Hannun, Y. A. (1989). *J. Biol. Chem.* **264,** 10976–19080.
Okazaki, T., Bielawska, A., Bell, R. M., and Hannun, Y. A. (1990). *J. Biol. Chem.* **265,** 15823–15831.
Ong, D. E., and Brady, R. N. (1973). *J. Biol. Chem.* **248,** 3884–3888.
Parenteau, H., Ho, T.-F. H., Eckel, L. A., and Carroll, K. K. (1992). *Nutr. Cancer* **17,** 235–241.
Patton, S. (1970). *J. Theor. Biol.* **29,** 489–491.
Pawelczyk, T., and Lowenstein, J. M. (1992). *Arch. Biochem. Biophys.* **297,** 328–333.
Pellkofer, R., and Sandoff, K. (1980). *J. Neurochem.* **34,** 988–992.
Perry, D. K., Hand, W. L., Edmondson, D. E., and Lambeth, J. D. (1992). *J. Immunol.* **149,** 2749–2758.
Petrova, D. H., Momchiliva-Pankova, A. B., Markovska, T. T., and Koumanov, K. S. (1988). *Exp. Gerontol.* **23,** 19–24.
Rando, R. R. (1988). *FASEB J.* **2,** 2348–2355.
Rao, B. G., and Spence, M. W. (1976). *J. Lipid Res.* **17,** 506–515.
Sabbadini, R. A., Betto, R., Teresi, A., Fachechi-Cassano, G., and Salviati, G. (1992). *J. Biol. Chem.* **267,** 15475–15484.
Sadahira, Y., Ruan, F., Hakomori, S.-I., and Igarashi, Y. (1992). *Proc. Natl. Acad. Sci. U.S.A.* **89,** 9686–9690.
Scheidl, H., Scita, G., Sampson, P. H., Park, H.-Y., and Wolf, G. (1992). *Biochim. Biophys. Acta* **1135,** 295–300.
Schneider, J. S., Pope, A., Simpson, K., Taggart, J., Smith, M. G., and DiStefano, L. (1992). *Science* **256,** 843–846.
Schneider, P. B., and Kennedy, E. P. (1967). *J. Lipid Res.* **8,** 202–209.
Schutze, S., Potthoff, K., Machleidt, T., Berkovič, D., Wiegmann, K., and Kronke, M. (1992). *Cell (Cambridge, Mass.)* **71,** 765–776.
Simons, K., and Wandinger-Ness, A. (1990). *Cell (Cambridge, Mass.)* **62,** 207–210.
Slife, C. W., Wang, E., Hunter, R., Wang, S., Burgess, C., Liotta, D. C., and Merrill, A. H., Jr. (1989). *J. Biol. Chem.* **264,** 10371–10377.
Slotte, J. P., Harmala, A.-S., Jansson, C., and Porn, M. I. (1990). *Biochim. Biophys. Acta* **1030,** 251–257.
Sohal, P. S., and Cornell, R. B. (1990). *J. Biol. Chem.* **265,** 11746–11750.
Soliven, B., Szuchet, S., Arnason, B. G. W., and Nelson, D. J. (1988). *J. Membr. Biol.* **105,** 177–186.

Sonderfeld, S., Conzelmann, E., Schwarzmann, G., Burg, J., Hinrichs, U., and Sandoff, K. (1985). *Eur. J. Biochem.* **149,** 247–255.
Song, W. X., Vacca, M. F., Welti, R., and Rintoul, D. A. (1991). *J. Biol. Chem.* **266,** 10174–10181.
Spence, M. W., and Burgess, J. K. (1976). *J. Neurochem.* **30,** 917–919.
Spence, M. W., Reed, S., and Cook, H. W. (1986). *Biochem. Cell Biol.* **64,** 400–404.
Spiegel, S., and Fishman, P. H. (1987). *Proc. Natl. Acad. Sci. U.S.A.* **84,** 141–145.
Spiegel, S., Fishman, P. H., and Weber, R. J. (1985). *Science* **230,** 1285–1287.
Stadler, J., Keenan, T. W., Bauer, G., and Derisch, G. (1989). *EMBO J.* **8,** 371–377.
Stein, O., Oette, K., Haratz, D., Halperin, G., and Stein, Y. (1988). *Biochim. Biophys. Acta* **960,** 322–333.
Stevens, V. L., Lambeth, J. D., and Merrill, A. H., Jr. (1986). *Biochemistry* **25,** 4287–4292.
Stevens, V. L., Winton, E. F., Smith, E. E., Owens, N. E., Kinkade, J. M., Jr., and Merrill, A. H., Jr. (1989). *Cancer Res.* **49,** 3229–3234.
Stevens, V. L., Nimkar, S., Jamison, W. C., Liotta, D. C., and Merrill, A. H., Jr. (1990a). *Biochim. Biophys. Acta* **1051,** 37–45.
Stevens, V. L., Owens, N. E., Winton, E. F., Kinkade, J. M., Jr., and Merrill, A. H., Jr. (1990b). *Cancer Res.* **50,** 222–226.
Sweeley, C. C. (1991). *In* "Biochemistry of Lipids, Lipoproteins, and Membranes" (D. E. Vance and J. E. Vance, eds.), pp. 327–361. Elsevier, Amsterdam.
Thompson, T. E., and Tillack, T. W. (1985). *Annu. Rev. Biophys. Biophys. Chem.* **14,** 361–386.
Trinchera, M., Wiesmann, U., Pitto, M., Acquotti, D., and Ghidoni, R. (1988). *Biochem. J.* **252,** 375–379.
Tsuji, S., Nakajima, U., Sasaki, T., and Nagai, Y. (1985). *J. Biochem. (Tokyo)* **97,** 969–972.
Tsuji, S., Yamashita, T., Tanaka, M., and Nagai, Y. (1988a). *J. Neurochem.* **50,** 414–423.
Tsuji, S., Yamashita, T., and Nagai, Y. (1988b). *J. Biochem. (Tokyo)* **104,** 498–503.
Usuki, S., Lyu, S.-C., and Sweeley, C. C. (1988a). *J. Biol. Chem.* **263,** 6847–6853.
Usuki, S., Hoops, P., and Sweeley, C. C. (1988b). *J. Biol. Chem.* **263,** 10595–10599.
Valsecchi, M., Chigormo, V., Sonnino, S., and Tettamanti, G. (1992). *Chem. Phys. Lipids* **60,** 247–252.
Van Veldhoven, P. P., Bishop, W. R., and Bell, R. M. (1989). *Anal. Biochem.* **183,** 177–189.
Van Veldhoven, P. P., Matthews, T. J., Bolognesi, D. P., and Bell, R. M. (1992). *Biochem. Biophys. Res. Commun.* **31,** 209–216.
Wang, E., and Merrill, A. H., Jr. (1986). *J. Biol. Chem.* **261,** 3674–3769.
Wang, E., Norred, W. P., Bacon, C. W., Riley, R. T., and Merrill, A. H., Jr. (1991). *J. Biol. Chem.* **266,** 14486–14490.
Wattenberg, B. W., and Silbert, D. F. (1983). *J. Biol. Chem.* **258,** 2284–2289.
Wedegaertner, P. B., and Gill, G. N. (1989). *J. Biol. Chem.* **264,** 11346–11353.
Wertz, P. W., and Downing, D. T. (1989). *Biochim. Biophys. Acta* **1002,** 213–217.
Williams, R. D., Sgoutas, D. S., Zaatari, G. S., and Santoianni, R. A. (1987). *J. Lipid Res.* **28,** 1478–1481.
Wilson, E., Olcott, M. C., Bell, R. M., Merrill, A. H., Jr., and Lambeth, J. D. (1986). *J. Biol. Chem.* **261,** 12616–12623.
Wilson, E., Wang, E., Mullins, R. E., Liotta, D. C., Lambeth, J. D., and Merrill, A. H., Jr. (1988). *J. Biol. Chem.* **263,** 9304–9309.
Winicov, I., and Gershengorn, M. C. (1988). *J. Biol. Chem.* **263,** 12179–12182.
Xu, X. X., and Tabas, I. (1991). *J. Biol. Chem.* **266,** 24849–24858.
Yeagle, R. L., and Young, J. E. (1986). *J. Biol. Chem.* **261,** 8175–8181.
Yung, B. Y.-M., Luo, K. J., and Hui, K.-W. (1992). *Cancer Res.* **52,** 3593–3597.
Zeller, C. B., and Marchase, R. B. (1992). *Am. J. Physiol.* **262** (Cell Physiol. 31), C1341–C1355.
Zhang, H., Buckley, N. E., Gibson, K., and Spiegel, S. (1990). *J. Biol. Chem.* **265,** 76–81.
Zhang, H., Desai, N. N., Olivera, A., Seki, T., Booker, G., and Spiegel, S. (1991). *J. Cell Biol.* **114,** 155–167.

Part I

FUNCTION OF SPHINGOLIPIDS AND SPHINGOLIPID-DERIVED PRODUCTS IN SIGNAL TRANSDUCTION

ADVANCES IN LIPID RESEARCH, VOL. 25

The Sphingomyelin Cycle: A Prototypic Sphingolipid Signaling Pathway

YUSUF A. HANNUN* AND ROBERT M. BELL†

*Departments of Medicine and Cell Biology
†Department of Biochemistry and Section of Cell Growth, Regulation, and Oncogenesis
Duke University Medical Center
Durham, North Carolina 27710

I. Introduction

A. Lipids as Bioregulators and Precursors to Second Messengers

Major insight has been achieved in our understanding of the functional complexity of membrane lipids beyond their critical role as structural components of biomembranes. The amphipathic character of membrane phospholipids is critical for bilayer formation, and the segregation of phospholipids into anionic and zwitterionic molecules is critical for maintaining polarity of biomembranes, with cytoplasmic membrane leaflets enriched in anionic phospholipids and external (or luminal) leaflets enriched in zwitterionic phospholipids. However, these essential functions do not appear to necessitate the vast structural complexity of phospholipids.

Initial insight into other fundamental functions for membrane phospholipids came with the pioneering work of Hokin and Hokin, who in the early 1950s discovered that the action of extracellular agents (acetylcholine) results in early turnover of inositol phospholipids (1). After four decades of intensive investigation, it has become obvious that this remodeling of membrane phospholipids in response to extracellular signals plays a critical role in signal transduction and cell regulation.

The regulated metabolism of inositol phospholipids (PI) constitutes a powerful paradigm for lipid-mediated signal transduction (PI cycle). According to this paradigm, the interaction of extracellular agents with specific membrane receptors results in activation of PI-specific phospholipases C (either through direct tyrosine phosphorylation or via heterotrimeric G proteins). In turn, phospholipase C acts on membrane PI (especially phosphatidylinositol-4,5-bisphosphate), resulting in the generation of the neutral lipid diacylglycerol and the soluble headgroup inositol trisphosphate. Both products play critical roles in transduction of the incoming extracellular signal: inositol trisphosphate interacts with specific intracellular membrane receptors which functionally cause release of stored calcium and a consequent elevation of free cytoplasmic calcium (2–4). In turn, calcium (primarily by interacting with calmodulin) regulates the activity and function of a number of cellular components, including protein kinases, ion channels, protein phosphatases, and other proteins. The other product of PI hydrolysis, diacylglycerol, was also found to function as a second messenger by activating a specific family of protein kinases (5–7). In pioneering studies, Nishizuka, Takai, and co-workers identified a novel cyclic nucleotide-independent protein kinase (later to be named protein kinase C) from rat brain. These investigators soon determined that diacylglycerol was a specific and potent activator of protein kinase C (8). The functional significance of protein kinase C has been highlighted by the realization that phorbol esters and related tumor promoters substitute for endogenous diacylglycerol in activating protein kinase C (9, 10), thus implicating protein kinase C in phorbol ester-mediated biology, which is vast and touches on most aspects of cell biology!

Of historical interest (and of special relevance to ceramide as discussed in this article), it is important to note that the appreciation and acceptance of a role for inositol trisphosphate as a second messenger was enhanced by the preceding established role of the soluble cyclic AMP as a second messenger. On the other hand, diacylglycerol provided the first example of a natural intermediary in lipid metabolism to achieve the status of a second messenger.

The PI cycle has served as a prototype for phospholipid-mediated signal transduction. Studies in the past decade have begun to define important roles for phosphatidylcholine (PC)-specific phospholipase C in causing the formation of diacylglycerol independent of inositol trisphosphate (11); phospholipase D, resulting in the formation of phosphatidic acid (12, 13); phospholipases A_2, resulting in the formation of arachidonic acid and bioactive eicosanoids (14–16); and other bioactive lipids such as platelet-activating factor, alkylacylglycerol, lysophosphatidic acid, and phosphatidic acid.

B. Overview of Biological and Biochemical Effects of Sphingosine

These studies on membrane phospholipids as precursors to bioactive molecules have provided the rationale(s) for the complexity of these lipids, the intricate pathways for their metabolism, and the significance of changes in levels of phospholipids and their derived products.

While the wealth of information accumulating on the role of membrane lipids in signal transduction has been primarily restricted to phospholipids, an equivalent role of the other major lipid constituents of biomembranes, sphingolipids, has been largely unexplored. Studies in the mid 1980s on regulation of protein kinase C led to the discovery of sphingosine and lysosphingolipids as potent inhibitors of activation of protein kinase C by phorbol esters or diacylglycerols (17,18). Since then, additional potential biochemical targets for the action of sphingosine have been identified (19,20), most notably phospholipase D (21,22), PA phosphohydrolase (23,24), and casein kinase II (25). Moreover, sphingosine has been shown to exert multiple biological effects, including inhibition of platelet and neutrophil activation (17,26,27), modulation of cell differentiation (28), and mitogenesis (29). These biochemical and biological activities of sphingosine are discussed in further detail in the introductory article of this volume (Merrill, Hannun, and Bell).

The discovery of inhibition of protein kinase C by sphingosine provided a conceptual bridge between phospholipid-dependent and potential sphingolipid-dependent signaling pathways. This work led to the development of a unifying hypothesis implicating the sphingolipids in "sphingolipid cycles," wherein minor sphingolipid metabolites would be produced in a transient fashion and participate in signal transduction (19). Although a physiological role for sphingosine has not yet been firmly established, the hypothesis developed heightened interest in signaling functions for sphingolipids.

C. Hypothesis: Sphingolipids as Precursors to Second Messengers and Cell Regulatory Molecules

Sphingolipids are complex membrane lipids structurally composed of a long-chain sphingoid base and an amide-linked fatty acid with different polar headgroups at the 1 position (30,31). The simplest sphingolipid, ceramide, which has a hydroxyl at the 1 position, serves as the precursor for more complex sphingolipids. These include sphingomyelin, which contains a phosphorylcholine headgroup, neutral glycolipids which have one or more glycose units, and more complex acidic glycosphingolipids which contain either sulfate (sulfatides) or sialic acid (the gangliosides).

Over the past few decades, vast and significant information has been obtained on the structural composition of sphingolipids and on potential multiple biological activities, including roles in cell transformation, tumor progression, cell differentiation, cell recognition, and receptor–ligand interaction (19,30). While these studies are beginning to elucidate the significance of sphingolipids in cell biology, they did not address the potential role of sphingolipids as precursors to second messengers. This may have been in part due to (1) the dogma that all sphingolipids are located in the external leaflet of the plasma membrane, (2) the only recent appreciation of the signaling function of phospholipids, (3) the absence of hormone-regulated short-term sphingolipid metabolism, and (4) the absence of information on biochemical and biological activities of sphingolipid-derived products. In this context, the demonstration of the effects of sphingosine on protein kinase C (and since then on other biochemical targets) suggested to us the more global hypothesis that sphingolipids may play a physiological role as precursors to bioactive molecules and that lipid-mediated signaling may not be restricted to the glycerol phospholipids (19). Investigation of this hypothesis required the demonstration of (1) regulated turnover of specific membrane sphingolipids in response to extracellular signals, (2) concomitant elevation of levels of sphingolipid-derived products, and (3) biochemical and/or biological activity of the generated molecules. Our pursuit of this hypothesis has led to the discovery of a sphingomyelin cycle of cell regulation and the potential role of ceramide as a second messenger/intracellular mediator. These studies are described below.

II. Sphingomyelin Cycle

A. Discovery of Sphingomyelin Turnover in HL-60 Cells

Since it was hypothesized that, in analogy with the PI cycle of transmembrane signaling, the action of certain extracellular agents may lead to the activation of cellular enzymes that would act on membrane sphingolipids to generate bioactive products, a search was initiated for hormone-mediated early sphingolipid metabolism. The first such experiments to yield substantial results were performed in human HL-60 leukemia cells. When these cells in tissue culture were labeled with sphingolipid precursors and stimulated with 1α,25-dihydroxyvitamin D_3, the levels of sphingomyelin changed transiently. Treatment with 1α,25-dihydroxyvitamin D_3 resulted in hydrolysis of approximately 25% of the cellular sphingomyelin; peak effects were achieved at 2 hours followed by a gradual return of sphingomyelin levels to baseline levels by 4–6 hours (32). No other significant changes were noted in the levels of glycosphingolipids or of phosphatidylcholine during the same time interval.

These initial results were very exciting on two accounts. First, they were the earliest known biochemical effects of vitamin D_3 in HL-60 cells, occurring well in advance of the effects of vitamin D_3 on c-*myc* protooncogene levels (which decrease by 4–8 hours) and well in advance of phenotypic changes of differentiation (which are detected 2–4 days following treatment with vitamin D_3). Second, the hormone-induced and reversible changes in sphingomyelin levels suggested that sphingomyelin may be hydrolyzed into a bioactive product providing a regulated intracellular signal. Therefore, it became important to define (1) the components of the sphingomyelin cycle, (2) the hormone-activated enzyme regulating sphingomyelin hydrolysis, (3) a role of sphingomyelin hydrolysis and the generated products in biological responses, and (4) extracellular inducers of sphingomyelin hydrolysis.

B. Elements of the Sphingomyelin Cycle

The turnover of sphingomyelin could result in the formation of ceramide, ceramide phosphate, lysosphingomyelin, or other products depending on the enzymatic pathways involved. In initial studies, it was determined that the action of vitamin D_3 resulted in the generation of choline phosphate and ceramide, with levels peaking at 2 hours and returning to baseline by 4 hours (32). These results suggested that vitamin D_3 caused the activation of a sphingomyelinase that resulted in the hydrolysis of sphingomyelin at the phosphodiesteric linkage of ceramide to choline phosphate (Fig. 1).

In an attempt to identify the sphingomyelinase responsible for this activity, HL-60 cells were treated with vitamin D_3, and membrane and cytosolic extracts were assayed for acid and neutral sphingomyelinase. The addition of vitamin D_3 resulted in elevation of neutral sphingomyelinase activity with kinetics paralleling the changes in ceramide levels and sphingomyelin hydrolysis (32). The main activity was cytosolic, neutral, and magnesium-independent (T. Okazaki, R. Bell, and Y. A. Hannun, unpublished observations). These features distinguish this hormone-activated sphingomyelinase from other previously described sphingomyelinases. The major neutral sphingomyelinase thus far described is a membrane-associated, magnesium-dependent sphingomyelinase which appears to reside on the outer leaflet of the plasma membrane (33). Ongoing studies aim at characterizing and purifying the hormone-activated sphingomyelinase and understanding its regulation.

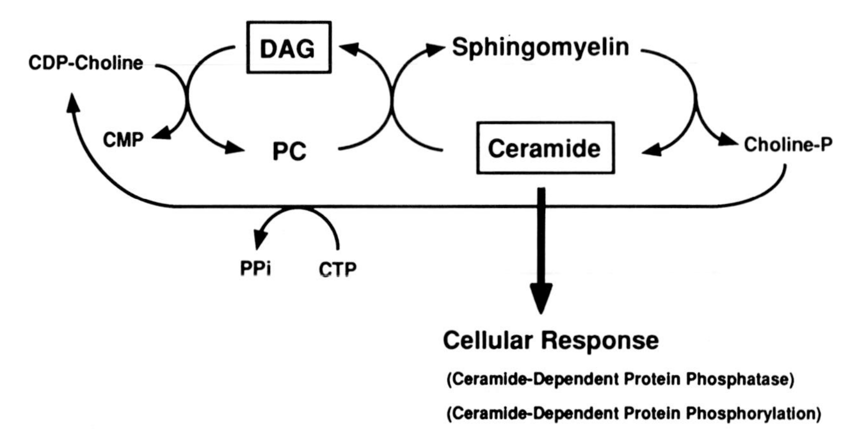

FIG. 1. The sphingomyelin cycle. Schematic representation of sphingomyelin hydrolysis and ceramide generation. Also shown is the possible metabolic interaction between sphingomyelin resynthesis and diacylglycerol (DAG) generation. PC, phosphatidylcholine.

Another major conclusion from these studies was based on the reciprocal changes in sphingomyelin and ceramide levels. Thus, it appears that ceramide is primarily reincorporated to regenerate sphingomyelin levels. Therefore, the "sphingomyelin cycle" serves to generate bioactive ceramide (see below) as well as to terminate the ceramide signal (Fig. 1).

The above features point to important similarities and differences between the sphingomyelin cycle and the well-characterized PI cycle of cell regulation. In both situations, the action of extracellular signals results in activation of a phospholipase of the C type which acts on membrane lipids causing the generation of bioactive products (diacylglycerol and inositol phosphates in one case, and ceramide in the other). Both diacylglycerol and ceramide are also ordinary intermediates in the biosynthesis of the glycerolipids and sphingolipids, respectively. Metabolism of the product and resynthesis of the precursor serve to terminate the signal and reset the pathway for subsequent modulation. The sphingomyelin cycle, however, demonstrates important differences in kinetics when compared to the PI cycle. Hydrolysis of PIP_2 and the generation of diacylglycerol occur within minutes of the action of extracellular agents and the diacylglycerol is usually short lived (a few minutes). On the other hand, sphingomyelin hydrolysis occurred more slowly; a prolonged phase of ceramide generation (approximately 1–2 hours with vitamin D_3) was observed, and the kinetics of regeneration required a few hours. While this undoubtedly reflects different metabolic processes between the two cycles, it may also underlie significant functional differences in that the sphingomyelin cycle may be more related to long-term cellular processes, such as regulation of cell proliferation and differentiation, whereas the PI cycle may be more relevant to short-term cellular activation, such as blood cell activation and hormone release. Further characterization of the sphingomyelin cycle and its role in biology may well provide important insights into the significance of these kinetic findings.

C. Inducers of Sphingomyelin Turnover—Role in Cell Regulation

Insight into the role of sphingomyelin turnover in cell regulation has come from studies employing bacterial sphingomyelinase, and by the determination of multiple important inducers of sphingomyelin hydrolysis, as well as from ongoing studies on the biological activity and function of the product of sphingomyelin hydrolysis, ceramide (see next section).

Initially, attempts were made to determine the biological consequences of sphingomyelin hydrolysis. For these studies, bacterial sphingomyelinase was added to HL-60 cells in concentrations sufficient to cause significant hydrolysis of membrane sphingomyelin without effects on other membrane phospholipids. While this approach did not result in significant biological activity, the addition of bacterial sphingomyelinase to subthreshold concentrations of 1α,25-dihydroxyvitamin D_3 caused remarkable potentiation of cell differentiation as well as inhibition of cell proliferation (32). These results strongly suggested that sphingomyelin hydrolysis provided an important signal contributing to the regulation of cell proliferation and differentiation. Therefore, the effects of vitamin D_3 on sphingomyelin hydrolysis appear to initiate a pathway of cell regulation mimicked by the addition of exogenous bacterial sphingomyelinase.

Depending on the particular inducer of differentiation, HL-60 cells differentiate into monocytes, granulocytes, eosinophils, or basophils. Therefore, further studies were conducted to determine whether sphingomyelin hydrolysis was peculiar to the action of vitamin D_3 in this cell line or whether it operated in response to other inducers. Such studies were of considerable importance since they could potentially define a broad role for sphingomyelin turnover in signal transduction. Indeed, this rationale led to the identification of tumor necrosis factor α (TNF-α) and γ-interferon as potent inducers of sphingomyelin turnover. With TNF-α, sphingomyelin hydrolysis occurred as early as 15 minutes and peaked at around 1 hour (34). On the other hand, neither retinoic acid nor dibutyrylcyclic AMP caused sphingomyelin hydrolysis (34). Since TNF-α, γ-interferon, and vitamin D_3 cause monocytic differentiation whereas retinoic acid and dibutyrylcyclic AMP cause neutrophilic differentiation, it appeared that sphingomyelin hydrolysis in HL-60 cells functions specifically to mediate monocytic rather than neutrophilic differentiation. This specificity was verified by studies with cell-permeable ceramide analogs (see below). Paradoxically, phorbol esters, which cause a macrophage-like differentiation of these cells, resulted in elevation of sphingomyelin levels rather than hydrolysis (34) perhaps as a result of negatively modulating the sphingomyelin cycle. These studies were in agreement with the results of Kiss *et al.* (35) and were confirmed by Dressler *et al.*, who also demonstrated a role for membrane sphingomyelin in cell adhesion (36).

The identification of TNF-α as a potent inducer of sphingomyelin hydrolysis has been of great significance in further studies on the regulation and function of sphingomyelin hydrolysis and ceramide generation. This cytokine, which interacts with two distinct membrane receptors, has pleiotropic activity (37); however, the signaling mechanisms involved in transducing the effects of TNF-α remain poorly understood. Therefore, sphingomyelin hydrolysis and ceramide generation may emerge as important transducers of some of the activities of TNF-α.

III. Role of Ceramide as a Second Messenger/Intracellular Mediator

A. Biological Activities of Ceramide

The ability of vitamin D_3 to cause a sustained and significant elevation in endogenous levels of ceramide suggested that this product of sphingomyelin hydrolysis may partake in a signal transduction pathway involved in mediating the effects of vitamin D_3 on cell growth and differentiation. The other product of sphingomyelin hydrolysis (Fig. 1), phosphorylcholine, was not investigated as a putative second messenger primarily because phosphorylcholine constitutes the major intracellular pool of choline and would be unlikely to play a signaling role, in contrast to inositol trisphosphate, for example, which exists in very low free levels in the cell and thus may be better suited for a signaling function.

To begin to investigate biological activities of ceramide, cell-permeable analogs were synthesized having either short-chain, amide-linked acyl groups or short-chain sphingosines. The most useful such analog was C_2-ceramide (Fig. 2) (*N*-acetylsphingosine) which, at low concentrations, acted in synergy with subthreshold concentrations of vitamin D_3 to cause monocytic differentiation and inhibition of growth of HL-60 cells. More significantly, addition of C_2-ceramide at slightly higher concentrations was able to induce differentiation and inhibit proliferation that was independent of the addition of other agonists (38). Cells treated with 1–10 μM C_2-ceramide displayed parameters indicating monocytic differentiation, including the ability to reduce nitroblue tetrazolium and expression of nonspecific esterase (a specific marker for monocytic differentiation).

Thus, the following criteria strongly suggest a physiological role for ceramide as a signal transducer/second messenger: (1) extracellular hormones and cytokines cause elevations in intracellular levels of ceramide; (2) the kinetics of ceramide generation are fast (relative to other activities of these agents such as

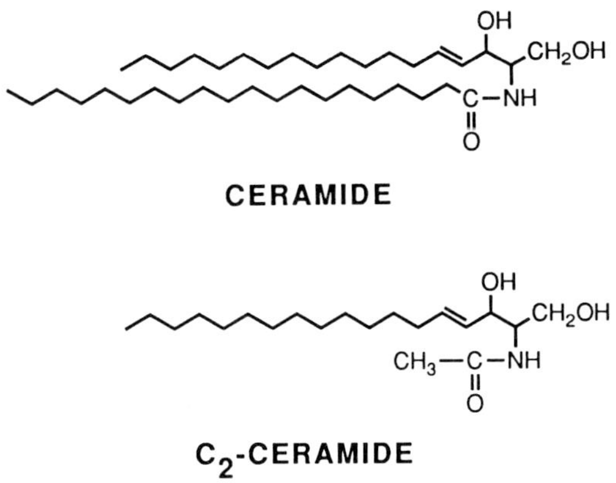

OH
CH_2OH
C—NH
O

CERAMIDE

OH
CH_2OH
CH_3—C—NH
O

C_2-CERAMIDE

FIG. 2. Structures of ceramide and C_2-ceramide.

their effects on gene expression, cell proliferation, and cell differentiation); (3) the increase in levels of ceramide are transient, with ceramide levels returning to baseline by 1–4 hours (depending on the extracellular inducer); (4) the demonstration of biological activity of exogenous ceramide at intracellular levels comparable to the changes in endogenous ceramide levels in response to extracellular agents; and (5) the exogenous ceramides need be present only 2–4 hours to induce their effects. While these results are not sufficient to establish conclusively that ceramide is the major mediator of the effects of vitamin D_3 or TNF-α on cell growth and differentiation, they nonetheless demonstrate the ability of ceramide to modulate cell function and to mediate the action of these extracellular agents. Therefore, according to these criteria, ceramide constitutes a new class of lipid second messengers.

B. Specificity of Action

As with other lipids and amphipathic molecules, it was essential to demonstrate specificity of action of ceramide such that the effects of ceramide would not be attributed to nonspecific toxicity due to membrane injury. Moreover, it was important to define whether the activity of ceramide was due to an action of ceramide itself or to some potential by-product of ceramide metabolism, especially sphingosine.

The potential role of sphingosine as a mediator of ceramide action was pursued along three lines of investigation (38). First, the action of vitamin D_3 failed to cause a measurable change in the levels of sphingosine, suggesting that ceramide was not further metabolized to sphingosine. Second, in studies using radiolabeled C_2-ceramide, it was found that this molecule was rather inert, with minor metabolism into sphingomyelin but no significant production of labeled sphingosine, suggesting that the action of exogenous C_2-ceramide was primarily due to the intact molecule and not a breakdown product. Finally, the ability of exogenous sphingosine to mimic the effects of ceramide was evaluated. Sphingosine was unable to act either alone or in synergy with subthreshold concentrations of vitamin D_3 to cause cell differentiation and inhibition of cell growth. Thus, these studies strongly suggested that ceramide was indeed the active species in the sphingomyelin cycle and that it probably interacts directly with an endogenous target (see below).

More recently, studies with phenyl-aminoalcohol derivatives as analogs of ceramide have pointed to important structural requirements and specificity for ceramide action (39). Using amide derivatives of 1-phenyl-2-amino-1,3-propanediol, it was shown that the activity of these compounds in cell differentiation and growth regulation required an optimal chain length, with the *N*-myristoyl derivative proving to be most effective. These studies also demonstrated the importance of the amide group and the hydrophobic character of the molecule. More significantly, the stereospecificity was established using *N*-myristoyl derivatives of 2-amino-1-phenyl-1-propanol of an enantiomeric pair: only D-*erythro*-*N*-myristoyl-aminophenylpropanol was active and mimicked the effects of C_2-ceramide (which has the D-erythro configuration in its natural form); L-*erythro*-*N*-myristoyl-aminophenylpropanol was largely inactive. Therefore, these studies demonstrate a high structural specificity for the action of ceramide analogs and strongly point to the existence of very specific targets for the action of ceramide (39).

C. Other Pathways of Ceramide Generation

Other studies on ceramide levels and cellular regulation do not fit neatly into the "sphingomyelin cycle" and effector paradigm described. These studies may be the initial descriptions of complex pathways both producing and attenuating cellular ceramide levels.

Investigations were undertaken on the effects of retinoic acid, which binds to a receptor of the thyroid–steroid hormone family of receptors that may modulate sphingolipid metabolism in a manner similar to vitamin D_3. Treatment of GH_4C_1 cells with 10 μM all-*trans* retinoic acid for 8 hours caused a 230% increase in cellular ceramide content (40). This concentration of retinoic acid also inhibited cell proliferation as measured by [^{3}H]thymidine incorporation. Under these conditions, no change in sphingomyelin, sphingosine, or phosphatidylcholine mass was observed. To determine the mechanism of increased ceramide production by 10 μM retinoic acid, cells were labeled with [^{3}H]palmitic acid. After a 2-hour period, a 4-fold increase in the incorporation of palmitate into ceramide was observed. Hydrolysis of the labeled ceramide to sphingosine and free fatty acid demonstrated that only 6% of the label was recovered in the sphingosine backbone of cells treated with retinoic acid, whereas 20% of the label was recovered in the sphingosine backbone of cells treated with vehicle alone. The data are consistent with retinoic acid causing an increase in cellular ceramide levels in GH_4C_1 cells through an increase in sphingosine *N*-acylation.

The ceramide levels of primary human T cells were found to be highest in nonproliferating cells (23 pmol/nmol phospholipid phosphate). When resting cells were treated with interleukin 2 (IL-2) (100 units/ml) or with phorbol myristate acetate (PMA) and ionomycin, cellular ceramide levels decreased 2.5 fold to 11 pmol/nmol phospholipid phosphate. These changes were not accompanied by changes in cellular sphingomyelin, phosphatidylcholine, or sphingosine levels. A correlation between ceramide levels and $[^3H]$thymidine incorporation was observed. However, addition of cell-permeable ceramide (C_2 and C_6-sphingosine) had no effect on $[^3H]$thymidine incorporation, whereas addition of D-erythrosphingosine (10 μM) decreased $[^3H]$thymidine incorporation by 40% (41). While these results do not show a direct correlation between T cell ceramide levels and growth, they raise the possibility of mitogen-dependent depletion of ceramide to some threshold level required for cell cycle progression. Growth arrest is also seen in response to D-*threo*-1-phenyl-2-(decanoylamino)-3-morpholino-1-propanol (PDMP), which increases T cell ceramide levels and reduces cellular glucocerebroside levels (42). Thus, a complex system regulating cellular ceramide levels may exist and function to modulate cellular growth.

IV. Mechanism of Action of Ceramide

The biological activity of sphingosine and specificity of action suggested that ceramide modulates an intracellular pathway of cell growth regulation. Therefore, it became very important to determine the mechanism of action of ceramide. Important insight into biochemical targets for action of ceramide came from examination of the interaction of ceramide with known critical pathways involved in regulation of cell growth and differentiation. This was primarily pursued at two levels: the effects of ceramide on the protooncogene c-*myc* and the effects of ceramide on protein phosphorylation.

A. Regulation of c-*myc* Levels

In HL-60 cells, a role for the c-*myc* protooncogene in regulation of cell proliferation has been proposed based on multiple studies correlating the level of c-*myc* expression and cell growth/differentiation. Thus, one of the earliest effects of TNF-α or vitamin D_3 in HL-60 cells is the downregulation of c-*myc* expression through a unique mechanism involving a block to elongation of transcription at the junction of the first exon and first intron. Since vitamin D_3 and TNF-α appear to induce an elevation in ceramide levels prior to their effects on c-*myc* expression, we evaluated whether ceramide is capable of mediating the effects of these agents on c-*myc* regulation. In initial studies, it was found that ceramide caused downregulation of c-*myc* protooncogene as early as 30 minutes after addition of ceramide, with the response peaking at 1–2 hours (34). Ongoing studies have demonstrated the specificity of action of ceramide in c-*myc* regulation as well as a similar mechanism of action for ceramide and TNF-α in inducing a block to elongation of transcription (R. Wolff and Y. A. Hannun, unpublished data). These studies carry two implications. First, the time course for effects of ceramide on c-*myc* suggests a temporal scheme whereby TNF-α (or vitamin D_3) induces an elevation in ceramide levels with subsequent downregulation of c-*myc*. Second, these studies strongly corroborate an important role for ceramide in modulating pathways of cell growth regulation.

B. Activation of Serine/Threonine Protein Phosphatase and Effects on Cellular Protein Phosphorylation

In another approach to determine the mechanism of action of ceramide, it was reasoned that protein phosphorylation as a fundamental pathway of posttranslational modification with important effects on cell growth regulation may be modulated by ceramide. Therefore, the effects of ceramide on protein phosphorylation were examined (Fig. 1). In initial studies, it was observed that ceramide may actually attenuate phosphorylation of a number of cellular proteins (43). Therefore, the effects of ceramide on protein phosphatases were examined (it should be noted that ceramide does not modulate the activity of protein kinase C). These studies led to the discovery of a *c*eramide-*a*ctivated *p*rotein *p*hosphatase (CAPP) that appears to be specifically activated by ceramide but not related sphingolipids such as sphingosine (44). Moreover, this ceramide-activated protein phosphatase appeared to have many features similar to those of one subgroup of serine/threonine protein phosphatases, PP2A, which are cation-independent and strongly inhibited by okadaic acid. Ongoing studies aim at characterizing and purifying this enzyme (see the article by Dobrowsky and Hannun in this volume).

In other studies, Kolesnick and co-workers showed that ceramide and sphingosine caused increased phosphorylation of the receptor for epidermal growth factor in membrane preparations of A431 cells and that the effects of sphingosine may be mediated by ceramide (45). Thus, one plausible scenario would be activation of CAPP by ceramide resulting in modulation of activity of other kinases and phosphatases with multiple effects on protein phosphorylation. These could constitute the earliest effects of ceramide on cell function with subsequent effects on nuclear events controlling cell proliferation and differentiation.

V. Major Questions and Future Directions

A. Physiological Roles of Ceramide

While at this point we can construct a tentative scheme for the sphingomyelin cycle, there remains a major question as to the physiological function of this cycle and primarily the activities mediated by ceramide as a second messenger/intracellular mediator. At this point, we may conclude that ceramide has important and specific effects on regulation of cell growth and differentiation. Ceramide is active in many leukemia cell lines and solid tumors (46). Ongoing studies aim at determining the effects of ceramide in nontransformed cell lines and tissues. In initial studies, ceramide did not effect short-term responses in human platelets, neutrophils, or macrophages. These preliminary results appear to indicate more long-term activities of ceramide and sphingomyelin hydrolysis rather than regulation of short-term responses. However, it is anticipated that examination of the effects of ceramide on multiple biological activities will be forthcoming. Such studies are instrumental in defining the spectrum of biological activity of ceramide as a prerequisite for defining the physiological function of the sphingomyelin cycle of cell regulation.

B. Role of CAPP

The identification of CAPP as an *in vitro* target for the action of ceramide provides an initial starting point for dissecting mechanisms of action of ceramide *in vivo*. The potent inhibition of CAPP by okadaic acid has provided an important tool for dissecting the role of this enzyme in mediating ceramide biology. In ongoing studies, it was shown that the ability of ceramide and TNF-α to downmodulate c-*myc* protooncogene levels in HL-60 cells is inhibited by the addition of low concentrations of okadaic acid, strongly pointing to a role for CAPP in mediating the effects of both ceramide and TNF-α (47). The purification and characterization of CAPP should allow insight as to its relation to other serine/threonine protein phosphatases and to the mechanism by which ceramide activates this phosphatase. In addition, a ceramide-activated protein kinase has been described and the potential role of this activity as a downstream target for ceramide action has been suggested.

C. Signaling through Sphingolipids

The identification of an intracellular signaling (cell regulatory) pathway involving sphingomyelin hydrolysis and the generation of ceramide may indicate a more generalized role for sphingolipids as precursors to second messengers/bioactive molecules with important roles in cell regulation. Thus, it is conceivable that other signaling pathways involving sphingolipids exist whereby sphingolipids such as ceramide, cerebroside, and other cellular glycolipids turn over, resulting in the generation of active molecules such as sphingosine, psychosine, lysosphingolipids, phosphosphingolipids, and others. The complexity of such signaling may rival that of glycerol phospholipids. Signaling through sphingolipids may explain, in part, the mechanism by which exogenous application of sphingolipids results in specific biological activities. This promises to be an exciting and productive area of investigation.

Acknowledgments

We thank Dr. Roy Borchardt for providing Fig. 1. We also thank Marsha Haigood for expert secretarial assistance. This work was supported in part by National Institutes of Health Grants GM-43825 and DK20205.

References

1. Hokin, M. R., and Hokin, L. E. (1953). *J. Biol. Chem.* **203,** 967–977.
2. Majerus, P. W., Connolly, T. M., Deckmyn, H., Ross, T. S., Bross, T. E., Ishii, H., Bansal, V., and Wilson, D. (1986). *Science* **234,** 1519–1526.
3. Berridge, M. J., and Irvine, R. F. (1989). *Nature (London)* **341,** 197–205.
4. Rhee, S. G., Suh, P. G., Ryu, S. H., and Lee, S. Y. (1989). *Science* **244,** 546–550.
5. Nishizuka, Y. (1989). *Cancer (Philadelphia)* **63,** 1892–1903.
6. Nishizuka, Y. (1988). *Nature (London)* **334,** 661–665.
7. Bell, R. M. (1986). *Cell (Cambridge, Mass.)* **45,** 631–632.
8. Kishimoto, A., Takai, Y., Mori, T., Kikkawa, U., and Nishizuka, Y. (1980). *J. Biol. Chem.* **255,** 2273–2276.
9. Castagna, M., Takai, Y., Kaibuchi, K., Sano, K., Kikkawa, U., and Nishizuka, Y. (1982). *J. Biol. Chem.* **257,** 7847–7851.
10. Niedel, J. E., Kuhn, L. J., and Vandenbark, G. R. (1983). *Proc. Natl. Acad. Sci. U.S..A.* **80,** 36–40.
11. Exton, J. H. (1990). *J. Biol. Chem.* **265,** 1–4.
12. Löffelholz, K. (1989). *Biochem. Pharmacol.* **38,** 1543–1549.
13. Billah, M. M., Pai, J.-K., Mullmann, T. J., Egan, R. W., and Siegel, M. I. (1989). *J. Biol. Chem.* **264,** 9069–9076.
14. Lapetina, E. G., and Crouch, M. F. (1989). *Ann. N.Y. Acad. Sci.* **559,** 153–157.
15. Dennis, E. A., Rhee, S. G., Billah, M. M., and Hannun, Y. A. (1991). *FASEB J.* **5,** 2068–2077.
16. Lin, L.-L., Lin, A. Y., and Knopf, J. L. (1992). *Proc. Natl. Acad. Sci. U.S.A.* **89,** 6147–6151.
17. Hannun, Y. A., Loomis, C. R., Merrill, A. H., Jr., and Bell, R. M. (1986). *J. Biol. Chem.* **261,** 12604–12609.
18. Hannun, Y. A., and Bell, R. M. (1987). *Science* **235,** 670–674.
19. Hannun, Y. A., and Bell, R. M. (1989). *Science* **243,** 500–507.
20. Merrill, A. H., Jr., and Stevens, V. L. (1989). *Biochim. Biophys. Acta* **1010,** 131–139.
21. Lavie, Y., and Liscovitch, M. (1990). *J. Biol. Chem.* **265,** 3868–3872.
22. Kiss, Z., and Anderson, W. B. (1990). *J. Biol. Chem.* **265,** 7345–7350.
23. Lavie, Y., Piterman, O., and Liscovitch, M. (1990). *FEBS Lett.* **277,** 7–10.

24. Mullmann, T. J., Siegel, M. I., Egan, R. W., and Billah, M. M. (1991). *J. Biol. Chem.* **266,** 2013–2016.
25. McDonald, O. B., Hannun, Y. A., Reynolds, C. H., and Sahyoun, N. (1991). *J. Biol. Chem.* **266** (in press).
26. Hannun, Y. A., Greenberg, C. S., and Bell, R. M. (1987). *J. Biol. Chem.* **262,** 13620–13626.
27. Wilson, E., Olcott, M. C., Bell, R. M., Merrill, A. H., Jr., and Lambeth, J. D. (1986). *J. Biol. Chem.* **261,** 12616–12623.
28. Merrill, A. H., Jr., Serenit, A., Stevens, V. L., Hannun, Y. A., Bell, R. M., and Kinkade, J. M., Jr. (1986). *J. Biol. Chem.* **261,** 12610–12615.
29. Zhang, H., Buckley, N. E., Gibson, K., and Spiegel, S. (1990). *J. Biol. Chem.* **265,** 76–81.
30. Hakomori, S. (1981). *Annu. Rev. Biochem.* **50,** 733–764.
31. Wiegandt, H. (1985). *In* "Glycolipids" (H. Wiegandt, ed.), pp. 199–259. Elsevier, New York.
32. Okazaki, T., Bell, R. M., and Hannun, Y. A. (1989). *J. Biol. Chem.* **264,** 19076–19080.
33. Sperker, E. R., and Spence, M. W. (1983). *J. Neurochem.* **40,** 1182–1184.
34. Kim, M.-Y., Linardič, C., Obeid, L., and Hannun, Y. A. (1991). *J. Biol. Chem.* **266,** 484–489.
35. Kiss, Z., Deli, E., and Kou, J. F. (1988). *Arch. Biochem. Biophys.* **265,** 38–42.
36. Dressler, K. A., Kan, C.-C., and Kolesnick, R. N. (1991). *J. Biol. Chem.* **266,** 11522–11527.
37. Loetscher, H. R., Brockhaus, M., Dembič, Z., Gentz, R., Gubler, U., Hohmann, H.-P., Lahm, H.-W., Van Loon, A. P. G. M., Pan, Y.-C. E., Schlaeger, E.-J., Steinmetz, M., Tabuchi, H., and Lesslauer, W. (1991). *Oxford Surv. Eukaryotic Genes* **7,** 119–142.
38. Okazaki, T., Bielawska, A., Bell, R. M., and Hannun, Y. A. (1990). *J. Biol. Chem.* **265,** 15823–15831.
39. Bielawska, A., Linardič, C. M., and Hannun, Y. A. (1992). *J. Biol. Chem.* **267,** 18493–18497.
40. Kalén, A., Borchardt, R. A., and Bell, R. M. (1992). *Biochim. Biophys. Acta, Lipids Lipid Metab.* **1125,** 90–96.
41. Borchardt, R., Kalén, A., Lee, T., and Bell, R. (1993). In preparation.
42. Felding-Habermann, B., Igarashi, Y., Fenderson, B. A., Park, L. S., Radin, N. S., Inokuchi, J., Strassman, G., Handa, K., and Hakomori, S. (1990). *Biochemistry* **29,** 6314–6322.
43. Dobrowsky, R. T., and Hannun, Y. A. (1993). In preparation.
44. Dobrowsky, R. T., and Hannun, Y. A. (1992). *J. Biol. Chem.* **267,** 5048–5051.
45. Goldkorn, T., Dressler, K. A., Muindi, J., Radin, N. S., Mendelsohn, J., Menaldino, D., Liotta, D., and Kolesnick, R. N. (1991). *J. Biol. Chem.* **266,** 16092–16097.
46. Werner, M., and Hannun, Y. A. (1993). In preparation.
47. Wolff, R. A., Obeid, L. M., and Hannun, Y. A. (1993). Submitted for publication.

ADVANCES IN LIPID RESEARCH, VOL. 25

The Novel Second Messenger Ceramide: Identification, Mechanism of Action, and Cellular Activity

YUSUF A. HANNUN,* LINA M. OBEID,[†] AND ROBERT A. WOLFF[†]

*Departments of Medicine and Cell Biology
[†]Department of Medicine
Duke University Medical Center
Durham, North Carolina 27710

I. Introduction

A. Sphingolipids in Signal Transduction: The Sphingomyelin Cycle

The discovery of the sphingomyelin cycle in HL-60 human myelocytic leukemia cells (reviewed by Hannun and Bell in this volume) has provided the clearest evidence yet of a role for membrane sphingolipids in signal transduction (1,2). The sphingomyelin cycle fulfulls the basic criteria of a signaling pathway. The action of extracellular agents and hormones [tumor necrosis factor α (TNF-α), γ-interferon, and 1α,25-dihydroxyvitamin D_3] results in early activation of a hormone-responsive enzyme (neutral sphingomyelinase) which acts on a precursor metabolite (sphingomyelin) resulting in the formation of a putative second messenger (ceramide) (1,3–5). In these aspects, the sphingomyelin cycle resembles the inositol phospholipid (PI) cycle of cell regulation and may provide a prototype for other sphingolipid signaling pathways.

B. CERAMIDE: A NOVEL SECOND MESSENGER

Three critical findings have begun to define a second messenger function for ceramide. First, the levels of cellular ceramide are acutely responsive to the action of certain extracellular agents. For example, the action of TNF-α results in elevation of ceramide levels within 5–10 minutes in U937 mononuclear cells (6) and within 15–60 minutes in HL-60 cells (4,7). Second, exogenous ceramide is able to mediate specifically some of the biological effects of those extracellular agents that induce formation of endogenous ceramide (3,4). Third, the discovery of biochemical targets for ceramide [notably, ceramide-activated protein phosphatase (8)] and the elucidation of cellular activities of ceramide provide strong evidence for the existence of a distinct ceramide pathway of signal transduction. In this article, ongoing studies which aim at elucidating the biological and biochemical activity of ceramide are presented, and evidence for ceramide as a second messenger is discussed.

II. Ceramide-Mediated Biology

A. INSIGHT FROM BIOLOGY OF 1α,25-DIHYDROXYVITAMIN D_3 AND TUMOR NECROSIS FACTOR α

The initial discovery of sphingomyelin turnover in response to 1α,25-dihydroxyvitamin D_3 (1), TNF-α, and γ-interferon (4) suggested that sphingomyelin hydrolysis may contribute to the mechanism of action of these extracellular agents on HL-60 cells. Therefore, it became important to determine whether the generated ceramide exerts significant biological and/or biochemical activities. Such effects would provide evidence for a functional role for the endogenously generated ceramide.

Two problems became evident at that point. First, natural ceramides are very hydrophobic molecules with very low aqueous solubility. Thus, their use is severely limited in tissue culture studies for biological evaluation. This problem was overcome by the design and synthesis of more soluble (cell-permeable) analogs of ceramide, including molecules with either a shorter sphingosine backbone or a shorter acyl group. The most useful of these molecules proved to be C_2-ceramide (*N*-acetylsphingosine) (Fig. 1) which appears to have a solubility of approximately 10 μ*M* in tissue culture media. Second, a useful approach was required that could result in identification of relevant biochemical and/or biological effects of ceramide. The guidelines for this approach were provided by assuming that, for ceramide to function as a second messenger/intracellular mediator, it should be able to mediate at least some of the effects of either vitamin D_3 or TNF-α on HL-60 cells and other cell types. The most obvious of these effects are the induction of differentiation and suppression of proliferation of

HL-60 cells. In particular, TNF-α proved to be extremely useful in this capacity since a large body of investigation has determined multiple biological and biochemical activities of TNF-α in various cell lines. TNF-α exerts cytocidal activities in many cell types, has antiproliferative effects in other cell types, and induces proliferation of still other cell types (9–11). These biological and biochemical effects are summarized in Table I (12–36).

The early effects of TNF-α on signal transduction events as well as on nuclear regulation have also provided the foundation for exploring the mechanism of action of ceramide. In various cell lines, TNF-α has been shown to cause early activation of protein kinase C (29,30), activation of phospholipase A_2 and the release of arachidonic acid (21,22), activation of the transcription factor NF-κB (35,37), and downregulation of the c-*myc* protooncogene (16) (Table I).

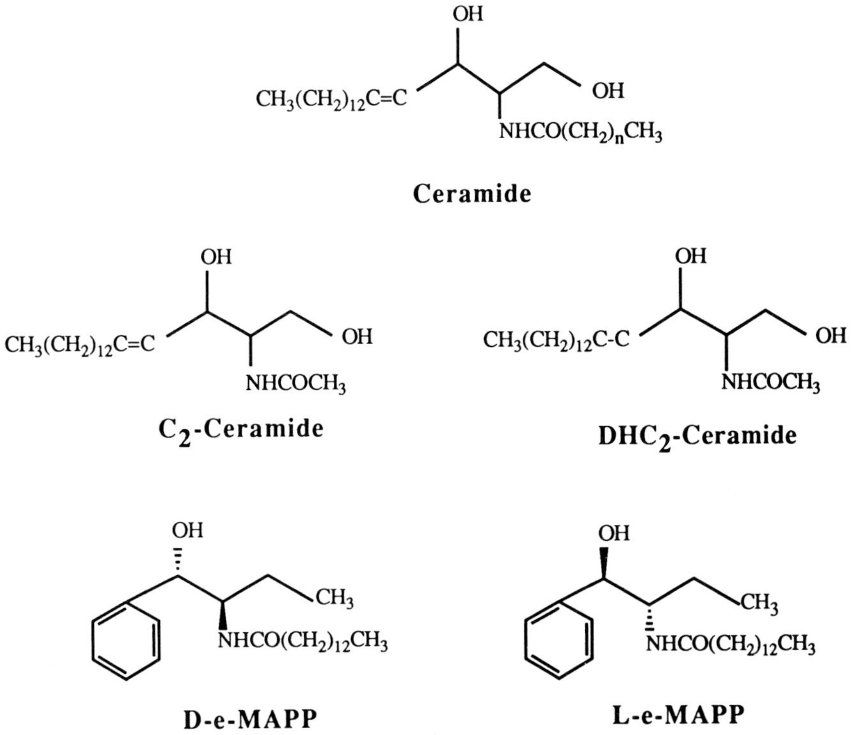

FIG. 1. Structures of ceramide and related molecules. Shown are the structures for natural ceramide, where the fatty acyl group in amide linkage may have a variable length of between 16 and 24 carbons (n = 14–22); C_2-ceramide, with an acetate in amide linkage; dihydro-C_2-ceramide (DHC_2-ceramide), which lacks the 4–5 *trans* double bond; D-*erythro-N*-myristoyl-amino-phenyl-1-propanol (D-*e*-MAPP); and L-*erythro-N*-myristoyl-amino-phenyl-1-propanol (L-*e*-MAPP).

Table I

BIOCHEMICAL AND BIOLOGICAL ACTIVITIES OF TUMOR NECROSIS FACTOR α[a]

Biochemical	Biological
HL-60 Cells	
↑ NBT and NSE positivity (12)	Monocytic differentiation (12)
↑ Tyrosine kinase activity (13)	Cytostatic and cytotoxic activity (14)
↑ GTPase activity (15)	↑ Phagocytic activity (14)
↓ c-*myc* gene expression (16–18)	
↓ c-*myb* gene expression (17)	
↑ HLA-A, -B, -C gene expression (18)	
Sphingomyelin hydrolysis (4)	
Other Cells	
PGE_2 production (19,20)	Mediator of immunity and inflammation (11)
PLA_2 activation (20–23)	Cytostatic and cytotoxic activity and apoptosis (26–28)
↑ GTPase activity (24,25)	
PKC activation and translocation (29,30)	Growth stimulatory activity (31)
Modulation of protein phosphorylation (32,33)	
↑ Collagenase gene expression (29)	
↑ *jun* gene expression (29)	
↑ *fos* gene expression (34)	
↑ c-*myc* gene expression (34)	
↑ NF-κB (35)	
Suppression of lipoprotein lipase (36)	

[a]Examples of cellular activities of TNF-α in HL-60 and other cell types are listed. More extensive discussion of these and other activities of TNF-α may be found in Refs. 9–11.

These have proved extremely useful in determining which actions of TNF-α are mediated by ceramide and have been instrumental in defining early biochemical targets for the action of ceramide.

B. GROWTH INHIBITION AND DIFFERENTIATION IN RESPONSE TO CERAMIDE

1. HL-60 Cells

The earliest studies on the effects of ceramide on HL-60 cells revealed that C_2-ceramide was capable of mimicking 1α,25-dihydroxyvitamin D_3 in inhibition of growth and induction of cell differentiation (3). The addition of C_2-ceramide at concentrations of between 1 and 10 μ*M* resulted in a dose-dependent inhibition of cell growth (Fig. 2A) such that a concentration of 3 μ*M* was cytostatic. Higher concentrations proved to be toxic.

HL-60 cells are pluripotent cells capable of differentiating along unique pathways depending on the nature of the extracellular agent. For example, 1α,25-dihydroxyvitamin D_3, TNF-α, and γ-interferon induce monocytic differentiation, whereas retinoic acid and dibutyryl cyclic AMP induce granulocytic differentiation, and phorbol esters induce a macrophage-like phenotype with significant adherence properties (12). C_2-Ceramide (1–5 μM) induced differentiation of HL-60 cells as assayed by the ability of differentiated cells to reduce nitro blue tetrazolium.

A specific role for sphingomyelin turnover and ceramide in monocytic differentiation of HL-60 cells became more apparent with further studies. Only inducers of monocytic differentiation of HL-60 cells (1α,25-dihydroxyvitamin D_3, TNF-α, and γ-interferon) were able to induce sphingomyelin turnover and cause elevation of endogenous ceramide levels. On the other hand, retinoic acid, dibutyryl cyclic AMP, dimethyl sulfoxide (DMSO), and phorbol esters were unable to induce sphingomyelin hydrolysis (4). Ceramide, in turn, mimicked the effects of D_3, TNF-α, and γ-interferon in inducing monocytic differentiation of HL-60 cells as determined by morphological and histochemical criteria. In addition, ceramide induced the expression of CD11b and CD14 in HL-60 cells as additional markers of monocytic differentiation (data not shown).

Ceramide behaves as a prototypic lipid second messenger in that its biological activities are modulated by surface dilution kinetics as well as by the presence of lipid-binding proteins in culture media (38). For example, in HL-60 cells, the effectiveness of ceramide on growth inhibition (measured as the concentration of ceramide required to produce 50% inhibition, IC_{50}) is directly proportional to cell density (Fig. 2B). However, if the concentration of ceramide is expressed relative to cell density (e.g., femtomoles of C_2-ceramide/10^5 cells), the effective concentrations of C_2-ceramide show little variation with cell density (Fig. 2C). Therefore, the cellular activities of ceramide appear to be primarily determined by the cellular rather than the molar concentrations of ceramide. Moreover, the presence of lipid-binding serum proteins in culture media results in a significant increase in the IC_{50} of ceramide. Consideration of these two important parameters (i.e., cell density and concentration of serum proteins) is critical in determining the effective concentration of ceramide.

2. *Other Cell Types*

Other ongoing studies have demonstrated the ability of ceramide to modulate cell proliferation in a number of cell systems. Thus, ceramide also inhibits the growth of U937 monocytic leukemia cells, human T cells, B cells, and human glioma cells. In all these cell lines, ceramide exerts potent antiproliferative effects similar to the results with HL-60 cells. In human T9 glioma cells, ceramide also induces differentiation into more mature astrocytic cells with neurites and cellular processes similar to those induced by nerve growth factor (39). Therefore, ceramide appears to be a more general regulator of cell growth and differentiation and it may play important roles in mediating the antiproliferative effects of various cytokines.

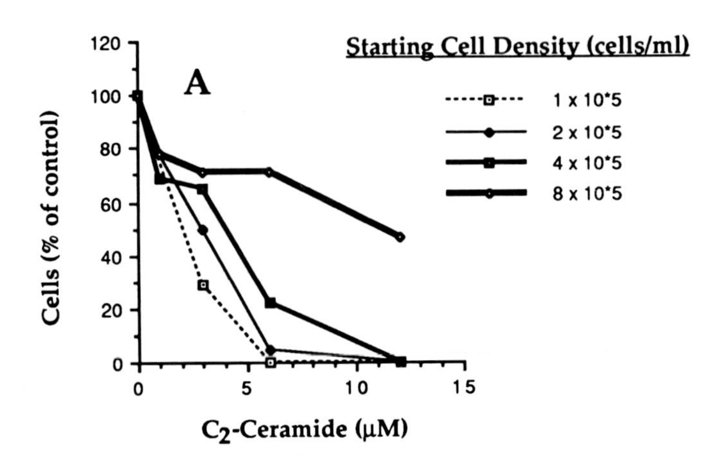
A
Starting Cell Density (cells/ml)
1 x 10*5
2 x 10*5
4 x 10*5
8 x 10*5
Cells (% of control)
C2-Ceramide (µM)
120
100
80
60
40
20
0
0
5
10
15

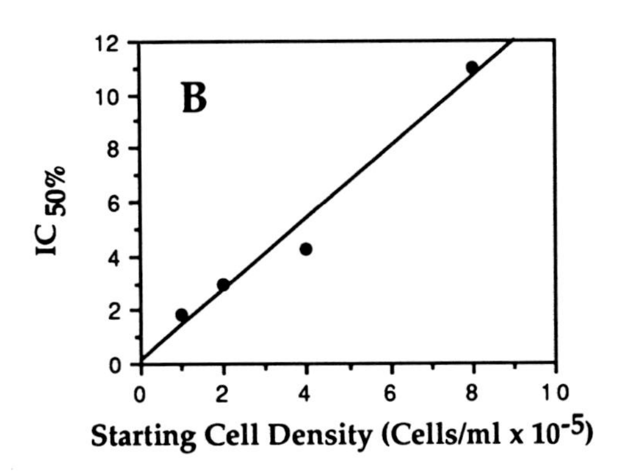
B
IC 50%
Starting Cell Density (Cells/ml x 10-5)
12
10
8
6
4
2
0
0
2
4
6
8
10

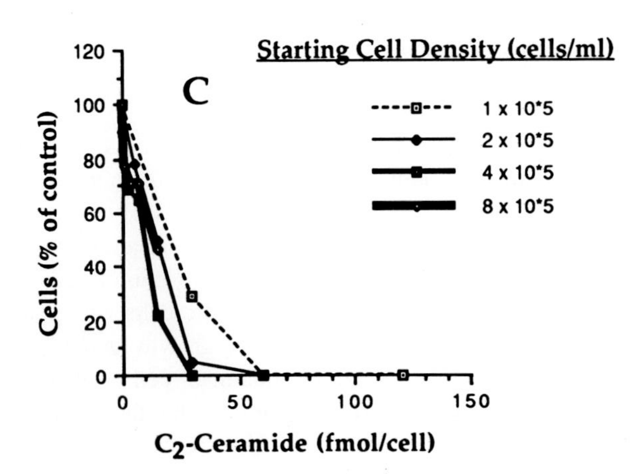
C
Starting Cell Density (cells/ml)
1 x 10*5
2 x 10*5
4 x 10*5
8 x 10*5
Cells (% of control)
C2-Ceramide (fmol/cell)
120
100
80
60
40
20
0
0
50
100
150

Recently, the biological effects of ceramide have been extended to yeast cells (40). In *Saccharomyces cerevisiae*, ceramide was found to induce potent inhibition of cell growth. This effect was specific to ceramide in that a number of other lipids and detergents (such as arachidonate, oleate, dioctanoylglycerol, and Triton X-100) were without effect. Maximal effects of ceramide occurred at a concentration of approximately 5–10 μ*M*. Similar to mammalian cells, the effects of ceramide were modulated by surface dilution as well as by the presence of lipid-binding proteins in culture media. In addition, yeast cells contained a ceramide-activated protein phosphatase (40) that is similar to one found in mammalian cells, which showed similar specificity in activation by ceramide (see Section IV). These studies document the conservation of a ceramide-activated pathway of growth regulation and suggest that ceramide may be a critical regulator of cell growth.

C. Apoptosis

In all cases examined, the antiproliferative effects of ceramide have been accompanied by cytotoxic effects that occur at higher concentrations of exogenous C_2-ceramide. While cytotoxicity may represent nonspecific lipid effects on membrane integrity, two observations suggested that the cytocidal/cytotoxic effects of ceramide may represent an integral part of ceramide-mediated biology. First, as with growth inhibition, the cytotoxic effects of ceramide were specific to ceramide and active analogs, such that inactive analogs were unable to inhibit growth or induce cytotoxicity (see Section III). For example, the L-*e*-MAPP inactive analog of ceramide (see Section III) did not demonstrate any significant cytotoxicity even at concentrations 10-fold higher than the concentrations of ceramide required to cause significant cytotoxicity. Second, the action of TNF-α on many cell types is closely associated with a significant cytotoxic/cytocidal activity. This latter observation suggested that ceramide may indeed mediate the effects of TNF-α on cell death. These considerations led us to speculate that ceramide may activate a specific program of cell death rather than nonspecific cell necrosis.

Programmed cell death or apoptosis has been defined as one or more discrete cellular mechanisms by which cells and tissues react to developmental and environmental signals to undergo cell death (41,42). A number of biochemical and morphological features distinguish apoptosis from necrosis (Table II). Notably, necrosis occurs in response to physical injury to cells, resulting in acute cytolysis

---

FIG. 2. Inhibition of cell growth by C_2-ceramide. (A) Dose response for growth inhibition of HL-60 cells treated with the indicated concentrations of C_2-ceramide at different starting cell densities. (B) Modulation of the effectiveness of ceramide by starting cell density. $IC_{50\%}$ is the concentration of ceramide that produces 50% inhibition of cell growth. (C) Data in A replotted as inhibition of cell growth in response to ceramide concentration measured as fmol/cell.

Table II
FEATURES DISTINGUISHING APOPTOSIS FROM NECROSIS

	Apoptosis	Necrosis
Inducers	Glucocorticoids, TNF-α, hypoxia, calcium ionophores, TGF-β	Physical injury, freeze-thawing, severe hyperthermia, acute hypoxia
Morphology	Apoptotic bodies, breakdown of nuclei and condensation of chromatin	Early swelling of cells and cytoplasmic organelles, lysis of nuclei, cellular disintegration
Membrane permeability	Unchanged	Increased
Lysosomal enzymes	Not released	Released
Tissue reaction	Phagocytosis by other cells	Inflammation and scar formation
DNA fragmentation (on agarose gel electrophoresis)	Characteristic ladder	Diffuse smearing

and release of intracellular contents. On the other hand, apoptosis occurs in response to discrete extracellular stimuli such as TNF-α (26,27), hypoxia (42,43), and glucocorticoids (44). The action of these agents results in activation of a calcium-dependent nuclease. This enzyme cleaves nuclear DNA at internucleosomal sites, resulting in a characteristic DNA ladder on gel electrophoresis, with DNA fragments evenly spaced by approximately 180 nucleotides (41). Morphologically, this is accompanied by nuclear fragmentation and the formation of membrane buds which eventually result in the formation of apoptotic bodies consisting of sealed membranes containing fragmented cytoplasm and nuclei. Eventually, these changes result in total cell destruction.

Although a number of "physiological" inducers of apoptosis such as TNF-α and glucocorticoids have been identified, little is known of the intracellular mediators of apoptosis (41). Because ceramide was identified as a mediator of the effects of TNF-α on growth inhibition, a role for ceramide in mediating apoptosis was considered. Indeed, addition of exogenous C_2-ceramide to human lymphoid and mononuclear cells resulted in early DNA fragmentation, with the characteristic DNA ladder (Fig. 3), considered a hallmark of apoptosis (41,44). Notably, the effects of ceramide were early (within 1–3 hours), potent (requiring 1–3 μM C_2-ceramide), and very specific such that other lipids, amphiphiles, and related sphingolipids were inactive (6). Remarkably, the closely related C_2-dihydroceramide was totally inactive (see Section III). The ability of ceramide to induce DNA fragmentation was opposed by the action of phorbol esters. This suggests that the diacylglycerol/protein kinase C pathway may act to prevent

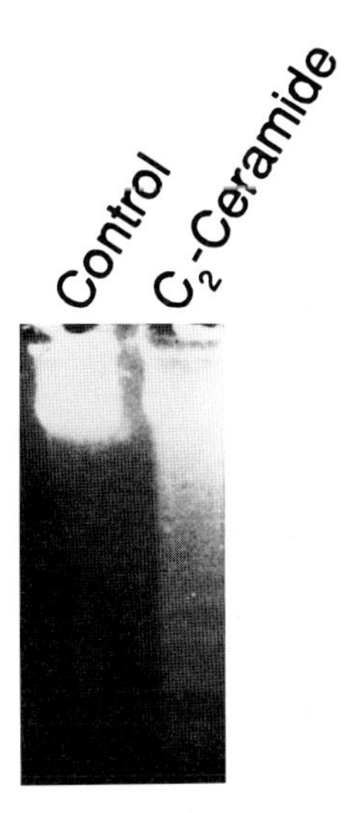

FIG. 3. DNA fragmentation in response to C_2-ceramide. U937 leukemia cells were treated with either ethanol or 3 μM C_2-ceramide, and DNA fragmentation was evaluated on agarose gel electrophoresis as described (6).

apoptosis while the ceramide/protein phosphatase pathway (see Section IV) may act in an opposite direction to induce apoptosis.

The putative role of ceramide in inducing apoptosis adds significant dimensions to the spectrum of biological function of ceramide. Apoptosis has been postulated to play important roles in development, differentiation, oncogenesis, and reaction of tissues to injury (41,45,46): (1) Apoptosis is the recognized mechanism by which the thymus undergoes involution. (2) Apoptosis is integral in various aspects of development and tissue regression and is also responsible for atrophy of hormone-dependent organs such as the prostate and adrenal glands upon hormone withdrawal. (3) Apoptosis may represent the mechanism by which mature white blood cells are programmed for self-death. (4) The ability of organs to adapt functionally to the demands of the organism appears to involve apoptosis whereby unnecessary differentiated cells undergo apoptosis, thus allowing the growth of other cells with different functions [for discussion and examples, see Michaelson (46)]. Thus, apoptosis appears to be a fundamental process of cell regulation operating at the level of individual cells, tissues, and whole organs.

Dysregulation of apoptotic pathways has been proposed as a novel mechanism for oncogenesis (47). This is illustrated with the *Bcl-2* protooncogene. *Bcl-2* appears to function normally in preventing apoptotic cell death; however, increased and unchecked activity of oncogenic *Bcl-2* may lead to unrestrained cell growth in some malignancies.

It is conceivable that ceramide may act as an intracellular sensor that regulates cell proliferation/death. At modest accumulations of intracellular ceramide, cells could be driven toward a cytostatic state with inhibition of cell growth. At higher intracellular concentrations of ceramide, cells initiate programmed cell death.

D. Downregulation of c-*myc*

The emerging evidence of the ability of ceramide to mimic at least part of the biological effects of TNF-α (especially on growth inhibition, differentiation, and apoptosis) strengthens the argument that ceramide may also mediate the effects of TNF-α on intracellular biochemical targets. One such important target is the c-*myc* protooncogene. In HL-60 cells, TNF-α causes early downregulation of c-*myc* mRNA at a transcriptional level. In this distinct mechanism (16), initiation of transcription is not perturbed by TNF-α, but elongation of the message is interrupted at the exon 1/intron 1 junction (Fig. 4).

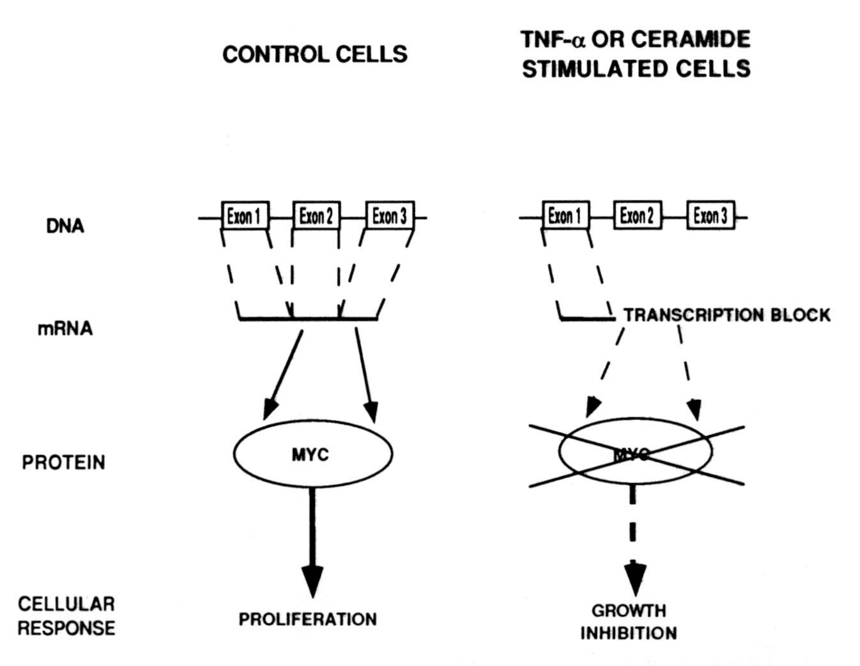

FIG. 4. Regulation of c-*myc* transcription by TNF-α and ceramide. In HL-60 cells, both TNF-α and ceramide induce a block in transcription elongation of c-*myc*, resulting in aborted formation of mRNA and functional protein.

The ability of ceramide to mediate the effects of TNF-α on c-*myc* regulation was investigated. The addition of C_2-ceramide to HL-60 cells resulted in early downregulation of c-*myc* mRNA within 30–120 minutes (4) at concentrations similar to those required for inhibition of cell growth. Importantly, C_2-ceramide exerted its effects on c-*myc* through a mechanism identical to that of TNF-α by inducing a block to transcription elongation with little effect on initiation of transcription (48). Thus, a role for ceramide in mediating the effects of TNF-α on c-*myc* transcription is supported by (1) the early effects of ceramide on c-*myc;* (2) the effective concentrations of exogenous ceramide, which approximate endogenous levels of ceramide in response to TNF-α; and (3) the similar mechanism by which TNF-α and ceramide modulate c-*myc*.

The effects of ceramide on c-*myc* also establish a strong link between the growth inhibitory effects of ceramide and downregulation of c-*myc* mRNA levels. The concentrations of C_2-ceramide required to inhibit growth were in close agreement with the concentrations required to downmodulate c-*myc* mRNA levels. Moreover, these effects demonstrated similar specificities (see Section III).

E. Regulation of NF-κB

While the above studies demonstrated the ability of ceramide to mimic many of the effects of TNF-α, other studies with the nuclear protein NF-κB suggested that ceramide was not sufficient to mediate all activities of TNF-α. NF-κB was initially described as a nuclear factor important in the regulation of transcription of immunoglobulin genes in B lymphocytes (49). Subsequently, it has become appreciated that NF-κB is a ubiquitous transcription factor involved in regulation of many genes in different cell types (49,50). In most of these non-B cells, NF-κB exists as an inactive cytosolic protein complexed to an inhibitory subunit, I-κB. The action of a number of extracellular agents such as TNF-α results in early (within minutes) activation of NF-κB. This appears to involve release of the NF-κB from the inhibitory I-κB subunit (51), translocation to the nucleus, and binding to NF-κB elements in responsive genes such as the genes for IL-2 and the IL-2 receptor (Fig. 5) (37,49,50,52).

The mechanisms regulating activation of NF-κB remain poorly understood. While it is known that activation of protein kinase C is sufficient to cause phosphorylation of I-κB and dissociation of this subunit (51,53), the protein kinase C pathway does not appear to be involved in transducing the effects of TNF-α on NF-κB (54,55). Therefore, it became important to determine whether ceramide was sufficient to mediate the effects of TNF-α on NF-κB regulation. Cells were treated with exogenous C_2-ceramide and the activity of nuclear NF-κB was monitored by gel-shift assays using oligonucleotides that contain NF-κB response elements. While TNF-α caused early and potent activation of NF-κB, C_2-ceramide was largely inactive over a wide concentration range and over prolonged periods of time (56). On the other hand, the addition of ceramide appeared to enhance activation of NF-κB by TNF-α. The significance of this latter finding remains undefined at this point.

It therefore appears that many, but not all, of the effects of TNF-α are mimicked by ceramide. Activation of NF-κB, which has been associated with increased proliferation of T lymphocytes in response to TNF-α, may be transduced by a mechanism not involving ceramide. It is known that TNF-α binds to two distinct plasma membrane receptors (57). The exact functions of these receptors and their signaling events remain largely unexplored. It is conceivable that only one of these receptors is coupled to ceramide while the other receptor is coupled to NF-κB. Further studies should help in clarifying signaling mechanisms for TNF-α and the role of ceramide as a transducer of TNF-α activities.

F. Insight from Cellular Activity of PDMP

Additional insight into potential biological consequences of changes in ceramide levels comes from studies with the sphingolipid analog D-*threo*-1-phenyl-2-(decanoylamino)-3-morpholino-1-propanol (PDMP). This compound is a potent inhibitor of glucosylceramide synthase *in vitro*. Inhibition of this enzyme in cells results primarily in reduced glycolipid synthesis and increased ceramide levels, as well as additional changes in phospholipid and diacylglycerol levels. These biochemical changes have been associated with multiple biological effects, including inhibition of cell proliferation (58,59), changes in receptor function (60), and inhibition of adherence during differentiation of human leukemia cells (61) (the biochemical and biological activities of this compound are detailed in this volume in the article by Radin, Shayman, and Inokuchi).

Because of the multiple biochemical effects of PDMP, it is difficult to distinguish the role of decreased glycolipid synthesis from other metabolic consequences. However, it is important to note that PDMP results in significant increases in ceramide levels that approach those achieved with TNF-α and 1α,25-dihydroxyvitamin D_3, suggesting that perhaps many of the biological activities of PDMP (such as growth inhibition) are mediated by increased cellular levels of ceramide. In this context, the accumulating wealth of information on the biological activities of PDMP may provide insight into potential cellular activities of ceramide. Therefore, with the use of cell-permeable ceramides, it becomes possible to investigate what effects of PDMP are mediated by ceramide rather than by the decrease in membrane glycosphingolipids.

III. Specificity of Action of Ceramide

An essential requirement for establishing second messenger function for ceramide necessitates the demonstration of specificity of action of ceramide. This is important for (1) demonstrating that the effects of ceramide are not due to nonspecific detergent-like actions resulting in injury to cell membranes, (2) defining the relevant endogenous active molecule, and (3) identifying the molecular pathway activated by ceramide. Thus, *in vitro* and cellular targets that share the same specificity for ceramide are more likely to participate in a related signaling pathway than effects that demonstrate distinct specificities. However, two caveats arise in this latter consideration. First, many molecules may be metabolically interrelated. For example, while sphingosine may not exert an effect on ceramide-activated protein phosphatase *in vitro*, it may be metabolically converted to ceramide *in vivo*. Second, distinct molecules may cause a similar biological effect through separate mechanisms. For example, both diacylglycerol and ceramide cause downregulation of c-*myc* which may be mediated by protein kinase C in the first case and by ceramide-activated protein phosphatase in the latter.

The specificity of the different biochemical and biological activities of ceramide have been investigated using three categories of analogs: (1) amphiphilic molecules and other bioactive lipids such as sphingosine, diacylglycerol, and oleate; (2) synthetic amides of phenyl-aminoalcohols as analogs of ceramide; and (3) the closely related dihydroceramide.

Of the different lipids tested, only ceramide appeared to cause monocytic differentiation of HL-60 cells, although sphingosine appeared to cause inhibition of proliferation but not significant monocytic differentiation (3). Ceramide could be added for only 2 hours then washed out and cells would undergo differentiation (3).

In another approach, amides of phenyl-aminoalcohols (Fig. 1) were synthesized and evaluated for their ability to mimic ceramide effects. Of the different amides studied, an enantiomeric pair was particularly informative. D-*erythro*-2-*N*-Myristoyl-amino-1-phenyl-1-propanol (D-*e*-MAPP) (Fig. 1) was nearly equally potent to ceramide, whereas its enantiomer L-*e*-MAPP was devoid of activity as assayed in cell proliferation and differentiation assays (62).

More recent studies have focused on evaluating the closely related D-*erythro*-dihydroceramide, which differs from ceramide only in lacking the 4–5-*trans* double bond (Fig. 1). In contrast to C_2-ceramide, C_2-dihydroceramide was inactive in yeast (40) and in proliferation and differentiation experiments in HL-60 cells. Thus, the biological end point of growth inhibition/differentiation appears to be specifically mediated by ceramide and the closely related D-*e*-MAPP but not by other sphingolipids and analogs.

Current studies also show that the effects of ceramide on apoptosis are very specific to D-*erythro*-ceramide with diacylglycerol, L-*e*-MAPP, and dihydroceramide being largely inactive. Thus, a close correspondence emerges between the effects of ceramide on apoptosis and growth inhibition, again supporting a related mechanism for the two effects. This would suggest that the ability of ceramide to induce apoptosis may be an exaggeration of its growth inhibitory effects (see Section II,C).

Downregulation of c-*myc* by ceramide does not show as clear-cut a specificity as the other biological effects primarily because phorbol esters and diacylglycerol are also able to downregulate c-*myc* in HL-60 cells. However, the action of ceramide appears to be specific in as much as neither L-*e*-MAPP nor dihydroceramide is active.

These studies clearly demonstrate a specific action for ceramide in inhibition of cell growth, apoptosis, and downregulation of the c-*myc* protooncogene. The most impressive specificity is that between ceramide and dihydroceramide. While other bioactive lipids (such as diacylglycerol) exert biochemical and biological activities that may overlap those of ceramide, dihydroceramide appears to be a totally inert molecule in the different assays utilized.

The studies with dihydroceramide carry three important implications. First, they establish conclusively the structural specificity of ceramide action since the two molecules differ only in the 4–5-*trans* double bond of ceramide. Second, this specificity for ceramide extends to downregulation of c-*myc,* apoptosis, growth inhibition, and cell differentiation. This specificity argues for a common pathway of cell regulation which is selectively activated by ceramide but not dihydroceramide. Third, the inactivity of dihydroceramide provides, for the first time, a rationale for the significance of the sphingolipid double bond. Thus, the introduction of the sphingolipid double bond (most probably at the level of dihydroceramide) imparts on ceramide the ability to modulate various biochemical and biological activities.

IV. Ceramide-Activated Protein Phosphatase: Role in Mediating Ceramide Biology

An important cornerstone in establishing a second messenger function for ceramide requires the identification of the direct cellular target for ceramide action. Modulation of protein phosphorylation is a near-universal mechanism of signal transduction occurring either as a direct or an indirect consequence of the action of most second messengers. Therefore, in an attempt to define such a target, the ability of ceramide to modulate protein phosphorylation was investigated. Preliminary studies failed to disclose any ability of ceramide to activate known protein kinases, but some observations suggested that ceramide may inhibit protein phosphorylation. Therefore, the effects of ceramide on protein phosphatases were investigated. This led to the identification of a *c*eramide-*a*ctivated *p*rotein *p*hosphatase (CAPP) that appears to be specifically activated by ceramide in the low micromolar range *in vitro* (see article by Dobrowsky and Hannun in this volume). This enzyme is potently inhibited by okadaic acid and appears to be a member of the heterotrimeric subgroup of the PP2A class of protein phosphatases. Since no endogenous substrates for CAPP have been identified so far, it has not been feasible to evaluate cellular activation of CAPP by ceramide directly. However, a role for CAPP in mediating ceramide biology is supported by two lines of evidence:

1. *Specificity of activation of CAPP. In vitro,* CAPP is activated potently by short-chain and natural D-*erythro*-ceramides, but it is not activated by D-*erythro*-dihydro-C_2-ceramide. This finding has two important implications. First, it underscores the specificity of activation of CAPP by ceramide and not by closely related lipids. Second, the specificity for D-*erythro*-ceramide is shared by all the relevant biochemical and biological activities of ceramide, including the effects on c-*myc* downregulation, apoptosis, growth inhibition, and induction of cell differentiation. That all these effects are induced by ceramide but not dihydroceramide provides strong evidence for a direct role for CAPP in mediating these effects.
2. *Effects of okadaic acid on ceramide-mediated biology.* CAPP is inhibited potently by okadaic acid *in vitro* over a concentration range of 1–10 n*M*. Thus, okadaic acid may present a useful tool for investigating *in vivo* effects of ceramide mediated by CAPP. In fact, in ongoing studies we have observed that okadaic acid in the nanomolar range inhibits the effects of ceramide on c-*myc* downregulation as well as on growth inhibition (unpublished observations). At these concentrations, okadaic acid selectively inhibits the PP2A class of protein phosphatases, including CAPP. Therefore, it may be concluded that these biological effects of ceramide require activation of a protein phosphatase, thus providing further evidence for the involvement of CAPP in ceramide-mediated biology.

The ability to distinguish specific effects of ceramide that are not shared by dihydroceramide and the inhibition of those effects by low concentrations of okadaic acid provide very important tools in determining the involvement of CAPP in ceramide biology. More importantly, these tools also should allow for the delineation of the role of CAPP in the action of a number of extracellular agents. For example, in initial studies, it was shown that okadaic acid inhibits the effects of TNF-α on c-*myc* downregulation, providing strong evidence for a role of CAPP in mediating the effects of TNF-α on c-*myc*.

V. Conclusions

A. Implications

These studies are defining a signaling function of ceramide whereby the action of extracellular agents results in activation of a neutral sphingomyelinase, causing sphingomyelin hydrolysis and ceramide generation. The resulting elevation in endogenous ceramide levels causes activation of a serine/threonine protein phosphatase which appears to be involved in mediating subsequent effects of ceramide on c-*myc* downregulation, growth inhibition, induction of cell differentiation, and apoptosis. These relationships are illustrated in a hypothetical scheme in Fig. 5 as it applies to signaling through the TNF-α receptor. Many of these interactions and relationships remain somewhat tentative, and their precise roles in signal transduction and cell regulation are not fully defined. Nonetheless, these studies have identified major players in this novel pathway of signal transduction.

At the very least, these studies provide a strong argument for a role for sphingolipids in signal transduction. It becomes obvious from Fig. 5 that the regulated metabolism of sphingomyelin in the sphingomyelin cycle and the existence of a ceramide/CAPP pathway of cell regulation share many of the basic features of the PI cycle and the diacylglycerol/PKC pathway of signal transduction. The similarities and differences of the sphingomyelin cycle to the PI cycle are discussed in detail by Hannun and Bell elsewhere in this volume. This can be taken as a prototype for other sphingolipid signaling pathways that may involve other sphingolipid precursors and/or other sphingolipid breakdown products.

The emerging features of the ceramide/CAPP pathway of cell regulation have introduced new implications as to our understanding of signal transduction/cell regulatory pathways. First, as with diacylglycerol, ceramide represents an essential intermediary in sphingolipid metabolism that appears to have specific cell regulatory activities in signal transduction. Ceramide is positioned at a focal point in the regulation of sphingolipid biosynthetic and catabolic pathways, analogous to the role of diacylglycerol in phospholipid metabolism. While there may be no connection between the metabolic and regulatory functions of diacylglycerol and ceramide, it is conceivable that, precisely because of their essential metabolic functions, they have assumed key regulatory roles. As such, a pattern is emerging whereby diacylglycerol may be involved in signal transduction in a mitogenic pathway through the activation of protein kinase C, while ceramide may be involved in an antiproliferative pathway through activation of a protein phosphatase. In addition, the two lipids are metabolically interconnected through the action of sphingomyelin synthase, which transfers a phosphorylcholine headgroup from phosphatidylcholine to ceramide:

$$\text{Phosphatidylcholine} + \text{ceramide} \rightleftharpoons \text{sphingomyelin} + \text{diacylglycerol}$$

This metabolic pathway may therefore regulate the relative levels of ceramide and diacylglycerol. Whether this plays a role in cell regulation remains to be investigated.

Second, the identification of ceramide as a specific activator of CAPP underscores a novel theme in signal transduction whereby regulation of protein phos-

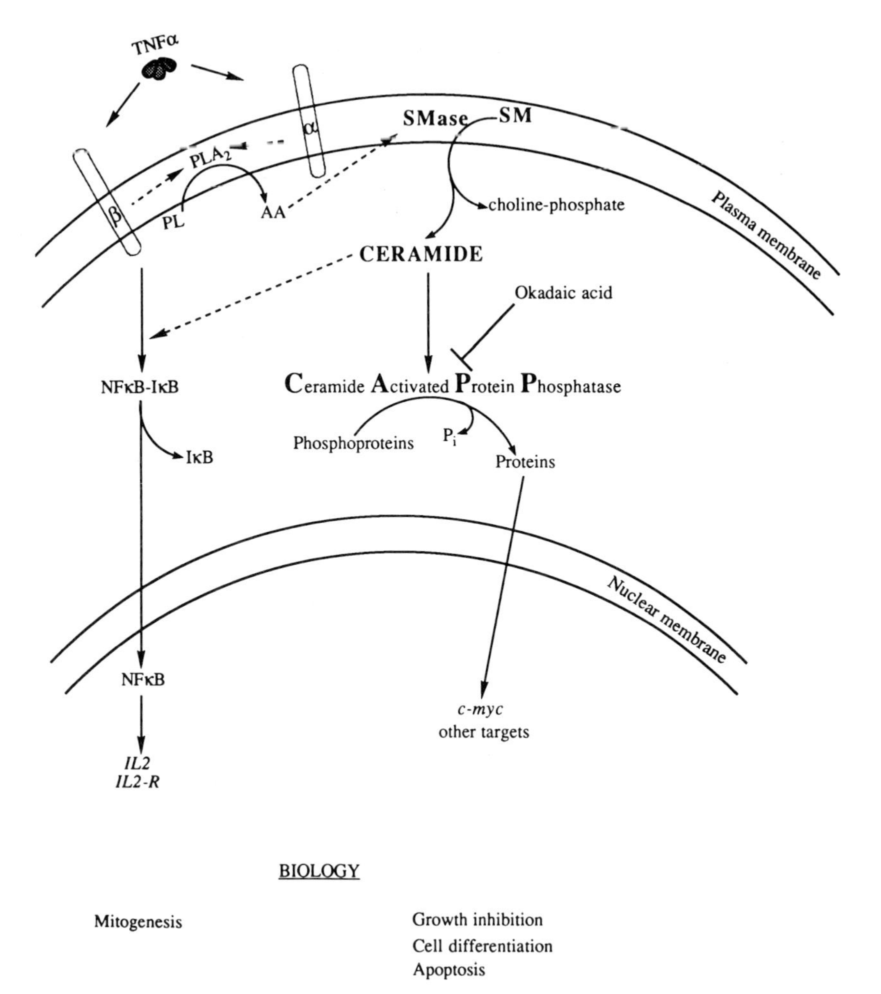

FIG. 5. Scheme of the proposed ceramide signaling pathway. Preliminary evidence indicates that the action of TNF-α results in early activation of phospholipase A_2 and the generation of arachidonic acid. Arachidonic acid is able to activate a sphingomyelinase, resulting in sphingomyelin (SM) hydrolysis and the formation of ceramide and choline phosphate. Ceramide in turn activates ceramide-activated protein phosphatase, which results in dephosphorylation of (unidentified) protein substrates. This pathway ultimately leads to suppression of c-*myc* transcription, growth inhibition, cell differentiation, and apoptosis. Ceramide is unable to activate NF-κB directly but may play a role in positive feedback regulation of NF-κB.

phatases may prove to be as critical as regulation of protein kinases. For a long time, protein phosphatases have been considered as constitutively active enzymes with more "housekeeping" regulatory roles in intermediary metabolism. It is becoming clear that protein phosphatases are also subject to direct regulation in a signaling modality. This is supported by the activation of PP2B (calcineurin) by calcium/calmodulin (63) and the activation of CAPP by ceramide.

Third, the ceramide/CAPP pathway may emerge as a critical signal transduction pathway involved in mediating the effects of an array of extracellular agents that belong to the subgroup of antimitogenic cytokines and chalones. While inhibitors of cell growth have not received as much attention as mitogenic stimuli, it is now increasingly appreciated that fine tuning of cell regulation involves both positive and negative stimuli. Indeed, an increasing number of such negative stimuli have been identified, including TNF-α, the interferons, TGF-β, NGF, and certain interleukins and other hormones that may have dual functions depending on the cell type. It is possible that the ceramide/CAPP pathway is critical in mediating the effects of many of these extracellular agents. Moreover, the ceramide/CAPP pathway may prove to be an essential link in the pathways of cell regulation initiated at the cell membrane and communicated to critical nuclear events such as the regulation of c-*myc*.

B. Questions

The scheme illustrated in Fig. 5 should serve as a focal point to identify major gaps in our knowledge concerning the regulation of the sphingomyelin cycle and the ceramide/CAPP pathway of cell regulation. Some of these important questions include

1. What are the different extracellular agents that activate sphingomyelin turnover? In addition to 1α,25-dihydroxyvitamin D_3, TNF-α, and γ-interferon, recent studies have identified interleukin 1 as an activator of sphingomyelin hydrolysis and ceramide generation, and that ceramide (or sphingosine) may play a role in transducing the effects of interleukin 1 (64). Our ongoing studies suggest that NGF may be another such important factor.
2. What are the proximal signaling events linking receptor occupancy to activation of the hormone-activate neutral sphingomyelinase? It should be noted that a common feature of many of the above identified activators of sphingomyelin hydrolysis is the absence of any known signaling function for their receptors. For example, with TNF-α, for which two distinct plasma membrane receptors exist, the cloned receptors do not demonstrate any intrinsic enzymatic activity (such as tyrosine kinase, tyrosine phosphatase, or guanylyl cyclase activities) nor do they appear to belong to the (seven-membrane-spanning) family of G protein-coupled receptors. It is tempting to speculate that these receptors may belong to a novel family of receptors with a unique and distinct signaling mechanism coupled to previously unidentified components. Whether neutral sphingomyelinase is an immediate proximal target for these receptors or a more indirect target has not been determined. In preliminary studies, we find that TNF-α may activate neutral sphingomyelinase through the generation of arachi-

donic acid following activation of phospholipase A_2. This area of signaling promises to be a novel and critical area for understanding regulation of sphingomyelinase and the generation of ceramide.

3. The cellular topologics of sphingomyelin hydrolysis and ceramide generation remain largely unknown. This is a particularly important problem since sphingomyelin has been located classically at the outer leaflet of the plasma membrane. However, the hormone-activated neutral sphingomyelinase appears as a cytosolic enzyme and therefore must act on accessible pools of sphingomyelin. In HL-60 cells, we find that only 40–60% of total cellular sphingomyelin is on the outer leaflet of the plasma membrane while the hormone-responsive pool of sphingomyelin is located in an intracellular membrane compartment (65). Notably, this pool of sphingomyelin is hydrolyzed in response to the action of Brefeldin A (66), a macrolide that induces collapse of the Golgi into the endoplasmic reticulum as well as having other effects on intracellular organelles and membranes (67). The identification of this membrane compartment is critical in understanding the cellular regulation of hormone-activated sphingomyelinase and in the subsequent activation of CAPP.

4. Are there direct targets for the action of ceramide other than CAPP? Ongoing studies support an important role for CAPP in mediating the effects of ceramide on the major activities defined for ceramide thus far. It is difficult to predict from existing paradigms of signal transduction whether CAPP will emerge as the sole mediator of ceramide effects or whether other targets will be identified. In the case of diacylglycerol, protein kinase C appears to be the major mediator of diacylglycerol effects, although other proteins with no discernable kinase activity such as *n*-chimaerin may prove to be additional targets for diacylglycerol. On the other hand, the existence of multiple targets for calcium/calmodulin is well recognized. In this context, Kolesnick and coworkers have demonstrated the ability of ceramide to increase protein phosphorylation of the EGF receptor, implicating a protein kinase in the action of ceramide (68,69). However, it is difficult to determine whether this is an immediate target for the action of ceramide or a downstream event. Further studies are required to determine the mechanisms of action of ceramide.

5. What is the spectrum of biochemical and biological activities of ceramide? At this point, a critical role for ceramide is increasingly appreciated in the regulation of cell growth, cell death, and cell differentiation. However, by activating a protein phosphatase, ceramide may have multiple other activities in cell function that could include effects on intermediary metabolism as well as effects on cell structure. This area of cell biology and biochemistry may prove to be extremely fruitful for the investigation of ceramide-mediated effects.

In conclusion, these investigations on the biological and biochemical activities of ceramide have provided strong evidence for a signal transduction function of ceramide and they are elucidating novel and complex pathways of signaling through sphingolipids that may play critical roles in the regulation of cell growth and function.

Acknowledgments

We thank Dr. Julie Fishbein and Corinne Linardic for careful and critical review of the manuscript, Supriya Jayadev for providing Fig. 5 and reviewing the manuscript, and Marsha Haigood for expert secretarial assistance.

References

1. Okazaki, T., Bell, R. M., and Hannun, Y. A. (1989). *J. Biol. Chem.* **264,** 19076–19080.
2. Hannun, Y. A., and Bell, R. M. (1993). This volume, Chapter 3.
3. Okazaki, T., Bielawska, A., Bell, R. M., and Hannun, Y. A. (1990). *J. Biol. Chem.* **265,** 15823–15831.
4. Kim, M.-Y., Linardič, C., Obeid, L., and Hannun, Y. (1991). *J. Biol. Chem.* **266,** 484–489.
5. Merrill, A. H., Jr. (1992). *Nutr. Rev.* **50,** 78–80.
6. Obeid, L. M., Linardič, C. M., Karolak, L. A., and Hannun, Y. A. (1993). *Science* **259,** 1769–1771.
7. Dressler, K. A., Mathias, S., and Kolesnick, R. M. (1992). *Science* **255,** 1715–1718.
8. Dobrowsky, R. T., and Hannun, Y. A. (1992). *J. Biol. Chem.* **267,** 5048–5051.
9. Beutler, B., and Cerami, A. (1988). *Biochemistry* **27,** 7575–7582.
10. Loetscher, H. R., Brockhaus, M., Dembič, Z., Gentz, R., Gubler, U., Hohmann, H.-P., Lahm, H.-W., Van Loon, A. P. G. M., Pan, Y.-C. E., Schlaeger, E.-J., Steinmetz, M., Tabuchi, H., and Lesslauer, W. (1991). *Oxford Surv. Eukaryotic Genes* **7,** 119–142.
11. Grunfeld, C., and Palladino, M. A. (1990). *Adv. Intern. Med.* **35,** 45–72.
12. Collins, S. J. (1987). *Blood* **70,** 1233–1244.
13. Glazer, R. I., Chapekar, M. S., Hartman, K. D., and Knode, M. C. (1986). *Biochem. Biophys. Res. Commun.* **140,** 908–915.
14. Darzynkiewicz, Z., Carter, S. P., and Old, L. J. (1987). *J. Cell. Physiol.* **130,** 328–335.
15. Imamura, K., Sherman, M. L., Spriggs, D., and Kufe, D. (1988). *J. Biol. Chem.* **263,** 10247–10253.
16. McCachren, S. S., Salehi, Z., Weinberg, J. B., and Niedel, J. E. (1988). *Biochem. Biophys. Res. Commun.* **151,** 574–582.
17. Schachner, J., Blick, M., Freireich, E., Gutterman, J., and Beran, M. (1988). *Leukemia* **2,** 749–753.
18. Kronke, M., Schluter, C., and Pfizenmaier, K. (1987). *Proc. Natl. Acad. Sci. U.S.A.* **84,** 469–473.
19. Haliday, E. M., Ramesha, C. S., and Ringold, G. (1991). *EMBO J.* **10,** 109–115.
20. Atkinson, Y. H., Murray, A. W., Krilis, S., Vadas, M. A., and Lopez, A. F. (1990). *Immunology* **70,** 82–87.
21. Neale, M. L., Fiera, R. A., and Matthews, N. (1988). *Immunology* **64,** 81–85.
22. Clark, M. A., Chen, M.-J., Crooke, S. T., and Bomalaski, J. S. (1988). *Biochem. J.* **250,** 125–132.
23. Godfrey, R. W., Johnson, W. J., and Hoffstein, S. T. (1987). *Biochem. Biophys. Res. Commun.* **142,** 235–241.
24. Yanaga, F., Abe, M., Koga, T., and Hirata, M. (1992). *J. Biol. Chem.* **267,** 5114–5121.

25. Brett, J., Gerlach, H., Nawroth, P., Steinberg, S., Godman, G., and Stern, D. (1989). *J. Exp. Med.* **169,** 1977–1991.
26. Schmid, D. S., Hornung, R., McGrath, K. M., Paul, N., and Ruddle, N. H. (1987). *Lymphokine Res.* **6,** 195–200.
27. Rubin, B. Y., Smith, L. J., Hellermann, G. R., Lunn, R. M., Richardson, N. K., and Anderson, S. L. (1988). *Cancer Res.* **48,** 6006–6010.
28. Flieger, D., Riethmuller, G., and Ziegler-Heitbrock, H. W. L. (1989). *Int. J. Cancer* **44,** 315–319.
29. Brenner, D. A., O'Hara, M., Angel, P., Chojkier, M., and Karin, M. (1989). *Nature (London)* **337,** 661–663.
30. Schütze, S., Nottrott, S., Pfizenmaier, K., and Krönke, M. (1990). *J. Immunol.* **144,** 2604–2608.
31. Sugarman, B. J., Aggarwal, B. B., Hass, P. E., Figari, I. S., Palladino, M. A., and Shepard, H. M. (1985). *Science* **230,** 943–945.
32. Marino, M. W., Pfeffer, L. M., Guidon, P. T., Jr., and Donner, D. B. (1989). *Proc. Natl. Acad. Sci. U.S.A.* **86,** 8417–8421.
33. Evans, J. P. M., Mire-Sluis, A. R., Hoffbrand, A. V., and Wickremasinghe, R. G. (1990). *Blood* **75,** 88–95.
34. Schutze, S., Scheurich, P., Schluter, C., Ucer, U., Pfizenmaier, K., and Kronke, M. (1988). *J. Immunol.* **140,** 3000–3005.
35. Osborn, L., Kunkel, W., and Nabel, G. (1989). *Proc. Natl. Acad. Sci. U.S.A.* **86,** 2336–2340.
36. Price, S. R., Olivecrona, T., and Pekälä, P. H. (1986). *Biochem. J.* **240,** 601–614.
37. Lowenthal, J. W., Ballard, D. W., Bogerd, H., Böhnlein, E., and Greene, W. C. (1989). *J. Immunol.* **142,** 3121–3128.
38. Bielawska, A., Linardič, C. M., and Hannun, Y. A. (1992). *FEBS Lett.* **307,** 211–214.
39. Werner, M. H., Bielawska, A. E., and Hannun, Y. A. (1993). Submitted for publication.
40. Fishbein, J. D., Dobrowsky, R. T., Bielawska, A., Garrett, S., and Hannun, Y. A. (1993). *J. Biol. Chem.* (in press).
41. Gerschenson, L. E., and Rotello, R. J. (1992). *FASEB J.* **6,** 2450–2455.
42. Kerr, J. F. R., and Searle, J. (1970). *J. Pathol.* **107,** 41–44.
43. Keler, T., Barker, C. S., and Sorof, S. (1992). *Proc. Natl. Acad. Sci. U.S.A.* **89,** 4830–4834.
44. Wyllie, A. H. (1980). *Nature (London)* **284,** 555–556.
45. Kerr, J. F. R., and Harmon, B. V. (1991). *In* "Apoptosis: The Molecular Basis of Cell Death" (L. D. Tomei and F. O. Cope, eds.), pp. 5–29. Cold Spring Harbor Lab. Press, Cold Spring Harbor, New York.
46. Michaelson, J. (1991). *In* "Apoptosis: The Molecular Basis of Cell Death" (L. D. Tomei and F. O. Cope, eds.), pp. 31–46. Cold Spring Harbor Lab. Press, Cold Spring Harbor, New York.
47. Williams. G. T. (1991). *Cell (Cambridge, Mass.)* **65,** 1097–1098.
48. Wolff, R. A., Obeid, L. M., and Hannun, Y. A. (1993). In preparation.
49. Lenardo, M. J., and Baltimore, D. (1989). *Cell (Cambridge, Mass.)* **58,** 227–229.
50. Molitor, J. A., Walker, W. H., Doerre, S., Ballard, D. W., and Greene, W. C. (1990). *Proc. Natl. Acad. Sci. U.S.A.* **87,** 10028–10032.
51. Ghosh, S., and Baltimore, D. (1990). *Nature (London)* **344,** 678–682.
52. Ruben, S. M., Dillon, P. J., Schreck, R., Henkel, T., Chen, C.-H., Maher, M., Baeuerle, P. A., and Rosen, C. A. (1991). *Science* **251,** 1490–1493.
53. Shirakawa, F., and Mizel, S. B. (1989). *Mol. Cell. Biol.* **9,** 2424–2430.
54. Meichle, A., Schütze, S., Hensel, G., Brunsing, D., and Krönke, M. (1990). *J. Biol. Chem.* **265,** 8339–8343.
55. Feuillard, J., Gouy, H., Bismuth, G., Lee, L. M., Debré, P., and Korner, M. (1991). *Cytokine* **3,** 257–265.

56. Dbaibo, G., Obeid, L. M., and Hannun, Y. A. (1993). Submitted for publication.
57. Vilcek, J., and Lee, T. H. (1991). *J. Biol. Chem.* **266,** 7313–7316.
58. Felding-Habermann, B., Igarashi, Y., Fenderson, B. A., Park, L. S., Radin, N. S., Inokuchi, J., Strassmann, G., Handa, K., and Hakomori, S. (1990). *Biochemistry* **29,** 6314–6322.
59. Shayman, J. A., Deshmukh, G. D., Mahdiyoun, S., Thomas, T. P., Wu, D., Barcelon, F. S., and Radin, N. S. (1991). *J. Biol. Chem.* **266,** 22968–22974.
60. Inokuchi, J., Momosaki, K., Shimeno, H., Nagamatsu, A., and Radin, N. S. (1989). *J. Cell. Physiol.* **141,** 573–583.
61. Kan, C.-C., and Kolesnick, R. N. (1992). *J. Biol. Chem.* **267,** 9663–9667.
62. Bielawska, A., Linardič, C. M., and Hannun, Y. A. (1992). *J. Biol. Chem.* **267,** 18493–18497.
63. Cohen, P., and Cohen, P. T. W. (1989). *J. Biol. Chem.* **264,** 21435–21438.
64. Ballou, L. R., Chao, C. P., Holness, M. A., Barker, S. C., and Raghow, R. (1992). *J. Biol. Chem.* **267,** 20044–20050.
65. Linardič, C. M., and Hannun, Y. A. (1993). In preparation.
66. Linardič, C. M., Jayadev, S., and Hannun, Y. A. (1992). *J. Biol. Chem.* **267,** 14909–14911.
67. Klausner, R. D., Donaldson, J. G., and Lippincott-Schwartz, J. (1992). *J. Cell Biol.* **116,** 1071–1080.
68. Goldkorn, T., Dressler, K. A., Muindi, J., Radin, N. S., Mendelsohn, J., Menaldino, D., Liotta, D., and Kolesnick, R. N. (1991). *J. Biol. Chem.* **266,** 16092–16097.
69. Mathias, S., Dressler, K. A., and Kolesnick, R. N. (1991). *Proc. Natl. Acad. Sci. U.S.A.* **88,** 10009–10013.

ADVANCES IN LIPID RESEARCH, VOL. 25

Ceramide: A Novel Second Messenger

SHALINI MATHIAS AND RICHARD KOLESNICK

Laboratory of Signal Transduction
Memorial Sloan-Kettering Cancer Center
New York, New York 10021

I. Introduction

Sphingomyelin (*N*-acylsphingosine-1-phosphocholine or ceramide phosphocholine) is preferentially concentrated in the outer leaflet of the plasma membrane of mammalian cells (Barenholz and Thompson, 1980). It was originally thought to be merely a structural component of membranes (Barenholz and Gatt, 1982). However, in 1986, while comparing effects of 1,2-diacylglycerol (DG) and phorbol ester, this laboratory found that DG activates a sphingomyelinase and that sphingomyelin is rapidly synthesized and degraded in response to various cellular signals (Kolesnick, 1987, 1989a,b; Kolesnick and Clegg, 1988). Since evidence was provided that sphingoid bases might serve as endogenous inhibitors of protein kinase C (PKC) (Hannun *et al.*, 1986; Merrill *et al.*, 1986; Wilson *et al.*, 1986), we suggested that DG might activate an inhibitory pathway for PKC involving sphingolipids. Subsequent studies resulted in the discovery of ceramide 1-phosphate (Dressler and Kolesnick, 1990) and helped define a specific pathway from sphingomyelin to ceramide 1-phosphate that bifurcated at the level of ceramide (Kolesnick and Hemer, 1990). Recent evidence suggests that ceramide plays a pivotal role in signal transduction (Okazaki *et al.*, 1989, 1990; Kim *et al.*, 1991; Goldkorn *et al.*, 1991; Mathias *et al.*, 1991; Dressler *et al.*, 1992). Thus, sphingomyelin and its derivatives appear to constitute a new pathway through which information may be transmitted.

The intent of this article is to focus on sphingomyelin and its derivatives as biomodulators and to define how signals may be transmitted through the sphingomyelin pathway during cellular activation. Emphasis is placed on the potential role of ceramide as a second messenger.

II. Sphingomyelin Hydrolysis and Signaling

A. Early Studies on Stimulated Sphingomyelin Degradation

Protein kinase C plays a central role in signal transduction via the phosphoinositide pathway (for review, see Nishizuka, 1988; Kikkawa *et al.*, 1989). In this paradigm, activation of certain cell surface receptors results in hydrolysis of plasma membrane phosphatidylinositol 4,5-bisphosphate by a phospholipase C to generate two intracellular messengers, inositol trisphosphate (IP_3) and 1,2-diacylglycerol (DG). IP_3 releases intracellular calcium stores, raising the cytoplasmic concentration of free Ca^{2+} to stimulate Ca^{2+}-dependent events (Berridge, 1989), while DG, the hydrophobic backbone of the cleaved phosphoinositide, remains within membranes and activates PKC (Nishizuka, 1989). This involves formation of complexes of PKC with Ca^{2+}, phospholipid, and DG. It results in a net redistribution of PKC from the inactive cytosolic pool to the membrane. PKC is also activated by the tumor-promoting phorbol esters, which presumably bind to the same site(s) as DG.

In addition to activation of PKC, the negative regulation of PKC by sphingoid bases was also investigated by Bell, Hannun, Merrill and co-workers. These investigators demonstrated that sphingosine possessed inhibitory properties *in vitro* and in intact cells (Hannun *et al.*, 1986; Merrill *et al.*, 1986; Wilson *et al.*, 1986). In a mixed micellar model, sphingosine exhibited competitive inhibition of DG- and phorbol ester-mediated activation of PKC. Similarly, sphingoid bases inhibited events in intact cells which were known to be mediated via PKC, including thrombin-stimulated phosphorylation of a 40-kDa protein in human platelets (Hannun *et al.*, 1986), phorbol ester-induced differentiation of human promyelocytic (HL-60) cells into macrophages (Merrill *et al.*, 1986), and the oxidative burst in human neutrophils (Wilson *et al.*, 1986).

While studying differences in the pattern of cellular activation by DG and phorbol esters, this laboratory observed that some of the effects of DG were independent of PKC. Early studies showed that DG, but not phorbol esters, stimulated sphingomyelin degradation via a sphingomyelinase (Kolesnick, 1987). A synthetic DG, 1,2-dioctanoylglycerol (diC_8), induced rapid reduction in sphingomyelin content and a concomitant and quantitative increase in the level of ceramide in GH_3 rat pituitary cells (Fig. 1). Subsequently, ceramide was deacylated to sphingoid bases. The level of sphingoid bases increased 1.9-fold from 24 to 46 pmol/10^6 cells after 10 minutes of diC_8 stimulation and remained ele-

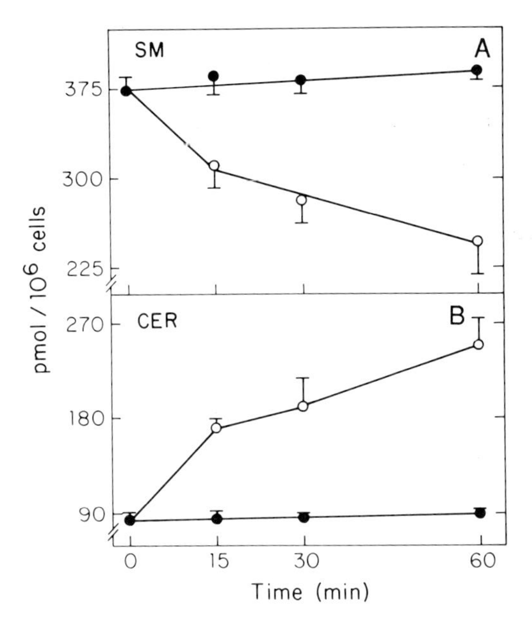

FIG. 1. Time course of the effect of dioctanoylglycerol (diC_8) on the levels of sphingomyelin and ceramide. (A) Effect on sphingomyelin (SM); (B) effect on ceramide (CER).

vated at 1.5-fold of control for at least 1 hour (Kolesnick and Clegg, 1988; Kolesnick and Hemer, 1989). Based on these data, this laboratory postulated that, in addition to stimulation of PKC, DG but not phorbol ester activated a potential inhibitory pathway for PKC involving a sphingomyelinase. These studies suggested that sphingomyelinase action might be involved in differences between physiological activation and tumor promotion via the PKC pathway.

The notion that differences in PKC activation by DG and phorbol ester were due to sphingomyelinase action was tested (Kolesnick and Clegg, 1988). Intact GH_3 cells were treated with maximal concentrations of diC_8 (Fig. 2, left-hand panels) or phorbol ester (Fig. 2, right-hand panels) and PKC was isolated by anion-exchange chromatography of cytosol and membrane fractions. Activity was determined by transfer of ^{32}P from the γ-position of ATP to histone as described by Nishizuka (1988; Kikkawa *et al.*, 1989). Stimulation resulted in quantitative translocation of PKC from the inactive cytosolic pool to the membrane. However, the effect of the diC_8 was smaller than that of phorbol ester and was transient. Hence, total PKC activity was conserved in response to diC_8. In contrast, after translocation to the membrane with phorbol ester, PKC activity was lost. This may represent the mechanism of downregulation of PKC in these cells.

Since DG activates a sphingomyelinase, studies were performed to determine whether the action of a sphingomyelinase was sufficient to inactivate PKC. It

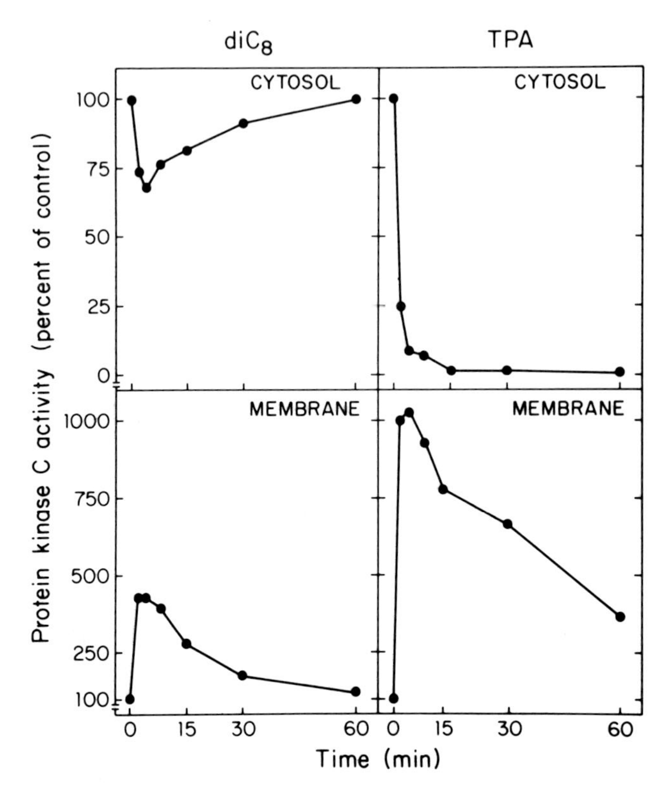

FIG. 2. Effect of diC_8 (left panels) and TPA (phorbol ester, right panels) on the distribution of protein kinase C (PKC) activity. PKC activity was measured in the cytosolic and particulate fractions.

was demonstrated that *Staphylococcus aureus* sphingomyelinase, added exogenously to GH_3 cells after PKC translocation was completed, inactivated PKC by inducing its redistribution back into the cytosol (Kolesnick and Clegg, 1988) (Fig. 3). Hence, the pattern of PKC activation/inactivation with phorbol ester plus sphingomyelinase mimicked that of DG alone. An identical effect was obtained using sphingoid bases. These studies were interpreted as evidence to support the concept that sphingomyelinase action, presumably by generation of sphingoid bases, was sufficient to inactivate PKC.

Subsequent studies were performed to assess whether sphingomyelinase action also blocked the biological consequences of signaling through the PKC pathway (Kolesnick, 1989a). Phorbol ester-induced differentiation of HL-60 cells into macrophages was used as a model for these studies, since Merrill *et al.* (1986) had demonstrated that sphingoid bases potently inhibited this event. Macrophage differentiation was measured by morphological changes, growth

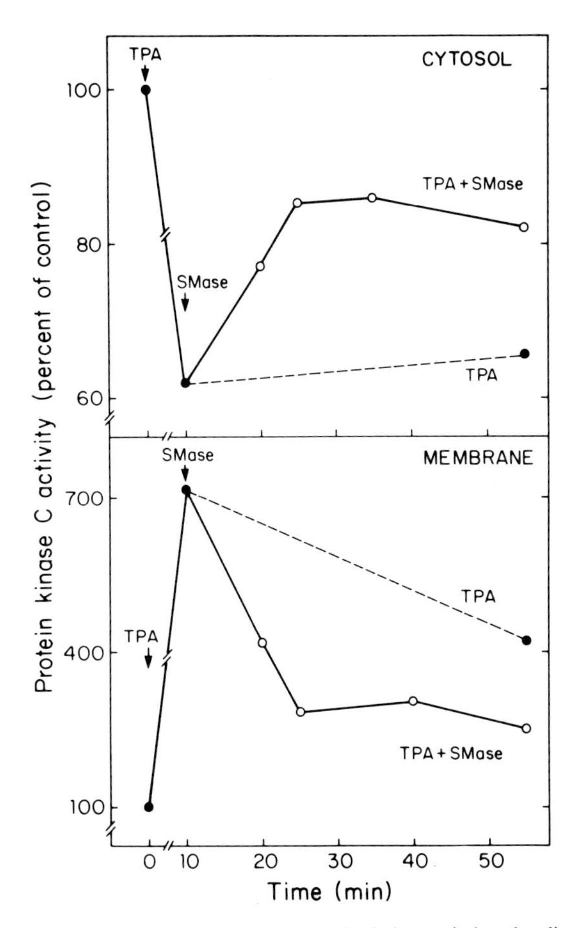

FIG. 3. Effect of sphingomyelinase (SMase) on phorbol ester-induced redistribution of PKC activity.

inhibition, adhesion to plastic tissue culture vessels, and by the enzymatic markers α-napthylacetate esterase and acid phosphatase (Kolesnick, 1989a). Exogenous sphingomyelinase elevated sphingoid base levels and inactivated PKC in HL-60 cells as in GH_3 cells. Further sphingomyelinase, like sphingoid bases, prevented macrophage differentiation by all the criteria noted above. These early studies showed that sphingomyelinase action was sufficient not only to generate free sphingoid bases and inactivate PKC but also to prevent the biological consequences of activation of this pathway.

It appeared from our initial studies that DG activated the acid sphingomyelinase in GH_3 cells (Kolesnick, 1987). Subsequently, Merrill and co-workers demonstrated that plasma membranes contained all of the components necessary for generation of sphingoid bases subsequent to degradation of sphingomyelin (Slife *et al.*, 1989). These investigators showed that activation of a neutral sphingomyelinase resulted in generation of ceramide and its deacylation to free sphingoid bases. This latter event presumably occurred by the action of a neutral ceramidase. Hence, these studies showed that the plasma membrane was capable of generating sphingoid bases at the presumed site of PKC translocation.

B. Identification of a Sphingomyelin Metabolic Pathway

In GH_3 cells, sphingomyelin hydrolysis resulted in the generation of relatively small quantities of sphingoid bases despite large increases in ceramide levels. For instance, an increase in the level of ceramide from 90 to 270 pmol/10^6 cells only resulted in an increase in sphingoid base levels from 24 to 48 pmol/10^6 cells (Kolesnick and Hemer, 1989). Thus, less than 10% of the ceramide generated was converted to sphingoid bases. Similar results were obtained with HL-60 cells (Kolesnick, 1989a) and human neutrophils (Wilson *et al.,* 1988). This prompted a search for other derivatives of ceramide.

Schneider and Kennedy (1973) had previously reported that *Escherichia coli* DG kinase utilized ceramide as substrate *in vitro,* although less efficiently than DG. Authentic ceramide 1-[^{32}P]phosphate was thus synthesized and used to determine whether HL-60 cells contained endogenous compound (Dressler and Kolesnick, 1990). Resting HL-60 cells were found to contain 30 pmol/10^6 cells of ceramide 1-phosphate and the level increased 2-fold after sphingomyelinase treatment (Fig. 4A). This was under the same conditions found sufficient for inactivation of PKC and inhibition of macrophage differentiation. An endoglycoceramidase generated a nearly equivalent elevation in ceramide levels from glycosphingolipids (Fig. 4B). However, ceramide derived from this source was not converted to ceramide 1-phosphate (Fig. 4C). These studies suggested the existence of a specific pathway from sphingomyelin to ceramide 1-phosphate via ceramide.

The discovery of a ceramide kinase distinct from DG kinase provided further support for the notion of a specific pathway from sphingomyelin to ceramide 1-phosphate. A Ca^{2+}-dependent ceramide kinase activity was first reported in rat brain synaptosomes by Bajjalieh *et al.* (1989). Subsequently, a similar activity was found in HL-60 cells (Kolesnick and Hemer, 1990). Figure 5 shows that cer-

FIG. 4. Effects of sphingomyelinase and endoglycoceramidase on ceramide and ceramide 1-phosphate. (A) Time course of the effect of sphingomyelinase on the level of ceramide 1-phosphate. (B) Measurement of ceramide levels after treatment with various concentrations of sphingomyelinase and endoglycoceramidase. (C) Measurement of ceramide 1-phosphate levels after treatment with sphingomyelinase (SMase) and endoglycoceramidase (EGC).

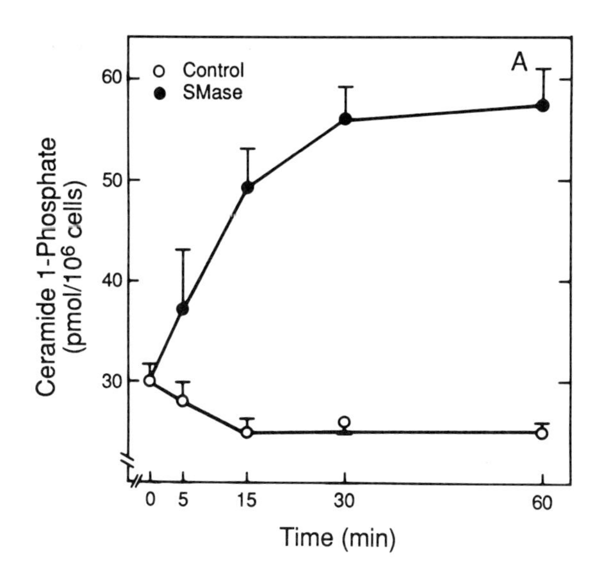
Control
SMase
A
Ceramide 1-Phosphate (pmol/10^6 cells)
60
50
40
30
0
5
15
30
60
Time (min)

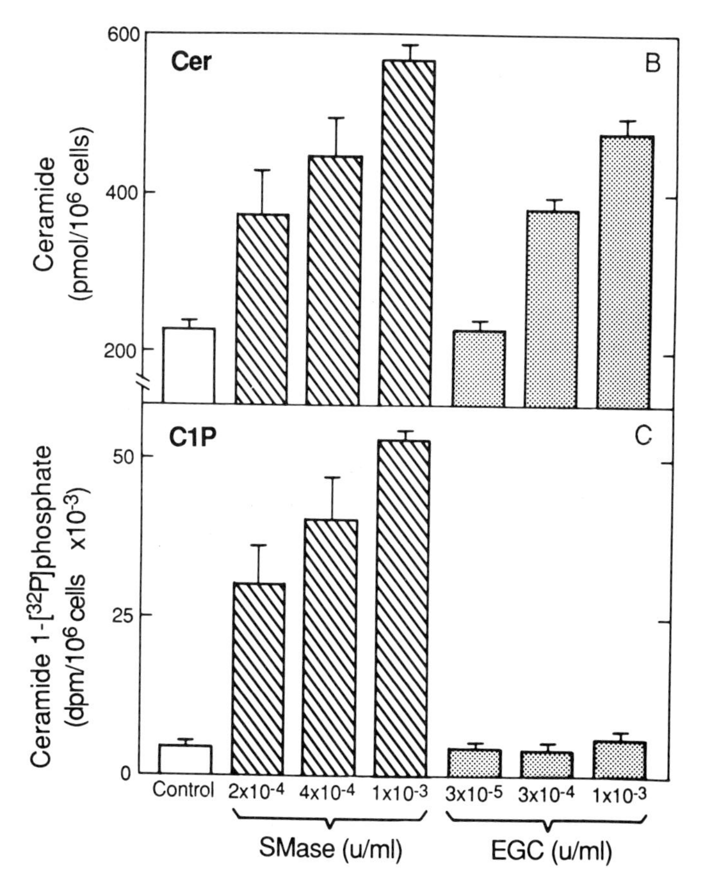
Cer
B
Ceramide (pmol/10^6 cells)
600
400
200
C1P
C
Ceramide 1-[32P]phosphate (dpm/10^6 cells x10^-3)
50
25
0
Control
2x10^-4
4x10^-4
1x10^-3
3x10^-5
3x10^-4
1x10^-3
SMase (u/ml)
EGC (u/ml)

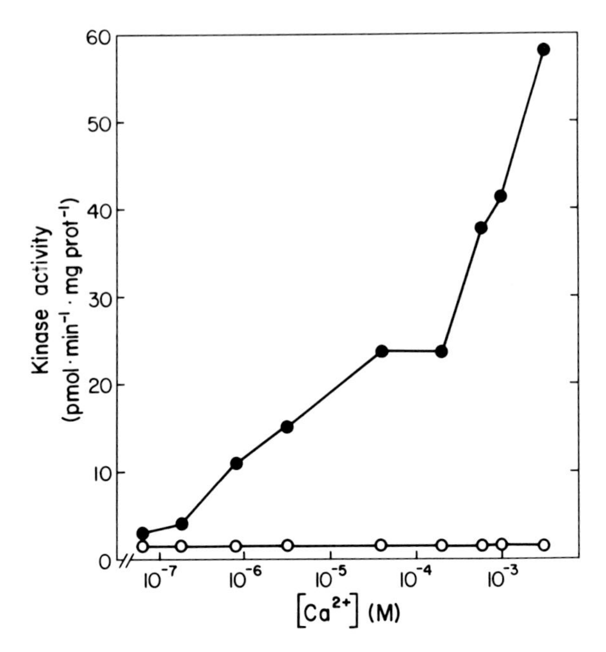

FIG. 5. Effect of calcium on phosphorylation of ceramide (●) and DG (○) by ceramide kinase.

amide but not DG was phosphorylated in the presence of Ca^{2+} by a microsomal preparation from HL-60 cells. Effective Ca^{2+} concentrations were within the physiological range and higher. This same microsomal preparation also contained Mg^{2+}-dependent DG kinase activity similar to that described in many systems (Kanoh *et al.,* 1983, 1986, 1990; Sakane *et al.,* 1990). The bulk of the DG kinase activity was separable from ceramide kinase activity by anion-exchange chromatography. Ceramide kinase was weakly anionic and appeared in the flow through of the column, whereas the major part of the DG kinase activity bound tightly to the column (Kolesnick and Hemer, 1990). These studies lent further credence to the proposal of a pathway from sphingomyelin to ceramide 1-phosphate.

The sphingomyelin pathway is in many ways similar to the phosphoinositide pathway (Fig. 6). Sphingomyelinase, the enzyme that initiates the sphingomyelin pathway, is a phospholipase C, and hence is similar to the enzyme that initiates the phosphoinositide pathway. Ceramide, the central molecule in the sphingomyelin pathway, also appears structurally similar to DG, the central molecule of the phosphoinositide pathway. This is evidenced by the fact that bacterial (but not human) DG kinase utilizes both ceramide and DG as substrates. A similar rationale suggests that the phosphorylated forms of DG and ceramide, phosphatidic acid and ceramide 1-phosphate, respectively, must also be structurally similar. To carry this analogy further, if DG activates a specific kinase, i.e., PKC, perhaps ceramide also possesses similar capabilities. In this regard, ceramide does not appear to activate PKC directly (Merrill *et al.*, 1989). Therefore, studies were initiated in this laboratory to determine whether ceramide activated a specific kinase.

C. Evidence That Ceramide Has Bioeffector Properties

Davis and co-workers had published a series of articles demonstrating that sphingoid bases stimulated phosphorylation of the epidermal growth factor receptor (EGFR) (Davis *et al.*, 1988; Faucher *et al.*, 1988; Northwood and Davis, 1988). This apparently occurred by a mechanism independent of PKC. Activation of PKC resulted in phosphorylation of a specific site, Thr^{654}, decreased receptor affinity for ligand, and inhibition of tyrosine kinase activity. In contrast, sphingosine enhanced EGFR phosphorylation on a different site, Thr^{669}. Site-directed mutagenesis of this locus showed it was involved in substrate selection and EGF-induced receptor downregulation (Heisermann *et al.*, 1990). These studies were interpreted as evidence that effects of sphingosine on the EGFR were not mediated by inhibition of PKC. Hence, this system provided the opportunity to compare effects of sphingosine to those of ceramide, independent of inhibition of PKC.

To circumvent the problem of the hydrophobicity of natural ceramide, a cell-permeable ceramide analog, *N*-octanoylsphingosine (termed C_8-Cer), was utilized (Goldkorn *et al.*, 1991). This molecule was designed on the same principle

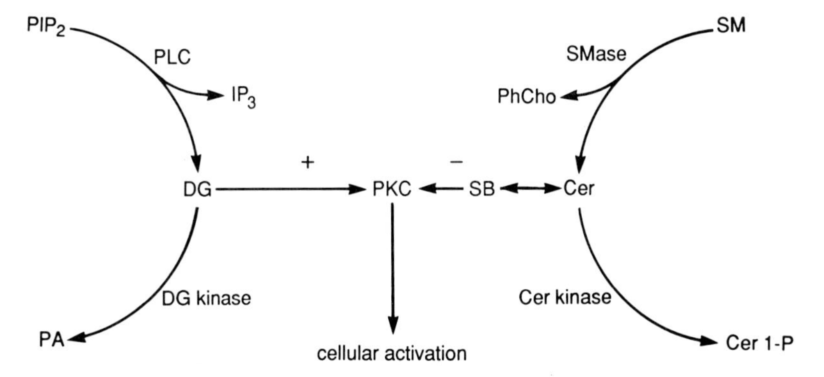

Fig. 6. Analogy between the phosphoinositide and sphingomyelin pathways. PIP_2, Phosphatidylinositol 4,5-bisphosphate; PLC, phospholipase C; IP_3, inositol trisphosphate; DG, diacylglycerol; PA, phosphatidic acid; PKC, protein kinase C; SB, sphingoid base; Cer, ceramide; SM, sphingomyelin; PhCho, phosphocholine; Cer 1-P, ceramide 1-phosphate.

as diC_8, the synthetic DG containing short-chain fatty acids, which has been extensively utilized to study activation of PKC in intact cells. Figure 7 demonstrates that C_8-Cer (0.1–10 μ*M*) induced concentration-dependent phosphorylation of the EGFR. The concentration dependence of this effect was similar to sphingosine in our laboratory. Phosphoamino acid analysis of total immunoprecipitated EGFR revealed that ceramide, like sphingosine, primarily induced Thr phosphorylation. Further, phosphopeptide maps of the EGFR from cells stimulated with sphingosine and ceramide were superimposable, as both agents enhanced phosphorylation of the Thr^{669}-containing major phosphopeptide peak resolved by high-performance liquid chromatography (HPLC).

Since sphingosine and ceramide are functionally interconvertible by acylation/deacylation reactions, studies were performed to determine which agent induced receptor phosphorylation. Table I shows that at concentrations (0.1–1 μ*M*) at which ceramide is effective to induce receptor phosphorylation, there was no conversion to sphingosine. Hence, ceramide appeared to act directly, not via conversion to sphingosine. In contrast, at concentrations (0.1–1 μ*M*) at which sphingosine was effective, exogenous sphingosine failed to increase its own cellular level. Rather, it was rapidly converted to ceramide, increasing the ceramide level 1.7-fold at 1 μ*M* sphingosine, a value comparable to that achieved by addition of a maximally effective concentration of exogenous ceramide. Concentrations of sphingosine above 1 μ*M*, which induced no further increase in ceramide levels, also induced no further increase in EGFR phosphorylation. At these concentrations, sphingosine appeared to overwhelm the acylation capacity of the cell and the level of free sphingosine increased markedly. These are the same concentrations found by most investigators to inhibit PKC activity in different systems (Merrill *et al.*, 1989). Hence, these studies provide an alternative path through which the biological effects of exogenous sphingosine may be mediated, i.e., by conversion to ceramide. These studies showed that ceramide had bioeffector properties and suggested the existence of a ceramide-activated threonine kinase since a specific site on the EGFR, Thr^{669}, was phosphorylated despite a multiplicity of other potential phosphorylation sites.

Subsequent investigations were aimed at identifying the kinase that mediated the effect of ceramide on EGFR phosphorylation (Mathias *et al.*, 1991). A synthetic peptide derived from the amino acid sequence around Thr^{669} of the EGFR was used as the substrate. This peptide represents the amino acid segment 663–681 and has a unique structure due to the presence of three proline residues within a span of nine amino acids. Phosphorylated peptide was usually resolved by HPLC, followed by Cerenkov counting of eluted fractions. Postnuclear supernates from A431 cells served as a source of enzyme activity. Incubation of the supernates with [γ-^{32}P]ATP and peptide resulted in time-dependent phosphorylation of the peptide (Fig. 8A). Kinase activity was detectable only at physiological pH (7–7.4) and was Mg^{2+}-dependent ($ED_{50} \approx 3.5$ m*M*). Other divalent

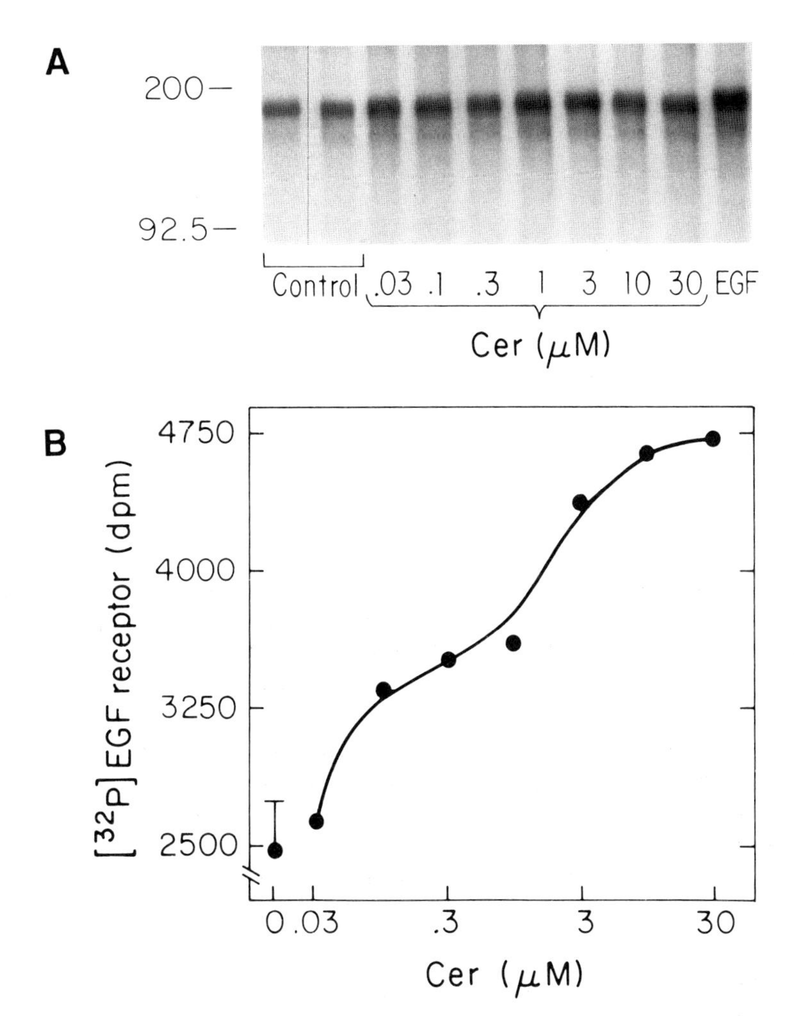

FIG. 7. Effect of ceramide on phosphorylation of the EGF receptor. (A) Autoradiogram of an individual experiment. (B) Compilation of data from four experiments.

cations such as Ca^{2+}, Mn^{2+}, and Zn^{2+} did not support activity. The activity was linearly related to substrate concentrations, with an apparent *K*m of 15 μ*M* for ATP and 0.25 mg/ml for the peptide. Cell fractionation studies revealed that kinase activity was equally distributed between membrane and cytosolic fractions. However, recent studies have suggested that the cytosolic activity is composed of other kinases that may recognize the same substrate (see below). The apparent V_{max} values for membrane activity ranged between 50 and 100 pmol/min/mg of protein.

Table I
EFFECT OF EXOGENOUS CERAMIDE AND SPHINGOSINE ON CELLULAR CERAMIDE AND SPHINGOSINE LEVELS

	Cellular level (pmol/10^6 cells)	
Stimulant	Ceramide	Sphingosine
Control	132	12
Ceramide (μ*M*)		
0.1	183	15
1	232	13
10	306	25
Sphingosine (μ*M*)		
0.1	145	12
0.3	162	12
1	227	20
10	232	326

Only membrane-associated activity was stimulated by ceramide. Ceramide stimulation of kinase activity was shown to be time- and concentration-dependent (Fig. 8). A maximal effect to 2.1-fold of control occurred with 1 μ*M* ceramide. Synthetic and natural ceramides were equally effective. Phosphoamino acid analysis showed that phosphorylation occurred exclusively on the Thr residue of the peptide, despite availability within the peptide of another potential phosphorylation site, Ser^{671}. Sphingosine also stimulated kinase activity, at levels comparable to those of ceramide. However, palmitic acid, the predominant fatty acid in natural ceramide, did not stimulate peptide phosphorylation, suggesting that the sphingoid base backbone is a critical determinant for stimulation.

D. SIGNAL TRANSDUCTION THROUGH THE SPHINGOMYELIN PATHWAY

Studies by Bell, Hannun, and co-workers provided evidence that vitamin D_3, tumor necrosis factor α (TNF-α), and γ-interferon stimulated a sphingomyelinase and increased cellular ceramide levels as an early event in monocytic differentiation of HL-60 cells (Okazaki *et al.*, 1989, 1990; Kim *et al.*, 1991). For vitamin D_3 and γ-interferon, peak activity occurred at 2 hours, and, for TNF-α, it occurred at 30–60 minutes. Synthetic ceramide enhanced or mimicked these agents. Exogenous sphingomyelinase acted similarly.

FIG. 8 (A) Time and (B) concentration dependence of ceramide-induced phosphorylation of the EGFR peptide.

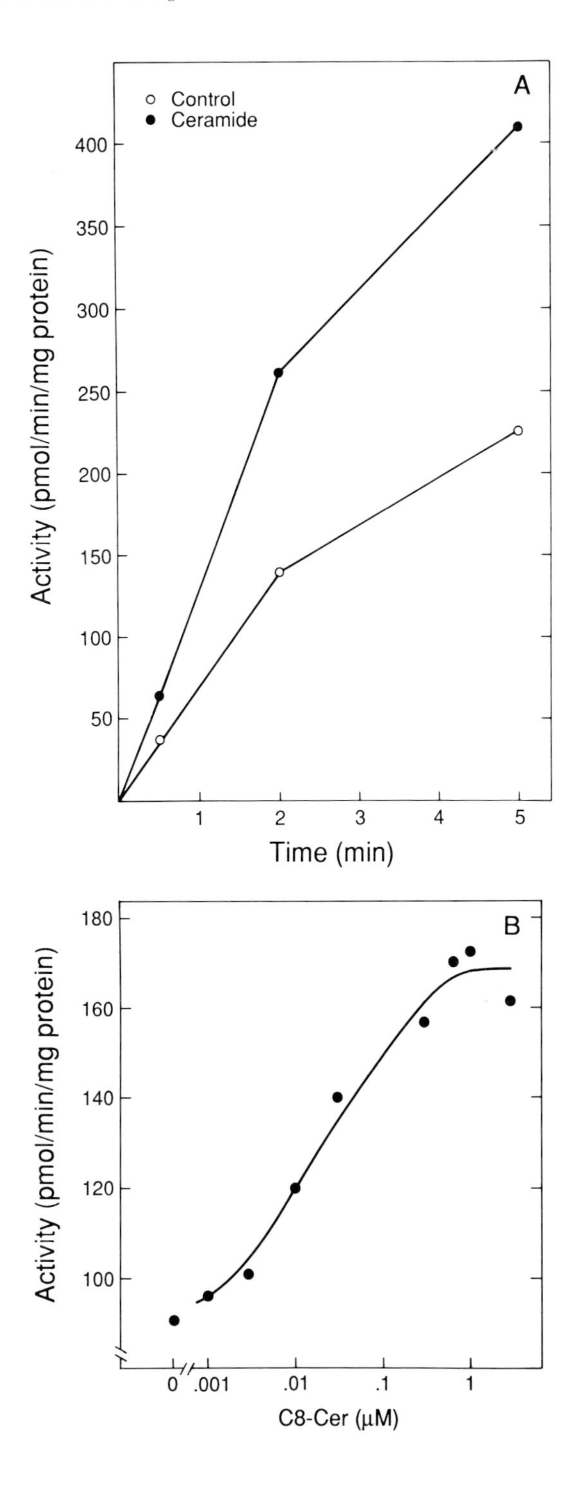
A
Control
Ceramide
Activity (pmol/min/mg protein)
400
350
300
250
200
150
100
50
1
2
3
4
5
Time (min)
B
Activity (pmol/min/mg protein)
180
160
140
120
100
0
.001
.01
.1
1
C8-Cer (μM)

Further investigations were conducted in this laboratory to determine whether TNF-α, by generating ceramide endogenously, activated a kinase in HL-60 cells similar to that detected in A431 cells (Mathias *et al.,* 1991). Initial studies demonstrated the existence of a ceramide-activated protein kinase activity in membranes from HL-60 cells. This activity was indistinguishable from that detected in A431 cells. Subsequent investigations from this laboratory extended the observations of Hannun and colleagues. These studies showed that stimulation of intact HL-60 cells with TNF-α resulted in very rapid degradation of sphingomyelin and a concomitant increase in ceramide (Fig. 9). Further, membrane-associated kinase activity was enhanced within 5 minutes and continued to increase in a concentration-dependent manner for as long as 2 hours to a maximum of 2.2-fold of control (Fig. 10). Treatment of intact HL-60 cells with synthetic cell-permeable ceramide or sphingosine also enhanced kinase activity in membranes derived from these cells. However, other agents, such as retinoic acid, cholera toxin, butyrate, and hexamethylene bisacetamide (Chaplinski and Niedel, 1982; Breitman *et al.,* 1980; Marks and Rifkind, 1984), which induce differentiation of HL-60 cells along a variety of lineages, did not have any effect on kinase activity (Fig. 11). This suggested that the pathway is specific for TNF-α (or perhaps monocytic generation in general). These studies demonstrated that HL-60 cells contain a ceramide-activated protein kinase and that TNF-α, which rapidly generates ceramide during cell stimulation, enhanced kinase activity.

Early kinetics do not necessarily designate a reaction as the most proximal event in a signal transduction cascade. To demonstrate that signal transduction for TNF-α is mediated via the sphingomyelin pathway, an attempt was made to reconstitute signaling in a cell-free system (Dressler *et al.,* 1992). To show direct evidence of the activation of the sphingomyelin pathway by TNF-α, postnuclear supernates from HL-60 cells were first incubated with TNF-α at 4°C to allow the formation of TNF–receptor complexes (Imamura *et al.,* 1988). The supernates were then warmed to room temperature under conditions which would allow for activation of neutral sphingomyelinase (Rao and Spence, 1976; Kolesnick, 1987). Under these conditions, TNF-α induced time- and concentration-dependent reduction in sphingomyelin content (Fig. 12a). The effect of TNF-α was evident at 1 minute and maximal by 7.5 minutes. Sphingomyelin concentrations decreased 27%, while ceramide increased quantitatively. As little as 300 p*M* was effective and a maximal effect occurred at 3 n*M* TNF-α; an $ED_{50} \approx 500$ p*M*. TNF-α did not affect the levels of DG or other choline-containing lipids, including phosphatidylcholine, lysophosphatidylcholine, and sphingosylphosphorylcholine. Thus, TNF-α activated a neutral sphingomyelinase in a cell-free system, resulting in generation of the potential second messenger ceramide.

The effect of TNF-α on ceramide-activated protein kinase activity was also assessed. TNF-α treatment enhanced kinase activity in a time- and concentration-dependent manner (Fig. 12b). Stimulation of kinase activity was detected by

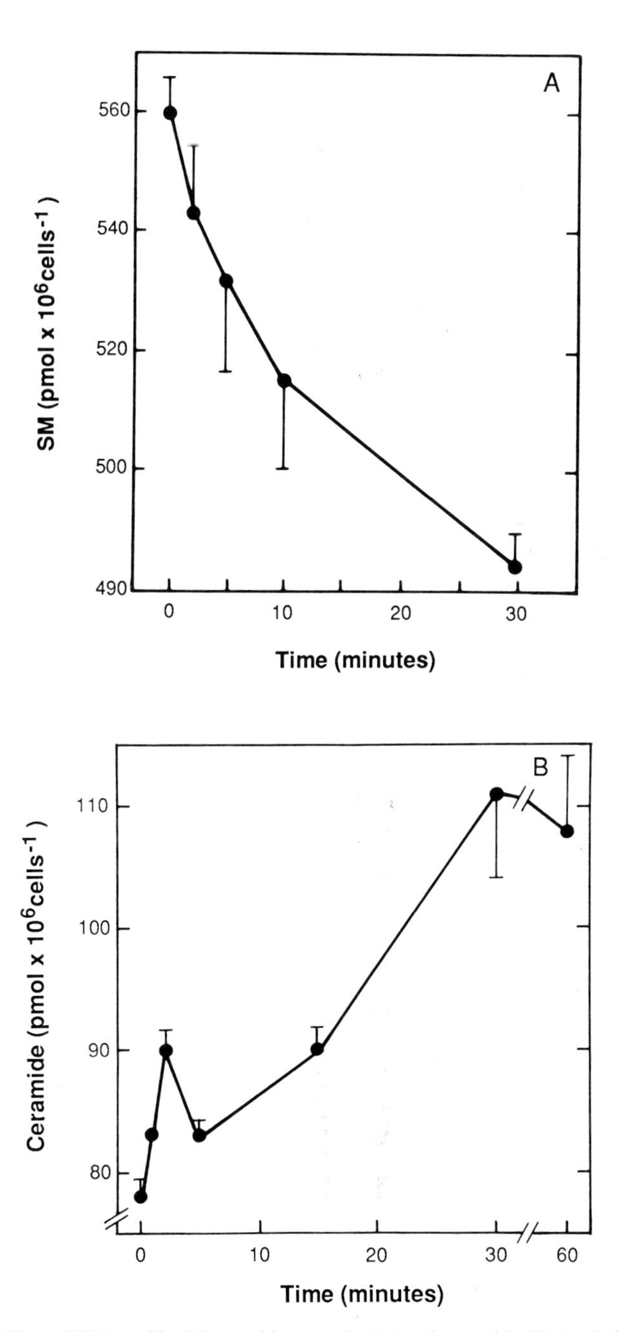

FIG. 9. Effect of TNF-α (30 n*M*) on sphingomyelin (A) and ceramide (B) levels in intact HL-60 cells.

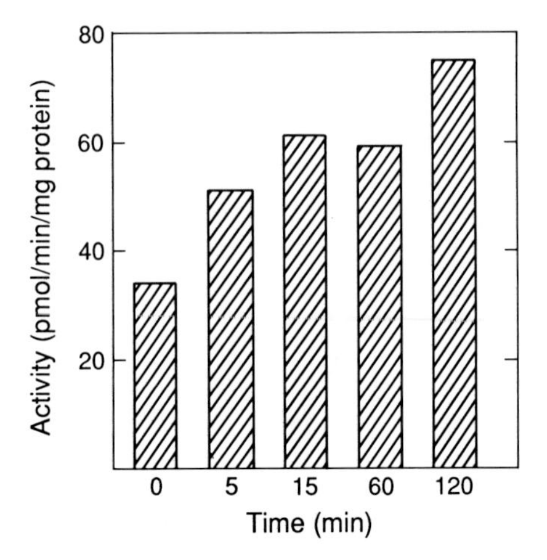

FIG. 10. Effect of TNF-α on ceramide-activated protein kinase activity in membranes derived from stimulated HL-60 cells.

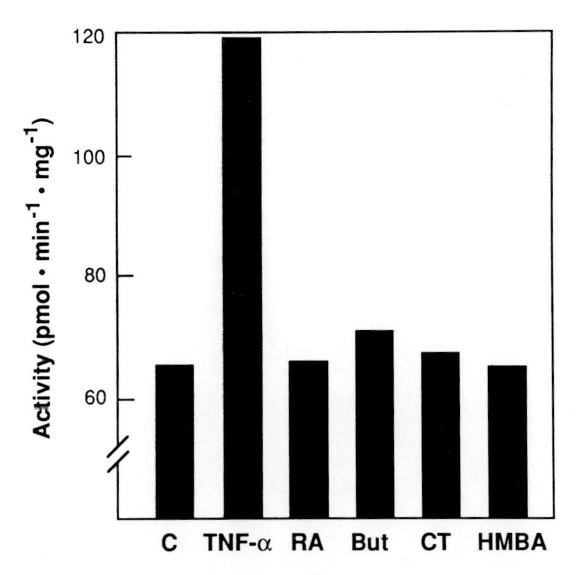

FIG. 11. Effect of differentiating agents on ceramide-activated kinase activity in membranes derived from stimulated HL-60 cells. C, Control; TNF-α, tumor necrosis factor α (30 n*M*); RA, retinoic acid (0.5 μ*M*); But, butyrate (0.5 m*M*); CT, cholera toxin (10 n*M*); HMBA, hexamethylene bisacetamide (1.5 m*M*).

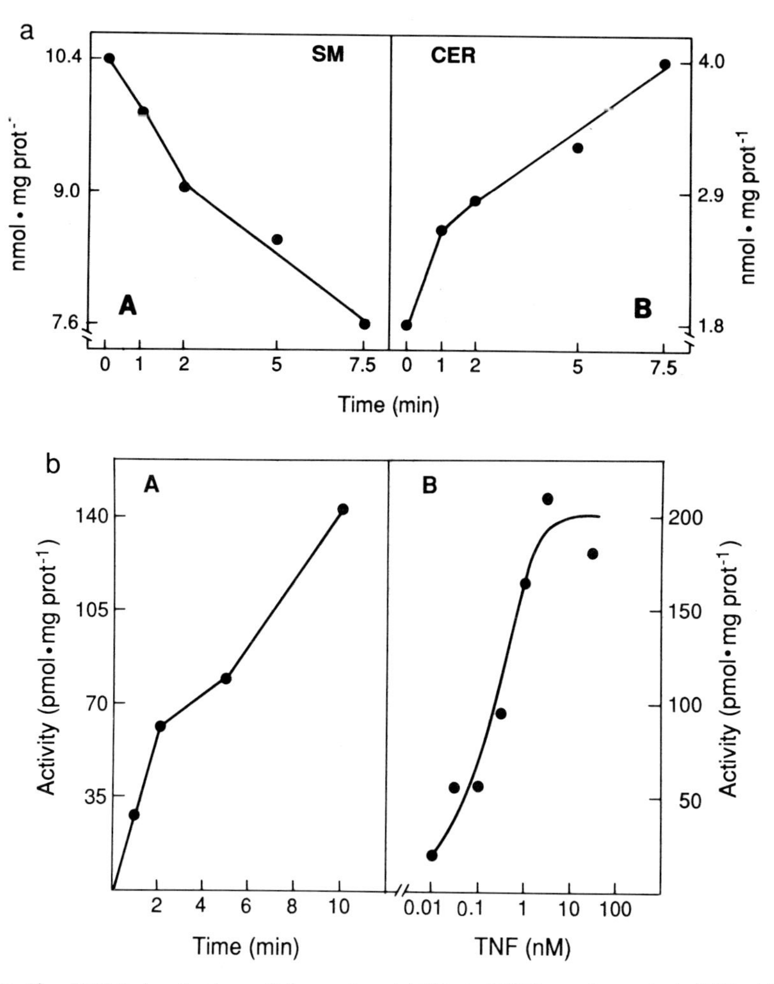

FIG. 12. TNF-α signaling in a cell-free system. (a) Effect of TNF-α on the content of (A) sphingomyelin and (B) ceramide. (b) Effect of TNF-α on ceramide-activated kinase activity: (A) kinetics, (B) concentration dependence.

1 minute and evident for at least 10 minutes. TNF-α was effective in the 10 p*M* to 3 n*M* range; an $ED_{50} \approx 300$ p*M* TNF-α. This is similar to the ED_{50} values of 100–300 p*M* TNF reported for stimulation of neutral sphingomyelinase (Kim *et al.*, 1991), ceramide-activated protein kinase activity (Mathias *et al.*, 1991), and monocyte differentiation (Kim *et al.*, 1991) in intact HL-60 cells. Hence, TNF-induced sphingomyelinase activation was accompanied by stimulation of ceramide-activated protein kinase in our cell-free system.

To demonstrate that this effect of TNF-α was mediated by sphingomyelin hydrolysis to ceramide, the cell-free supernates were treated with sphingomyelinase (1×10^{-3} U/ml) from *Staphylococcus aureus*. This concentration of sphingomyelinase stimulated a 2-fold elevation in ceramide levels in intact HL-60 cells (Kolesnick, 1989a; Dressler and Kolesnick, 1990). Under these conditions, enhancement in kinase activity was comparable to that induced by TNF-α (1 n*M*). In contrast, phospholipases A_2, C, and D, at concentrations 40 to 400-fold higher than that of sphingomyelinase, did not enhance kinase activity. Hence, the effect of TNF-α in broken cell preparations was mimicked by sphingomyelinase but not by other phospholipases.

The mechanism of coupling of the TNF receptor to sphingomyelinase has not yet been investigated. The TNF receptor exists in two isoforms of 55 and 75 kDa (Loetscher *et al.*, 1990; Schall *et al.*, 1990; Smith *et al.*, 1990; Dembič *et al.*, 1990). It is not known which receptor form is linked to the sphingomyelin pathway, although both forms are found in HL-60 cells (Brockhaus *et al.*, 1990; Thoma *et al.*, 1990). The use of activating and blocking receptor antibodies and cell lines containing a single receptor form (Brockhaus *et al.*, 1990; Thoma *et al.*, 1990; Hohmann *et al.*, 1989; Espevik *et al.*, 1990; Engelmann *et al.*, 1990; Ryffel *et al.*, 1991) may resolve this issue. Neutral sphingomyelinase appears to be ubiquitous in mammalian cells and is externally oriented in the plasma membrane (Das *et al.*, 1984). Similarly, sphingomyelin is preferentially localized to the outer leaflet of the plasma membrane (Barenholz and Thompson, 1980). This colocalization of receptor, phospholipase, and substrate at the plasma membrane suggests that ceramide is generated at this site. The exact intracellular site of the ceramide-activated protein kinase has not yet been investigated but preliminary evidence suggests it is an intrinsic membrane protein. However, the kinase would not have to be present in the outer leaflet of the plasma membrane in order for signaling to occur, as ceramide can redistribute across a membrane bilayer (Lipsky and Pagano, 1983).

Ceramide-activated protein kinase may be a member of an emerging family of serine/threonine protein kinases that includes microtubule-associated protein 2 (MAP2) kinases/extracellular signal-regulated kinases (ERK) (Seger *et al.*, 1991; Boulton *et al.*, 1991; Northwood *et al.*, 1991), glycogen synthase kinase-3 (Aitken *et al.*, 1984), and $p34^{cdc2}$ kinases (Hall *et al.*, 1991). These kinases are cytosolic whereas ceramide-activated protein kinase appears membrane associated. Furthermore, ceramide-activated protein kinase is not anionic, whereas the MAP/ERK kinases have been reported as strongly anionic. The substrates for these kinases appear to have a minimal recognition sequence, X-Ser/Thr-Pro-X, in which the phosphorylated site is flanked by a carboxyl-terminal proline residue (Hall *et al.*, 1991; Alvarez *et al.*, 1991) and X can be any amino acid. The proline residue is required for recognition of the substrate by the kinase. Evidence has been presented that the residues around the minimal recognition sequence also modulate substrate recognition (Gonzalez *et al.*, 1991). Substrates for this class of kinases include EGFR, protooncogene products c-*jun* and c-*myc*, tyrosine hydroxylase, histone H1, glycogen synthase, synapsin I, and protein phosphatase inhibitor II (Northwood *et al.*, 1991; Alvarez *et al.*, 1991; Hall *et al.*, 1991; Aitken *et al.*, 1984). TNF-induced proline-directed phosphorylation of these proteins has not yet been demonstrated. The X-Ser/Thr-Pro-X sequence is different from consensus substrate sequences for other major serine/threonine kinases (Heisermann and Gill, 1988). In fact, PKC, cyclic adenosine monophosphate (cAMP)- and cyclic guanosine monophosphate (cGMP)-dependent protein kinases, Ca^{2+}/calmodulin-dependent protein kinase, and ribosomal S6 protein kinase have only limited activity toward this proline-containing sequence (Heisermann *et al.*, 1990).

Various distinct signaling systems, including protein kinases A and C, phospholipases A_2 and C, the EGFR tyrosine kinase, and a novel serine kinase, have been reported to mediate TNF-α action (Vilcek and Lee, 1991). However, no single second messenger pathway accounts for all of the reported biological effects of TNF-α. The role of the sphingomyelin pathway in events other than monocytic differentiation has not been investigated, nor has its relationship to these other signaling systems been determined. However, the rapid kinetics of activation of the sphingomyelin pathway by TNF-α in intact cells, the ability of cell-permeable ceramide analogs to bypass receptor activation and mimic TNF-α action, and the reconstitution of this cascade in a cell-free system provide strong support for the notion that this pathway serves to couple TNF receptor activation to cellular stimulation (Fig. 13).

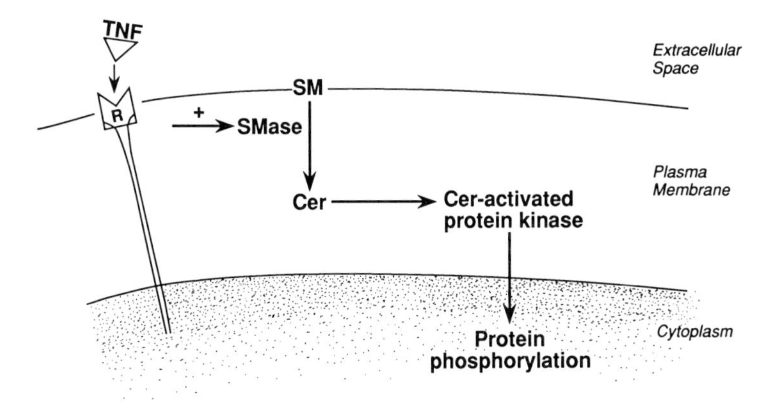

FIG. 13. Activation of the sphingomyelin pathway by TNF-α. R, Receptor; SM, sphingomyeline; Cer, ceramide.

III. Sphingolipid Synthesis and Signaling

A. Sphingomyelin Synthesis and Signaling

Sphingomyelin synthesis is mediated by the enzyme phosphatidylcholine:ceramide cholinephosphotransferase and occurs by transfer of phosphocholine from phosphatidylcholine to ceramide (Diringer *et al.*, 1972; Marggraf *et al.*, 1972; Ullman and Radin, 1974). It has been demonstrated that this reaction occurs chiefly in the *cis* and medial cisternae of the Golgi apparatus (Pagano, 1990a,b). Subsequently, the majority of sphingomyelin is transferred to the plasma membrane via a vesicular default pathway.

There is limited information concerning receptor-mediated sphingomyelin synthesis or the role of sphingomyelin synthesis in signal transduction. Nelson and co-workers first reported that dexamethasone stimulated a selective increase in the sphingomyelin content of rat epididymal fat cell ghosts obtained from adrenalectomized animals (Murray *et al.*, 1979). This event correlated with inhibition of glucose transport. Subsequently, the stimulatory effect of dexamethasone on sphingomyelin synthesis has also been demonstrated in various other cell types (Johnston *et al.*, 1980; Nelson and Murray, 1982; Nelson *et al.*, 1982).

Stimulation of sphingomyelin synthesis through a cell surface receptor was first reported by this laboratory for thyrotropin-releasing hormone (TRH) in GH_3 pituitary cells (Kolesnick, 1989b). This effect appeared to occur subsequent to phosphatidylcholine synthesis and was mediated via PKC. In contrast, TRH-induced phosphatidylinositol synthesis occurred independently of PKC. The role of stimulated sphingomyelin synthesis in TRH action was not investigated.

In contrast to the effect of TNF (Kim *et al.*, 1991; Mathias *et al.*, 1991; Dressler *et al.*, 1992) and vitamin D_3 (Okazaki *et al.*, 1989, 1990) on sphingomyelin degradation, Kiss *et al.* (1988) showed that phorbol esters stimulated early sphingomyelin synthesis in HL-60 cells. This was measured by enhanced incorporation of [^{3}H]choline into sphingomyelin in short-term radiolabeled cells. Effects on sphingomyelin mass were not measured in these studies.

Recent studies from this laboratory have extended these observations and demonstrated that early sphingomyelin synthesis defines a population of promyelocytes destined to become adherent macrophages (Dressler *et al.*, 1991). Phorbol esters increased the mass of sphingomyelin within the first hour of stimulation. After 14 hours, an adherent macrophage population developed. Table II shows that this population contained the entire elevation of sphingomyelin content detected in the total cell population. Further, direct elevation of cellular sphingomyelin levels with sphingomyelin vesicles (Gatt and Bierman, 1980) or with sphingosylphosphorylcholine, which may be acylated to sphingomyelin (Sugiyama *et al.*, 1990), potentiated the effect of a submaximal concentration of

Table II
EFFECT OF PHORBOL ESTER ON SPHINGOMYELIN AND PHOSPHATIDYLCHOLINE LEVELS IN NONADHERENT AND ADHERENT CELLS

	Level (pmol/10^6 cells)	
	Sphingomyelin	Phosphatidylcholine
Control	560 ± 62	5040 ± 403
TPA		
Nonadherent	644 ± 106	4486 ± 605
Adherent	907 ± 78[a]	5685 ± 554

[a] $p < 0.001$ versus control.

phorbol ester on the development of the adherent macrophage population. Hence, sphingomyelin synthesis correlated closely with the development of an adherent macrophage population during differentiation.

Studies were also performed to assess the role of sphingomyelin in the adherence process. For these studies, cells were differentiated into adherent macrophages by a 48-hour incubation with a maximal concentration of phorbol ester. Thereafter, the medium was replaced with medium containing a variety of phospholipases or sphingoid bases and the ability of these agents to induce detachment of adherent cells was determined. Figure 14 demonstrates that only sphingomyelinase was sufficient to induce detachment. These studies support an as-yet-undefined role of sphingomyelin in the adherence process. In this regard, phorbol esters also induced sphingomyelin synthesis and adherence of hairy cell leukemia cells in culture (Lockney *et al.*, 1984). Since the lineage of this B cell leukemia is separate from that of HL-60 cells, the association of adherence and sphingomyelin synthesis may be a general property of PKC activation.

Pagano (1988) has suggested that stimulated sphingomyelin synthesis might also serve another purpose, i.e., as a mechanism for selective generation of DG at the Golgi apparatus. This is particularly relevant in light of the observations that PKC translocates not only to the plasma membrane but to intracellular membranes as well. In fact, in a rat neuronal cell line, the βII subspecies has recently been localized to the Golgi apparatus (Nishizuka, 1989).

B. ROLE OF CERAMIDE IN GLYCOSPHINGOLIPID SYNTHESIS AND SIGNALING

Glycosphingolipids are known to influence cellular activities such as differentiation and adhesion (Hakomori, 1990). However, the mechanisms involved are not yet well understood. Since ceramide forms the core structure of all glycosphingolipids, an analog of ceramide synthesized by Radin and co-workers

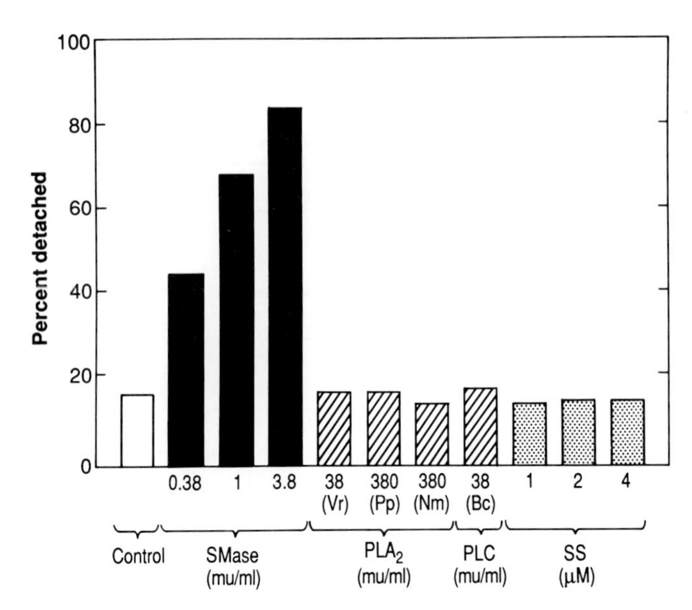

FIG. 14. Effect of phospholipases and sphingosine on detachment of adherent macrophages. Vr, *Vipera ruselli;* Pp, porcine pancreas; Nm, *Naja mocambique mocambique;* Bc, *Bacillus cereus;* SMase, sphingomyelinase; PLA_2, phospholipase A_2; PLC, phospholipase C; SS.

(Vunnam and Radin, 1980; Radin and Vunnam, 1981), D-*threo*-1-phenyl-2-decanoylamino-3-morpholino-1-propanol (PDMP), was used to investigate the biological role(s) of glycosphingolipids. PDMP is a competitive inhibitor of UDPglucose:*N*-acylsphingosine glucosyltransferase (EC 2.4.1.80), an enzyme which regulates conversion of ceramide to glucosylceramide (GlcCer). This is the first step in the synthesis of a large number of glycosphingolipids. During macrophage differentiation, phorbol esters stimulate a 6- to 12-fold increase in the levels of GlcCer and sialosyllactosylceramide (G_{M3}), specifically in the adherent population. Based on this information, PDMP was used to determine whether glycosphingolipids play a general role in the differentiation process itself or selectively affect adherence. Several biological events which correlate with differentiation were studied. Initial studies demonstrated that PDMP markedly reduced glycosphingolipid levels in resting cells and blocked phorbol ester-induced synthesis. PDMP also caused an 80% inhibition of adherence induced by phorbol esters (Fig. 15A). However, other aspects of phorbol ester-induced differentiation, such as inhibition of growth (Fig. 15B), transcription of the protooncogene c-*fos* (Fig. 15C), appearance of enzymatic markers, and changes in morphology were unaffected by PDMP. These studies suggest a specific role for glycosphingolipid synthesis in the adherence process during macrophage differentiation.

IV. Summary

The data discussed in the preceding sections suggest that information may be transmitted both through synthesis and through degradation of sphingomyelin. Although the sphingomyelin pathway holds promise as a new signaling system coupling TNF receptor activation to cellular stimulation (Fig. 5), the work is still

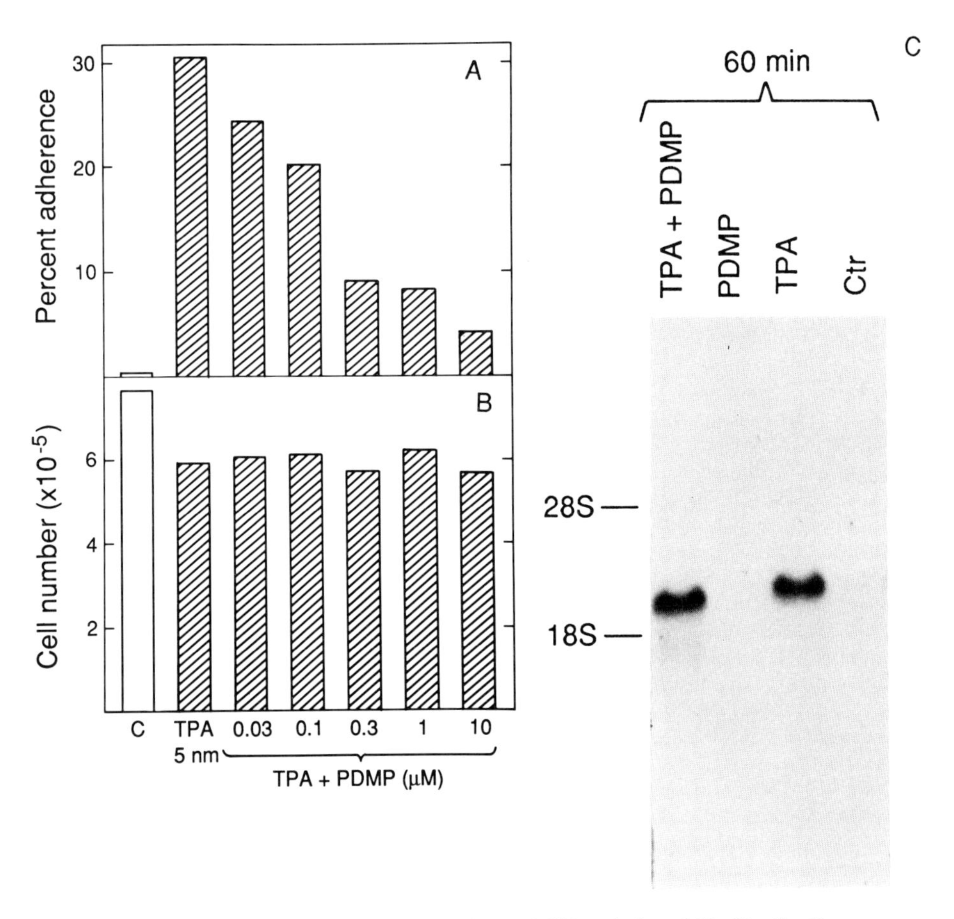

FIG. 15. Effect of PDMP on phorbol ester-induced differentiation of HL-60 cells. Shown are the effects on (A) adherence, (B) growth, (C) expression of c-*fos*.

at a preliminary stage. A physical association of receptors with neutral sphingomyelinase has yet to be established and the ceramide-activated protein kinase has yet to be isolated. Endogenous substrates for the kinase have also to be identified. Furthermore, the exact role of this pathway has not been defined. It is unclear whether this pathway is specific to monocyte differentiation or cytokine action. It also seems unlikely that a single signal transduction mechanism can account for all the diverse effects of TNF-α in different systems (Vilcek and Lee, 1991). Interactions with other signaling systems are sure to complicate elucidation of the exact role(s) of this pathway. Nevertheless, availability of cell permeable analogs of ceramide, localization of many components of the system at the cell surface, and recent development of anti-TNF receptor antibodies to receptor isotypes may allow for greater definition of the sphingomyelin pathway in the near future.

Acknowledgments

RNK is supported by grants RO-1-CA-42385 from the National Institutes of Health and FRA-345 from the American Cancer Society.

References

Aitken, A., Holmes, C. F. B., Campbell, D. G., Resnick, T. J., Cohen, P., Leung, C. T. W., and Williams, D. H. (1984). *Biochim. Biophys. Acta* **790,** 288–291.

Alvarez, E., Northwood, I. C., Gonzalez, F. A., Latour, D. A., Seth, A., Abate, C., Curran, T., and Davis, R. J. (1991). *J. Biol. Chem.* **266,** 15277–15285.

Bajjalieh, S. M., Martin, T. F. J., and Floor, E. (1989). *J. Biol. Chem.* **264,** 14354–14360.

Barenholz, Y., and Gatt, S. (1982). *In* "Phospholipids" (J. N. Hawthorne and G. B. Ansel, eds.), pp. 129–177. Elsevier, Amsterdam.

Barenholz, Y., and Thompson, T. E. (1980). *Biochim. Biophys. Acta* **604,** 129–158.

Berridge, M. J. (1989). *JAMA, J. Am. Med. Assoc.* **262,** 1834–1841.

Boulton, T. G., Nye, S. H., Robbins, D. J., Ip, N. Y., Radziejewska, E., Morgenbesser, S. D., DePinho, R. A., Panayotatos, N., Cobb, M. H., and Yancopoulos, G. D. (1991). *Cell (Cambridge, Mass.)* **65,** 663–675.

Breitman, T. R., Selonick, S. E., and Collins, S. J. (1980). *Proc. Natl. Acad. Sci. U.S.A.* **77,** 2936–2940.

Brockhaus, M., Schoenfeld, H. J., Schlaeger, E. J., Hunziker, W., Lesslauer, W., and Loetscher, H. (1990). *Proc. Natl. Acad. Sci. U.S.A.* **87,** 3127–3131.

Chaplinski, T. J., and Niedel, J. E. (1982). *J. Clin. Invest.* **70,** 953–964.

Das, D. V. M., Cook, H. W., and Spence, M. W. (1984). *Biochim. Biophys. Acta* **777,** 339–342.

Davis, R. J., Girones, N., and Faucher, M. (1988). *J. Biol. Chem.* **263,** 5373–5379.

Dembič, Z., Loetscher, H., Gubler, U., Pan, Y.-C. E., Lahm, H.-W., Gentz, R., Brockhaus, M., and Lerner, W. (1990). *Cytokine* **2,** 231–237.

Diringer, H., Marggraf, W. D., Koch, M. A., and Anderer, F. A. (1972). *Biochem. Biophys. Res. Commun.* **47,** 1345–1352.

Dressler, K. A., and Kolesnick, R. N. (1990). *J. Biol. Chem.* **265,** 14917–14921.

Dressler, K. A., Kan, C.-C., and Kolesnick, R. N. (1991). *J. Biol. Chem.* **266,** 11522–11527.
Dressler, K. A., Mathias, S., and Kolesnick, R. N. (1992). *Science* **255,** 1715–1718.
Engelmann, H., Holtmann, H., Brakebusch, C., Avni, Y. S., Sarov, I., Nophar, Y., Hadas, E., Leitner, O., and Wallach, D. (1990). *J. Biol. Chem.* **265,** 14497–14504.
Espevik, T., Brockhaus, M., Loetscher, H. R., Nonstad, U., and Shalaby, R. (1990). *J. Exp. Med.* **171,** 415–426.
Faucher, M., Girones, N., Hannun, Y. A., Bell, R. M., and Davis, R. J. (1988). *J. Biol. Chem.* **263,** 5319–5327.
Gatt, S., and Bierman, E. L. (1980). *J. Biol. Chem.* **255,** 3371–3376.
Goldkorn, T., Dressler, K. A., Muindi, J., Radin, N. S., Mendelsohn, J., Menaldino, D., Liotta, D., and Kolesnick, R. N. (1991). *J. Biol. Chem.* **266,** 16092–16097.
Gonzalez, F. A., Raden, D. L., and Davis, R. J. (1991). *J. Biol. Chem.* **266,** 22159–22163.
Hakomori, S.-I. (1990). *J. Biol. Chem.* **265,** 18713–18716.
Hall, F. L., Braun, R. K., Mihara, K., Fung, Y.-K. T., Berndt, N., Carbonaro-Hall, D. A., and Vulliet, P. R. (1991). *J. Biol. Chem.* **266,** 17430–17440.
Hannun, Y. A., Loomis, C., Merrill, A. H., Jr., and Bell, R. M. (1986). *J. Biol. Chem.* **261,** 12604–12609.
Heisermann, G. J., and Gill, G. N. (1988). *J. Biol. Chem.* **263,** 13152–13158.
Heisermann, G. J., Wiley, H. S., Walsh, B. J., Ingraham, H. A., Fiol, C. J., and Gill, G. N. (1990). *J. Biol. Chem.* **265,** 12820–12827.
Hohmann, H. P., Remy, R., Brockhaus, M., and van Loon, A. P. G. M. (1989). *J. Biol. Chem.* **264,** 14927–14934.
Imamura, K., Sherman, M. L., Spriggs, D., and Kufe, D. (1988). *J. Biol. Chem.* **263,** 10247–10253.
Johnston, D., Matthews, E. D., and Melnkovych, G. (1980). *Endocrinology (Baltimore)* **107,** 1482–1488.
Kan, C.-C., and Kolesnick, R. N. (1992). *J. Biol. Chem.* **267,** 9663–9667.
Kanoh, H., Kondoh, H., and Ono, T. (1983). *J. Biol. Chem.* **258,** 1767–1774.
Kanoh, H., Iwata, T., Ono, T., and Suzuki, T. (1986). *J. Biol. Chem.* **261,** 5597–5602.
Kanoh, H., Yamada, K., and Sakane, F. (1990). *Trends Biochem. Sci.* **15,** 47–50.
Kikkawa, U., Kishimoto, A., and Nishizuka, Y. (1989). *Annu. Rev. Biochem.* **58,** 31–44.
Kim, M.-Y., Linardič, C., Obeid, L., and Hannun, Y. (1991). *J. Biol. Chem.* **266,** 484–489.
Kiss, Z., Deli, E., and Kuo, J. F. (1988). *Arch. Biochem. Biophys.* **265,** 38–42.
Kolesnick, R. N. (1987). *J. Biol. Chem.* **262,** 16759–16762.
Kolesnick, R. N. (1989a). *J. Biol. Chem.* **264,** 7617–7623.
Kolesnick, R. N. (1989b). *J. Biol. Chem.* **264,** 11688–11692.
Kolesnick, R. N., and Clegg, S. (1988). *J. Biol. Chem.* **263,** 6534–6537.
Kolesnick, R. N., and Hemer, M. (1989). *J. Biol. Chem.* **264,** 14057–14061.
Kolesnick, R. N., and Hemer, M. (1990). *J. Biol. Chem.* **265,** 18803–18808.
Lipsky, N. G., and Pagano, R. E. (1983). *Cell Biol.* **80,** 2608.
Lockney, M. W., Golomb, H. M., and Dawson, G. (1984). *Biochim. Biophys. Acta* **796,** 384–392.
Loetscher, H., Pan, Y.-C. E., Lahm, H.-W., Gentz, R., Brockhaus, M., and Tabuchi, H. (1990). *Cell (Cambridge, Mass.)* **61,** 351–359.
Marggraf, W. D., Diringer, H., Koch, M. A., and Anderer, F. A. (1972). *Hoppe-Seyler's Z. Physiol. Chem.* **353,** 1761–1768.
Marks, P. A., and Rifkind, R. A. (1984). *Cancer (Philadelphia)* **54,** 2766–2769.
Mathias, S., Dressler, K. A., and Kolesnick, R. N. (1991). *Proc. Natl. Acad. Sci. U.S.A.* **88,** 10009–10013.
Merrill, A. H., Jr., Sereni, A. M., Stevens, V. L., Hannun, Y. A., and Bell, R. M. (1986). *J. Biol. Chem.* **261,** 12610–12615.

Merrill, A. H., Jr., Nimkar, S., Menaldino, D., Hannun, Y. A., Loomis, C., Bell, R. M., Tyagi, S. R., Lambeth, J. D., Stevens, V. L., Hunter, R., and Liotta, D. C. (1989). *Biochemistry* **28,** 3138–3145.

Murray, D. K., Ruhmann-Wennhold, A., and Nelson, D. H. (1979). *Endocrinology (Baltimore)* **105,** 774–777.

Nelson, D. H., and Murray, D. K. (1982). *Proc. Natl. Acad. Sci. U.S.A.* **79,** 6690–6692.

Nelson, D. H., Murray, D. K., and Brady, R. O. (1982). *J. Clin. Endocrinol. Metab.* **54,** 292–295.

Nishizuka, Y. (1988). *Nature (London)* **334,** 661–665.

Nishizuka, Y. (1989). *JAMA, J. Am. Med. Assoc.* **262,** 1826–1833.

Northwood, I. C., and Davis, R. J. (1988). *J. Biol. Chem.* **263,** 7540–7543.

Northwood, I. C., Gonzalez, F. A., Wartmann, M., Raden, D. L., and Davis, R. J. (1991). *J. Biol. Chem.* **266,** 15266–15276.

Okazaki, T., Bell, R. M., and Hannun, Y. A. (1989). *J. Biol. Chem.* **264,** 19076–19080.

Okazaki, T., Bielawaska, A., Bell, R. M., and Hannun, Y. A. (1990). *J. Biol. Chem.* **265,** 15823–15831.

Pagano, R. E. (1988). *Trends Biochem. Sci.* **13,** 202–205.

Pagano, R. E. (1990a). *Curr. Opin. Cell Biol.* **2,** 652–663.

Pagano, R. E. (1990b). *Biochem. Soc. Trans.* **18,** 361–366.

Radin, N. S., and Vunnam, R. R. (1981). *In* "Methods in Enzymology" (J. Lowenstein, ed.), vol. 72, pp. 673–684. Academic Press, New York.

Rao, B. G., and Spence, M. W. (1976). *J. Lipid Res.* **17,** 506–515.

Ryffel, B., Brockhaus, M., Durmuller, U., and Gudat, F. (1991). *Am. J. Pathol.* **139,** 7–15.

Sakane, F., Yamada, K., Kanoh, H., Yokoyama, C., and Tanabe, T. (1990). *Nature (London)* **344,** 345–348.

Schall, T., Lewis, M., Koller, K. J., Lee, A., Rice, G. C., Wong, G. H. W., Gatanaga, T., Granger, G. A., Lentz, R., Raab, H., Kohr, W. J., and Goedell, (1990). *Cell (Cambridge, Mass.)* **61,** 361–370.

Schneider, E. G., and Kennedy, E. P. (1973). *J. Biol. Chem.* **248,** 3739–3741.

Seger, R., Ahn, N. G., Boulton, T. G., Yancopoulos, G. D., Panayotatos, N., Radziejewska, E., Ericsson, L., Bratlien, R. L., Cobb, M. H., and Krebs, E. G. (1991). *Proc. Natl. Acad. Sci. U.S.A.* **88,** 6142–6146.

Slife, C. W., Wang, E., Hunter, R., Wang, S., Burgess, C., Liotta, D. C., and Merrill, A. H., Jr. (1989). *J. Biol. Chem.* **264,** 10371–10377.

Smith, C. A., Davis, T., Anderson, D., Solam, L., Beckmann, M. P., Jerzy, R., Dower, S. K., Cosman, D., and Goodwin, R. G. (1990). *Science* **248,** 1019–1023.

Sugiyama, E., Uemura, K.-I., Hara, A., and Taketomi, T. (1990). *Biochem. Biophys. Res. Commun.* **169,** 673–679.

Thoma, B., Grell, M., Pfizenmaier, K., and Scheurich, P. (1990). *J. Exp. Med.* **172,** 1019–1023.

Ullman, M. D., and Radin, N. S. (1974). *J. Biol. Chem.* **249,** 1506–1513.

Vilcek, J., and Lee, T. H. (1991). *J. Biol. Chem.* **266,** 7313–7316.

Vunnam, R. R., and Radin, N. S. (1980). *Chem. Phys. Lipids* **26,** 265–278.

Wilson, E., Olcott, M. C., Bell, R. M., Merrill, A. H., Jr., and Lambeth, J. D. (1986). *J. Biol. Chem.* **261,** 12616–12623.

Wilson, E., Wang, E., Mullins, R. E., Uhlinger, D. J., Liotta, D. C., Lambeth, J. D., and Merrill, A. H., Jr. (1988). *J. Biol. Chem.* **263,** 9302–9309.

Ceramide-Activated Protein Phosphatase: Partial Purification and Relationship to Protein Phosphatase 2A

RICK T. DOBROWSKY* AND YUSUF A. HANNUN†

*Department of Medicine
†Departments of Medicine and Cell Biology
Duke University Medical Center
Durham, North Carolina 27710

I. Introduction

A. Intracellular Generation and Bioactivity of Ceramide

Ceramide has been identified as a direct metabolic product of agonist-induced sphingolipid metabolism. In response to the cytokines TNF-α or interferon-γ (Kim *et al.*, 1991), 1α,25-dihydroxycholecalciferol (vitamin D_3) (Okazaki *et al.*, 1989), or growth factors, i.e., NGF (M. Werner and Y. A. Hannun, unpublished results) the activation of a cellular sphingomyelinase generates ceramide and phosphocholine. These products are then recycled back to sphingomyelin, defining a sphingomyelin cycle (Okazaki *et al.*, 1989).

Utilizing ceramide analogs, biological studies have indicated that ceramide possesses bioactivity and can directly affect cell growth and differentiation. For example, addition of the ceramide analog C_2-ceramide (1,3-dihydroxy-2-acetamido-4-*trans*-octadecene) to human HL-60 leukemia cells results in a dose-dependent inhibition of cell proliferation with concomitant induction of the monocytic phenotype (Kim *et al.*, 1991; Okazaki *et al.*, 1991). This ceramide-induced biology mimics that produced by addition of vitamin D_3 to these cells. Furthermore, C_2-ceramide induces transcriptional downregulation of the c-*myc* protooncogene, in part through induction of an elongation block at the juncture of the first exon and first intron of the c-*myc* gene (Wolfe *et al.*, 1993). These data suggest that ceramide-induced effects on cell proliferation and differentiation may involve control of c-*myc* mRNA levels (Kim *et al.*, 1991; Wolfe *et al.*, 1993). As such, agonist-induced ceramide generation may, in part, account for the effects of vitamin D_3, TNF-α, and NGF on cell proliferation and differentiation. These studies have established a potential signaling pathway involving sphingomyelin hydrolysis with generation of a putative second messenger molecule, ceramide.

B. Potential Biochemical Targets for Intracellular Ceramide

Although ceramide generation has been observed in response to extracellular messengers, a proximal target for the direct interaction of ceramide has not been described. Recently, two cellular proteins have been identified as potential targets for mediating the *in vivo* downstream effects of ceramide. The first is a ceramide-responsive serine/threonine protein kinase activity present in the membrane fraction of A431 cells (Goldkorn *et al.*, 1991; Mathias *et al.*, 1991). In response to ceramide, this kinase activity induced a rapid increase in threonine phosphorylation of the epidermal growth factor receptor present in the membrane of the A431 cells (Goldkorn *et al.*, 1991). However, it was not determined whether ceramide directly activated this kinase or whether kinase activation was a secondary effect for the action of ceramide in this crude membrane preparation. Interestingly, this kinase was equally stimulated by sphingosine, presumably due to its rapid conversion to ceramide. Of considerable importance is the need for further biochemical and biological data to determine the relevance of this sphingolipid-activated kinase to ceramide-mediated biology.

A second potential protein target for ceramide has been identified as a serine/threonine protein phosphatase activity. We have recently reported on the presence of a ceramide-activated protein phosphatase (CAPP) activity isolated from crude cytosol of rat T9 glioblastoma cells (Dobrowsky and Hannun, 1992). CAPP has been partially purified from cytosol of both rat T9 glioma cells (Dobrowsky and Hannun, 1992) and rat brain (Dobrowsky and Hannun, 1993). CAPP activity is directly activated by ceramide, cation independent, and inhibited by low concentrations of the serine/threonine protein prosphatase inhibitor okadaic acid (Dobrowsky and Hannun, 1992).

C. Overview of Serine/Threonine Protein Phosphatases

Serine/threonine protein phosphatases comprise a family of enzymes defined by a broad *in vitro* substrate specificity, variable requirements for cations, and sensitivity to various protein and lipid inhibitors (Cohen, 1989; Cohen and

Cohen, 1989). All serine/threonine protein phosphatases have been found to fall within two general classifications. Type 1 protein phosphatases (PP1) can be isolated in high yield from the glycogen pellet of striated muscle as a cation-independent activity which is potently inhibited by either of two thermostable proteins, inhibitor 1 and inhibitor 2.

Type 2 serine/threonine protein phosphatases comprise three distinct enzymes distinguished primarily by their cation requirements (PP2A, PP2B, PP2C). PP2A activity is cation independent, while the phosphatase activities of PP2B and PP2C require Ca^{2+}/calmodulin and Mg^{2+} ions, respectively (Cohen, 1989). The PP2A protein phosphatases further comprise three distinct subclasses, based on their elution from DEAE-Sephadex chromatography (Tung *et al.,* 1985). $PP2A_0$ and $PP2A_1$ are heterotrimeric proteins of 181 and 202 kDa, respectively. They are composed of two regulatory subunits (termed A and B) and a single catalytic subunit (C). A third form, $PP2A_2$, is a heterodimer of molecular mass 107 kDa and possesses one regulatory subunit (A) bound to the catalytic subunit (C). The catalytic subunit may be readily dissociated from the regulatory subunits and purified as an active monomer, but does not exist as a free monomer *in vivo* (Tung *et al.,* 1985).

In contrast to PP1, type 2 protein phosphatases are resistant to inhibition by either inhibitor 1 or inhibitor 2. However, okadaic acid has recently been identified as a natural product capable of inhibiting both classes of phosphatases, albeit with differential efficacy. Okadaic acid is a polyether carboxylic acid isolated from marine sponges (Bialojan and Takai, 1988). PP2A activity is most sensitive to inhibition by okadaic acid with an IC_{50} of 1 n*M,* while inhibition of PP1 activity can require 10- to 40-fold greater concentrations to achieve 50% inhibition (Cohen and Cohen, 1989). The remaining type 2 phosphatases are either totally unaffected by okadaic acid, i.e., PP2C, or, in the case of PP2B, require micromolar concentrations to achieve 50% inhibition. Therefore, establishing both the response to specific inhibitors and ionic requirements can facilitate in the identification and quantitation of serine/threonine protein phosphatases (Cohen and Cohen, 1989).

D. Potential Role of CAPP in Ceramide-Mediated Biology

Okadaic acid is cell permeable and has proved useful in elucidating many biological activities mediated via modulation of serine/threonine protein phosphatase activity (Hardie *et al.,* 1991). As such, the sensitivity of CAPP to okadaic acid provides a powerful tool for determining if ceramide may exert its bioactive effects via modulation of an okadaic-sensitive serine/threonine protein phosphatase *in vivo.* For example, 10 n*M* okadaic acid completely inhibits ceramide-induced downregulation of c-*myc* mRNA in HL-60 cells (Wolfe *et al.,*

1993). Additionally, in T9 glioma cells, ceramide induces the dephosphorylation of several ^{32}P-labeled phosphoproteins of relative molecular masses 66, 32 and 28 kDa. Pretreatment of the cells with 0.1 μM okadaic acid inhibits the ceramide-induced dephosphorylation of these proteins (R. T. Dobrowsky and Y. A. Hannun, unpublished results). Taken together, these studies provide preliminary evidence to support the hypothesis that a serine/threonine protein phosphatase may be a proximal target for ceramide and may provide the molecular switch for initiating downstream biological effects of this sphingolipid via modulation of cellular phosphoprotein metabolism.

The following studies have been conducted with the aim of elucidating some of the biochemical characteristics of CAPP and establishing its relationship to other known serine/threonine protein phosphatases.

II. Results

A. Properties of CAPP from Rat T9 Glioblastoma Cells

CAPP was initially identified as a ceramide-stimulated protein phosphatase activity derived from cytosolic extracts of rat T9 glioma cells (Dobrowsky and Hannun, 1992). Protein phosphatase activity increased in a dose-dependent manner in response to addition of C_2-ceramide, utilizing protein kinase C phosphorylated histone as substrate (Fig. 1A). Increased dephosphorylation was apparent at C_2-ceramide concentrations as low as 1 μM. However, because of the high background phosphatase activity in crude cytosol, statistically relevant differences occurred at C_2-ceramide concentrations of ≥ 5 μM. Since C_2-ceramide is a biologically active, but nonphysiological structural analog of naturally occurring ceramides, more hydrophobic ceramide species were tested for their ability to stimulate cytosolic phosphatase activity. Both C_6-ceramide and natural ceramide (from bovine brain sphingomyelin) stimulated cytosolic phosphatase activity with a similar concentration dependence as C_2-ceramide (Fig. 1B).

Since the effect of ceramide may be due to nonspecific lipid–protein interaction, other structurally related sphingolipids were tested for their ability to stimulate cytosolic phosphatase activity. Neither sphingosine nor sphingomyelin, both precursor/products of ceramide metabolism, were capable of stimulating cytosolic phosphatase activity (Dobrowsky and Hannun, 1992). Moreover, neither phosphatidylcholine nor dioleoylglycerol showed dose-dependent stimulation of phosphatase activity at similar concentrations. These studies indicate that enhancement of cytosolic phosphatase activity requires the specific structure of ceramide, and is not due to nonspecific lipid–protein interaction.

To determine the relationship of CAPP to known protein phosphatases, the effects of cations and protein phosphatase inhibitors on CAPP activity were

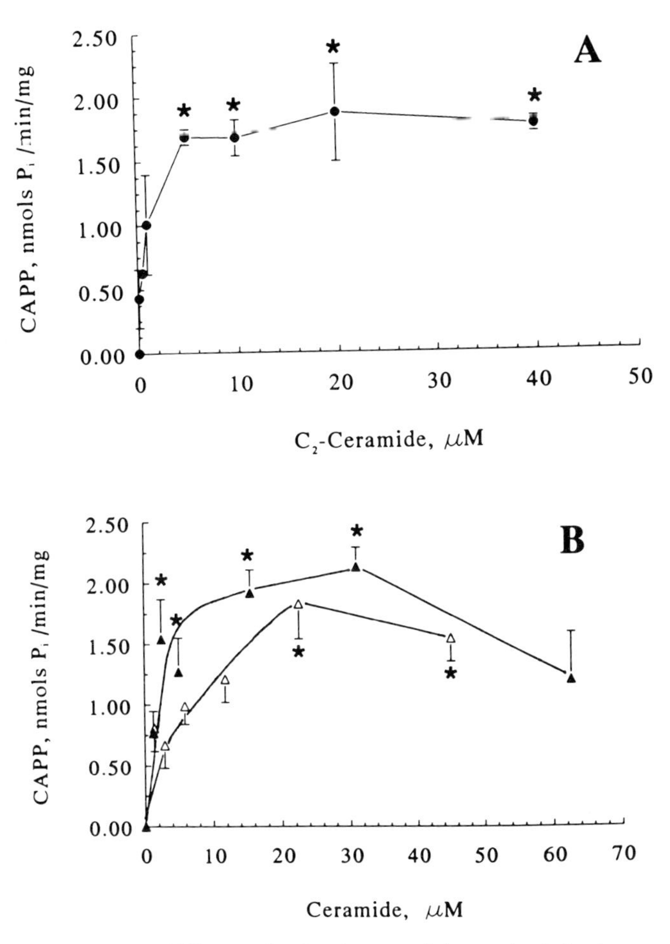

FIG. 1. Dose response of T9 cytosolic phosphatase activity to stimulation by ceramides. (A) CAPP activity in the presence of C_2-ceramide or (B) C_6-ceramide (▲) and natural ceramide (△). CAPP activity was determined after subtraction of activity in the absence of ceramide. Control activity was (A) 4.75 ± 1.25 nmol P_i/min/mg, (B) 5.89 ± 0.21 nmol P_i/min/mg. Asterisks indicate $p < 0.05$ versus control. Reprinted from Dobrowsky and Hannun (1992) with permission of the copyright holder, The American Society for Biochemistry and Molecular Biology.

assessed. CAPP activity was found to be unaffected by addition of either Mn^{2+}, Mg^{2+}, or Ca^{2+} ions. This cation independence is similar to that recognized for protein phosphatases type 1 (PP1) and type 2A (PP2A).

The phosphatase activities of PP1 and PP2A may be readily discerned based on their sensitivity to the phosphatase inhibitor okadaic acid (Bialojan and Takai, 1988). Figure 2 shows the sensitivity of CAPP to inhibition by low concentrations

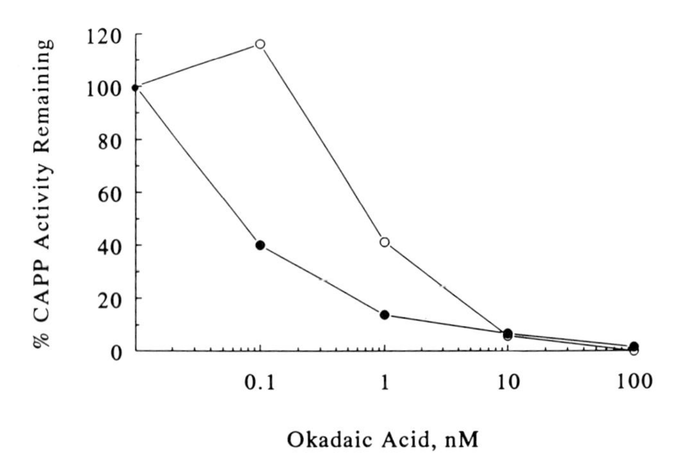

FIG. 2. Okadaic acid inhibition of T9 cytosolic CAPP activity. CAPP activity was determined at 5 μ*M* (●) and 20 μ*M* C_2-ceramide (○) in the presence of various concentrations of okadaic acid. CAPP activity at each okadaic acid concentration was determined after subtraction of activity in the absence of ceramide. Reprinted from Dobrowsky and Hannun (1992) with permission of the copyright holder, The American Society for Biochemistry and Molecular Biology.

of okadaic acid. Depending on the ceramide concentration, the IC_{50} of okadaic acid was 0.1–1 n*M*. This concentration range for inhibition of CAPP is consistent with that for the inhibition of PP2A by okadaic acid (IC_{50} = 1 n*M*). By comparison, the IC_{50} for okadaic acid inhibition of PP1 is 15–40 n*M* (Cohen, 1989). These results suggest that CAPP is more related to PP2A than to PP1 activity.

B. Partial Purification of CAPP from Rat Brain

CAPP has been partially purified from rat brain by sequential ion exchange and affinity fast protein liquid chromatography. Frozen rat brain (15–20 g) is homogenized in 5 volumes of 20 m*M* Tris-HCl, 1 m*M* ethylenediaminetetraacetic acid (EDTA), 1 m*M* phenylmethylsulfony fluoride (PMSF), 1 m*M* benzamidine, 1 m*M* dithiothreitol (DTT), 2.5 μg/ml leupeptin, and 2.5 μg/ml pepstatin A, pH 7.4. The homogenate is centrifuged at 5000 *g* for 20 minutes at 4°C, the supernatant centrifuged at 100,000 *g* for 1 hour to obtain cytosol, and the cytosolic protein adsorbed to DEAE-Sephahacel at 15–20 mg protein/ml packed gel for 1 hour at 4°C. CAPP is recovered by sequentially washing the resin with 200 ml of 20 m*M* Tris-HCl, 1 m*M* EDTA, 0.1 m*M* EGTA, 10% glycerol, 1 m*M* benzamidine, 1 m*M* DTT, pH 7.4 (buffer A) containing 100 m*M* NaCl, 100 ml of 200 m*M* NaCl in buffer A, and finally with 50 ml of 300 m*M* NaCl in buffer A. CAPP activity is assayed as described (Dobrowsky and Hannun, 1992) and primarily elutes in the 300 m*M* salt fraction. The 300 m*M* fraction from DEAE chromatography is diluted 3-fold with buffer A and applied to a column of poly (L-lysine) agarose at 3–4 mg protein/ml gel. Protein is initially eluted with a linear gradient from 100–220 m*M* NaCl in buffer A over 20 ml and the column washed with an additional 20 ml of 220 m*M* NaCl. Remaining protein is then eluted with a linear gradient to 800 m*M* NaCl over 60 ml. Fractions of 2 ml are collected and assayed for CAPP, which elutes at about 630–650 m*M* NaCl. Poly(L-lysine) chromatography separates the bulk of protein from CAPP activity, which migrates as the trailing edge of phosphatase activity (Fig. 3). Active fractions are stored at –70°C after adding 10 μg/ml leupeptin and are stable for about 1 week. All phosphatase assays are performed with appropriately diluted enzyme so that no more than 15–20% of the substrate is hydrolyzed.

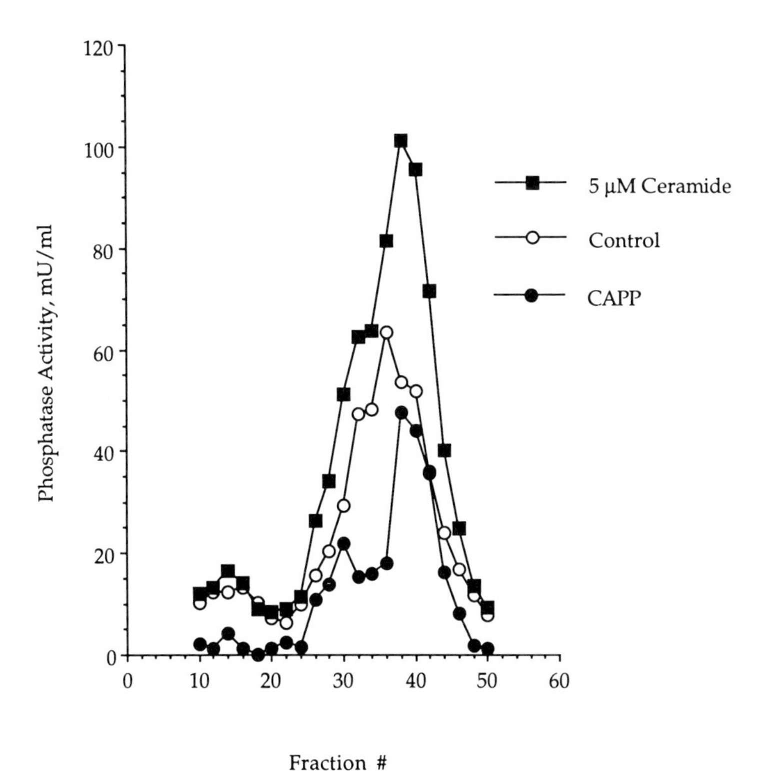

FIG. 3. Partial purification of CAPP from rat brain by poly(L-lysine) agarose fast protein liquid chromatography. Cytosol was fractionated on DEAE-Sephacel as described in the text. The 300 m*M* NaCl fraction was diluted 3-fold and chromatographed on a column of poly(L-lysine) agarose. Fractions of 2 ml were collected and 2 μl assayed for CAPP activity for 4 minutes at 37°C (Dobrowsky and Hannun, 1992). CAPP activity was determined by subtraction of activity in the absence of C_2-ceramide. Activity is expressed as milliunits per milliliter; 1 U activity is defined as 1 nmol P_i hydrolyzed/minute.

Figure 4 shows the ceramide dependence of CAPP activity of peak fractions from poly(L-lysine) chromatography. Dose dependence is similar to that seen in crude cytosol (Fig. 1) except that partial purification results in a 2- to 3-fold increase in ceramide-activated activity. This is in contrast to the 30% increase typically seen in crude cytosol (Dobrowsky and Hannun, 1992). Additionally, partially purified CAPP retains its insensitivity to cations (data not shown). As such, CAPP preparations purified through poly(L-lysine) chromatography are useful for biochemical studies, since ceramide-activated phosphatase is readily distinguishable from baseline phosphatase activity.

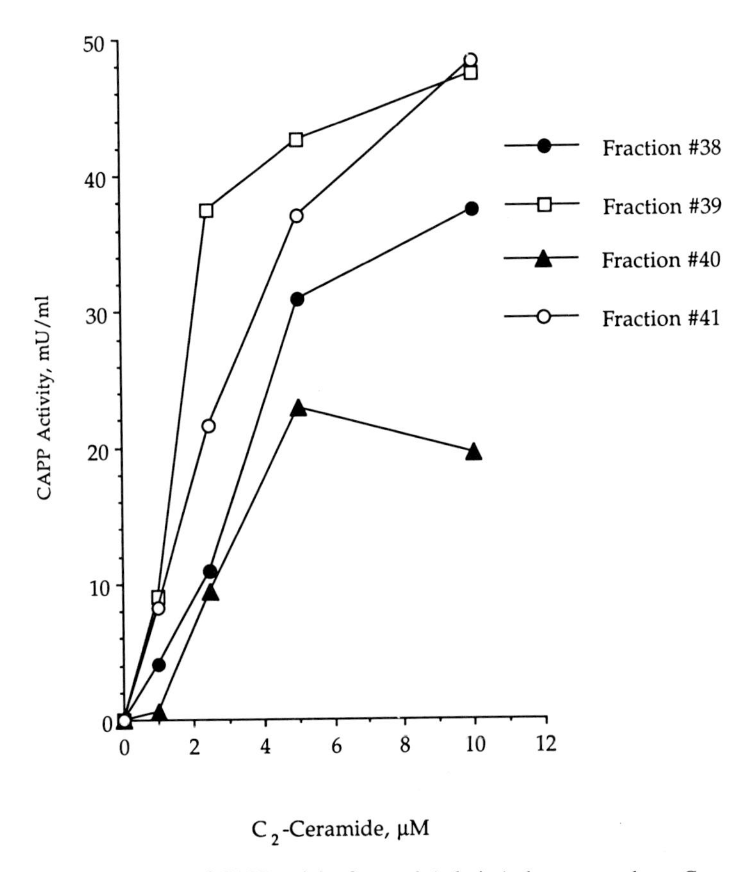

FIG. 4. Dose response of CAPP activity from poly(L-lysine) chromatography to C_2-ceramide. Fractions 38–41 were assayed for CAPP activity in the presence of 0–10 μ*M* C_2-ceramide. Assays were performed as previously described with 2 μl of each fraction for 4 minutes at 37°C. CAPP activity was determined after subtraction of activity in the absence of C_2-ceramide. Baseline activity was 38–42 mU/ml for fractions 38–40 and 22 mU/ml for fraction 41.

C. Effect of Ceramide and Okadaic Acid on CAPP, $PP2A_1$, and $PP2A_c$

Since cytosolic preparations of CAPP share some biochemical characteristics with PP2A, it was of interest to determine the effect of ceramide and okadaic acid on partially purified CAPP and two forms of PP2A. $PP2A_1$ was partially purified from bovine brain through aminohexyl-Sepharose chromatography essentially as described by Cohen *et al.* (1988). The catalytic subunit of PP2A was prepared by ethanol precipitation and isolated as described (Cohen *et al.*, 1988). Figure 5 shows the effect of C_2-ceramide and okadaic acid on partially purified CAPP. The IC_{50} for okadaic acid in the absence of ceramide was about 0.6 n*M*. However, in the presence of 2.5 or 10 μ*M* C_2-ceramide the IC_{50} is shifted to the left to 0.25–0.3 n*M*.

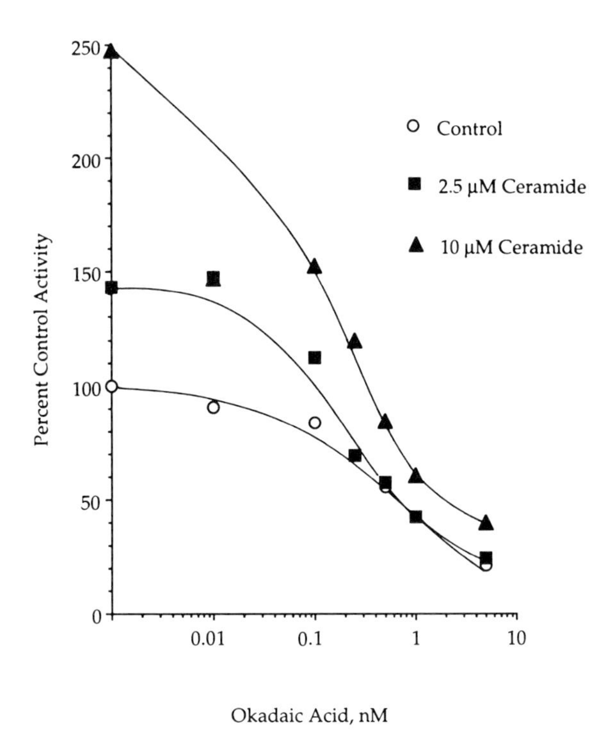

Fig. 5. Effect of C_2-ceramide and okadaic acid on partially purified CAPP. Partially purified CAPP was incubated in the presence of 0, 2.5, or 10 μ*M* C_2-ceramide at various okadaic acid concentrations. Assays were performed as previously described with 2 μl of poly(L-lysine) purified enzyme for 4 minutes at 37°C. Data are expressed as percent activity in the absence of C_2-ceramide and okadaic acid.

In contrast, bovine brain $PP2A_1$ is not stimulated by up to 10 μM ceramide (Fig. 6A) and the IC_{50} for okadaic acid inhibition remains unchanged in the presence of 2.5 or 5 μM C_2-ceramide (Fig. 6B). C_2-ceramide also fails to stimulate the catalytic subunit of PP2A and has no effect on the okadaic acid inhibition of this form of PP2A (Fig. 6C). These data indicate that ceramide modestly affects the okadaic acid inhibition of CAPP but not that of $PP2A_1$ or $PP2A_c$. Since ceramide shifts the IC_{50} of okadaic acid for CAPP to the left, it is unlikely that ceramide and okadaic acid act at the same site on the catalytic subunit (Bialojan and Takai, 1988). This conclusion is also supported by the lack of effect of up to a 500,000-fold excess of ceramide versus okadaic acid when using only the catalytic subunit of PP2A. Alternatively, ceramide may be interacting with some regulatory subunit and indirectly affecting catalytic activity. However, a detailed understanding of the interaction of ceramide with CAPP will require isolation of homogeneous holoenzyme.

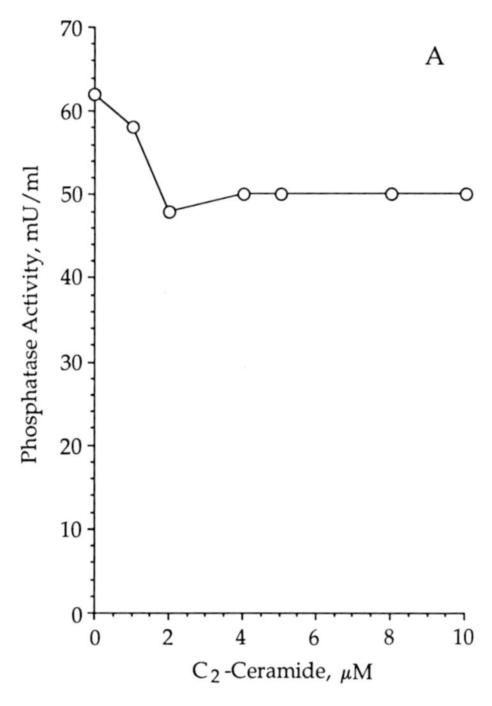

FIG. 6. Effect of C_2-ceramide and okadaic acid on partially purified $PP2A_1$ and $PP2A_c$. (A) Partially purified $PP2A_1$ (2 μl) was assayed in the presence of 0–10 μM C_2-ceramide for 5 minutes at 37°C. (B) Partially purified $PP2A_1$ was incubated in the presence of 0, 2.5, or 5 μM C_2-ceramide at various okadaic acid concentrations. Assays were performed with 2 μl of aminohexyl-Sepharose purified enzyme for 5 minutes at 37°C. (C) Partially purified $PP2A_c$ was incubated with 2.5 or 5 μM

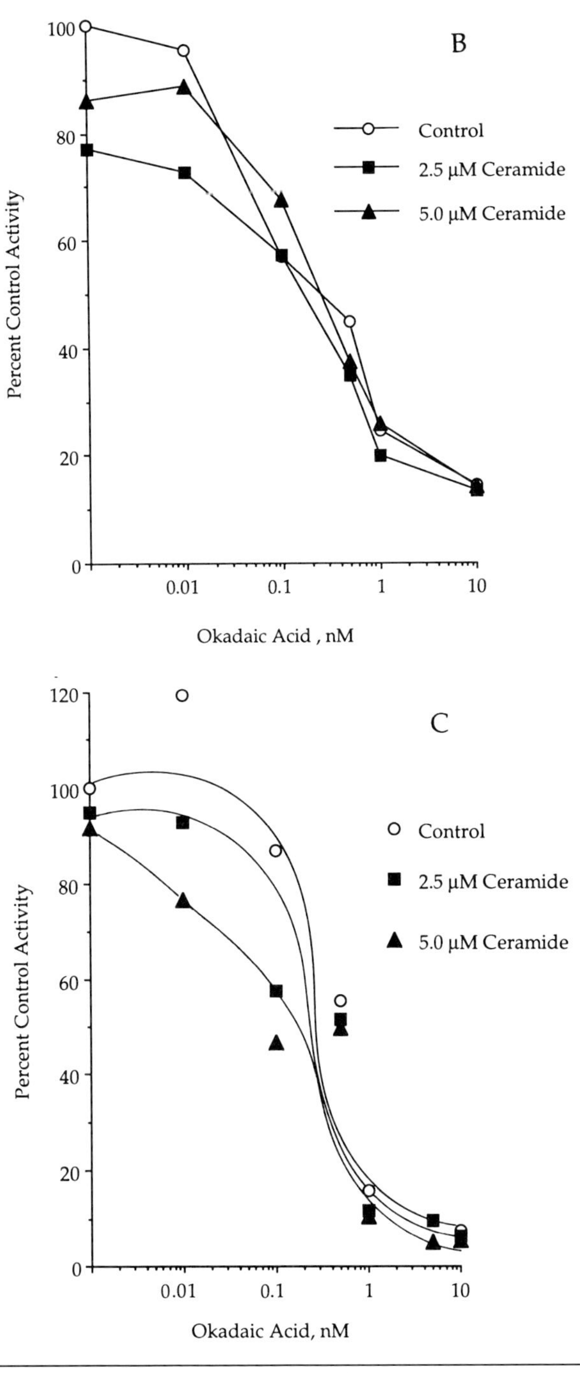

C_2-ceramide in the presence of various concentrations of okadaic acid. Assays were performed with 2 μl of enzyme for 5 minutes at 37°C. Data in B and C expressed as percent activity in the absence of C_2-ceramide and okadaic acid.

III. Discussion

The data presented provide evidence for the existence of a cytosolic protein phosphatase activity which is specifically activated by ceramide. CAPP activity is readily separable from ceramide-unresponsive serine/threonine protein phosphatases by conventional chromatographic procedures and shares some biochemical characteristics with PP2A. As described above, CAPP activity is cation independent and sensitive to inhibition by okadaic acid in a similar concentration range apparent for bovine brain $PP2A_1$. Additionally, CAPP migrates as a high-molecular-weight complex by gel filtration (M_r = 203,000) and is resistant to inhibition by inhibitor 2 (data not shown). Further, preliminary evidence indicates that antibodies directed against purified $PP2A_c$ can immunoprecipitate CAPP activity (data not shown).

Taken together, these data suggest that CAPP is related to PP2A type phosphatases. However, the absence of any effect of ceramide on partially purified $PP2A_1$ and on the sensitivity of $PP2A_1$ to okadaic acid suggests that ceramide activation may be associated with a distinct subpopulation of PP2A activity. Indeed, recent evidence indicates that a specific subfraction of PP2A activity exists in *Xenopus* extracts which is responsible for inactivation of maturation-promoting factor activity ($p34^{cdc2}$–cyclin B complex) (Lee *et al.*, 1991). This PP2A subfraction also appears to be biochemically and antigenically similar to bovine cardiac PP2A. However, bovine cardiac $PP2A_1$ holoenzyme was inactive in dephosphorylating maturation-promoting factor (Lee *et al.*, 1991). Thus, although this PP2A-like activity shares biochemical and antigenic properties with authentic $PP2A_1$, distinct subpopulations of PP2A activity may exist *in vivo* which differ based on the association of various regulatory subunits (see below). It is these novel regulatory subunits which may impart distinct substrate specificities to these unique PP2A subpopulations.

The subunit composition of PP2A is well recognized as being a heterotrimer composed of two regulatory subunits (A and B) and a catalytic subunit (C) (Cohen, 1989). The basic subunit structure of PP2A holoenzyme (ABC) isolated from a variety of tissues is composed of a M_r = 38,000 C subunit, M_r = 60,000–63,000 A subunit, and several distinct B subunits of M_r = 55,000 (B), M_r = 54,000 (B′), or M_r = 74,000 (B″) (Kamibayashi *et al.*, 1991). Although the subunit composition of CAPP remains to be determined, its relationship to PP2A would suggest that a similar heterotrimeric composition may exist. Waelkens *et al.* (1987) have noted that putative regulatory noncatalytic subunits may readily dissociate during purification and be present in substoichiometric amounts in purified phosphatase preparations. Since $PP2A_1$ is not responsive to ceramide, the presence of an additional regulatory subunit may impart ceramide responsiveness. It has been previously suggested that a large native holoenzyme may indeed exist *in vivo*. This complex contains a tight association of three main subunits to each other, with additional regulatory subunits of lesser affinities loosely bound to this core heterotrimer (Waelkens *et al.*, 1987). During purification, CAPP has been found to be extremely labile. This phenomenon may be indicative of a loosely associated ceramide-responsive subunit(s) which is readily lost during the purification procedure and storage.

IV. Summary and Future Directions

Presently, our data suggest that CAPP may be a novel serine/threonine protein phosphatase activity with biochemical and antigenic properties similar to the PP2A class of these enzymes. CAPP may prove to be the proximal molecular target for initiating further downstream effects of ceramide, i.e., c-*myc* downregulation, but additional work is needed to establish this connection. Figure 7 summarizes a hypothetical scheme for the molecular mechanism of ceramide-mediated bioactivity. Agonist-induced activation of a membrane sphingomyelinase activity generates ceramide and phosphocholine. Although these sphingomyelin metabolites can be recycled back to the parent compound (Okazaki *et al.*, 1989), a pool of ceramide is produced, providing a second messenger activator for CAPP. Once activated by ceramide, CAPP may then dephosphorylate critical substrates involved in the regulation of c-*myc* levels (Wolfe *et al.*, 1993), leading to growth arrest and cell differentiation. It is also of interest to determine if CAPP may be involved in modulating the phosphorylation of other phosphoproteins involved in cell cycle regulation, i.e., retinoblastoma, $p34^{cdc2}$, p53.

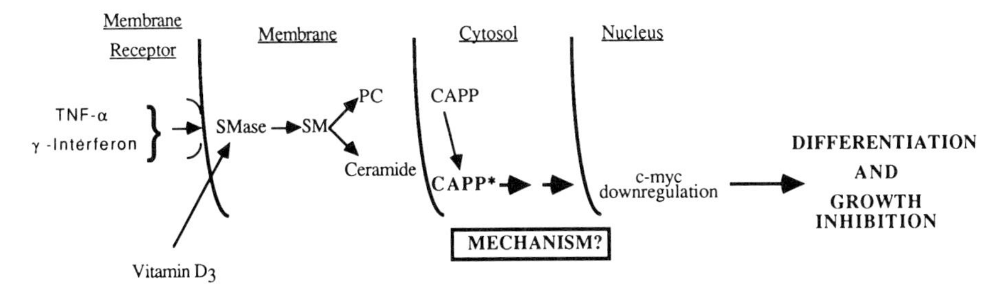

FIG. 7. Hypothetical molecular mechanism for ceramide-mediated biology. Agonist-induced stimulation results in activation of a neutral sphingomyelinase producing ceramide and phosphocholine. Ceramide becomes available to activate CAPP (CAPP*) which may then dephosphorylate critical substrates regulating c-*myc* mRNA levels as well as other potential phosphoproteins involved in cell cycle regulation. Changes in c-*myc* levels and alteration of cellular phosphoprotein metabolism by ceramide then lead to growth arrest and cell differentiation.

Future biochemical studies are being directed primarily at purifying CAPP and defining the specific subunit structure which imparts ceramide responsiveness. Related studies are aimed at identifying potential cellular substrates for CAPP activity and to determine if these proteins may be relevant to ceramide-induced biology.

ACKNOWLEDGMENTS

We would like to thank Dr. W. Kahn for the protein kinase C preparations and Dr. J. Fishbein and C. Linardic for critical reading of the manuscript. This work was supported in part by National Institutes of Health Grant GM 43825 and by a scholarship from the PEW Foundation. YAH is a Mallinckrodt Scholar. RTD is recipient of National Research Service Award F32 AG 05531.

References

Bialojan, C., and Takai, A. (1988). *Biochem. J.* **256,** 283–290.

Cohen, P. (1989). *Annu. Rev. Biochem.* **58,** 453–508.

Cohen, P., and Cohen, P. T. W. (1989). *J. Biol. Chem.* **264,** 21435–21438.

Cohen, P., Alemany, S., Hemmings, B., Resink, T. J., Stralfors, P., and Tung, H. Y. L. (1988). *In* "Methods in Enzymology" (J. Corbin and R. Johnson, eds.), vol. 159, pp. 390–408. Academic Press, Orlando, FL.

Dobrowsky, R. T., and Hannun, Y. A. (1992). *J. Biol. Chem.* **267,** 5048–5051.

Dobrowsky, R. T., and Hannun, Y. A. (1993). *J. Biol. Chem.* (in press).

Goldkorn, T., Dressler, K. A., Muindi, J., Radin, N. S., Mendelsohn, J., Menaldino, D., Liotta, D., and Kolesnick, R. N. (1991). *J. Biol. Chem.* **266,** 16092–16097.

Hardie, D. G., Haystead, T. A. J., and Sim, A. T. R. (1991). *In* "Methods in Enzymology" (T. Hunter and B. Sefton, eds.), vol. 201, pp. 469–476. Academic Press, San Diego.

Kamibayashi, C., Estes, R., Slaughter, C., and Mumby, M. (1991). *J. Biol. Chem.* **266,** 13251–13260.

Kim, M.-Y., Linardic, C., Obeid, L., and Hannun, Y. A. (1991). *J. Biol. Chem.* **266,** 484–489.

Lee, T. H., Solomon, M. J., Mumby, M. C., and Kirschner, M. W. (1991). *Cell (Cambridge, Mass.)* **64,** 415–423.

Mathias, S., Dressler, K. A., and Kolesnick, R. N. (1991). *Proc. Natl. Acad. Sci. U.S.A.* **88,** 10009–10013.

Okazaki, T., Bell, R. M., and Hannun, Y. A. (1989). *J. Biol. Chem.* **264,** 19076–19080.

Okazaki, T., Bielawska, A., Bell, R. M., and Hannun, Y. A. (1990). *J. Biol. Chem.* **265,** 15823–15831.

Tung, H. Y., Alemany, S., and Cohen, P. (1985). *Eur. J. Biochem.* **148,** 253–263.

Waelkens, E., Goris, J., and Merlevede, W. (1987). *J. Biol. Chem.* **262,** 1049–1059.

Wolfe, R. A., Obeid, L. M., and Hannun, Y. A. (1993). Submitted for publication.

The Role of Sphingosine in Cell Growth Regulation and Transmembrane Signaling

SARAH SPIEGEL, ANA OLIVERA, AND ROBERT O. CARLSON

Department of Biochemistry
Georgetown University Medical Center
Washington, D.C. 20007

I. Regulation of Cellular Functions by Sphingolipids

A. A Sphingolipid Cycle May Provide Intracellular Messengers

Over the past decade, it has become clear that phospholipids, in addition to acting as primary structural constituents of cell membranes and as regulators of protein activities through direct interactions, also serve as sources of intracellular signaling molecules, such as diacylglycerol (DAG), inositol trisphosphate (IP_3), phosphatidic acid (PA) and arachidonic acid (AA), which are critical regulatory elements in many fundamental cellular processes (Berridge and Irvine, 1989; Dennis *et al.*, 1991; Exton, 1990b). Within the past few years, there has been a dramatic increase in attention to a similar role of sphingolipids in the regulation of cellular events (Hannun and Bell, 1987, 1989; Merrill, 1991).

Complex sphingolipids, particularly gangliosides, have long been known to play a role in the regulation of cellular functions, including transformation, proliferation, differentiation, and modulation of receptor function (Sweeley, 1985; Hakomori, 1990). However, the mechanisms by which sphingolipids act have remained poorly understood. Complex sphingolipids are known to alter activity of protein kinases, which may explain some sphingolipid-mediated effects (Hakomori, 1990; Weis and Davis, 1990). Recently, certain metabolites of complex sphingolipids, such as sphingosine and ceramide, have been implicated in signal transduction processes, possibly acting as intracellular messengers (Hannun and Bell, 1989, 1990; Merrill and Stevens, 1989; Kolesnick, 1991; Merrill, 1991). Therefore, similar to phospholipids, complex sphingolipids may function in part as sources of metabolites which have regulatory functions.

In seminal studies, Hannun and colleagues demonstrated that activation of neutral sphingomyelinase activity may be receptor-mediated in HL-60 cells (Okazaki *et al.*, 1989, 1990; Kim *et al.*, 1991). Although in those studies, ceramide was the major product of sphingomyelin breakdown that was detected, free sphingosine can potentially be produced through any ceramide-generating pathway via ceramidase-catalyzed deacylation of ceramide (Al *et al.*, 1989; Sweeley, 1985). Glycosphingolipids, such as gangliosides, could also serve as sources of ceramide and sphingosine, although a receptor-mediated breakdown of gangliosides has yet to be described.

In addition to providing pathways for production of sphingosine and ceramide, the metabolism of complex sphingolipids provides a potential mechanism for removal and reutilization of these bioactive metabolites. Ceramide is converted to sphingomyelin primarily through phosphorylcholine transfer from phosphatidylcholine (Pagano, 1988), a pathway which could serve as the major means of reducing elevated levels of ceramide. Ceramide levels could also be lowered by conversion to glycosphingolipids, initially involving glycosyltransferase activities (Schwarzmann and Sandhoff, 1990). Since sphingosine is known to undergo rapid acylation to ceramide in cells (Sweeley, 1985), free sphingosine could enter synthetic pathways for complex sphingolipids through conversion to ceramide. Alternatively, both ceramide and sphingosine levels can be reduced through degradative pathways, involving conversion to sphingosine-1-phosphate (SPP), which can subsequently be degraded to phosphoethanolamine and hexadecanal (Sweeley, 1985).

Rapid turnover is a critical feature of any metabolic cycle responsible for regulating levels of intracellular messengers. A metabolic cycle with a high turnover rate allows for rapid production of intracellular messengers in response to effector stimulation and equally rapid removal of the intracellular messenger. Rapid turnover has recently been demonstrated for sphingolipids in pulse-chase experiments with [^{14}C]serine (Medlock and Merrill, 1988). Therefore, the metabolism of complex sphingolipids has the elements which are essential for the production of intracellular messengers and such a sphingolipid cycle could function in previously unappreciated and profoundly important ways to regulate cellular processes.

B. Sphingosine as an Intracellular Signaling Molecule

In a series of elegant and thorough studies, Bell and co-workers revealed the potential for sphingosine to act as a negative intracellular effector through inhibition of protein kinase C (PKC) (Hannun *et al.*, 1986). This group originally demonstrated that the naturally occurring stereoisomer, D-*erythro*-sphingosine, inhibited PKC *in vitro* through competitive displacement of diacylglycerol or phorbol ester binding with an IC_{50} of 100 μ*M* in a mixed micellar assay (Hannun *et al.*, 1986). *In vitro* PKC inhibition was later demonstrated for many analogs of sphingosine, including the L-*erythro*, D-*threo*, and L-*threo* stereoisomers, *N*-methyl derivatives, sphingosines of varying acyl chain lengths, stearylamine and other long-chain aliphatic amines, and a variety of lysosphingolipids (Hannun and Bell, 1987, 1989; Igarashi *et al.*, 1990; Merrill, 1991).

Sphingosine and other sphingoid bases have also been found to regulate the activity of other enzymes. Those that are inhibited by sphingoid bases in the effective dose range for *in vitro* inhibition of PKC include CTP:phosphocholine cytidyltransferase (Sohal and Cornell, 1990), Na/K-ATPase (Oishi *et al.*, 1990), and insulin receptor tyrosine kinase (Arnold and Newton, 1991). Sphingosine inhibited purified calcium/calmodulin-dependent protein kinase (CAM kinase) with an IC_{50} of 4.8 μ*M*, and calcium/calmodulin-dependent phosphodiesterase and myosin light chain kinase each with an IC_{50} of about 35 μ*M* (Jefferson and Schulman, 1988). Increasing the concentration of calmodulin decreased the extent of sphingosine-induced inhibition for these calcium/calmodulin-dependent enzymes, suggesting that sphingosine may interfere with the binding of calmodulin (Jefferson and Schulman, 1988). Also, sphingosine differentially affected two distinct diacylglycerol kinases purified from porcine thymus, markedly inhibiting the 80-kDa form while activating the 150-kDa form with a maximal effect for each at 20 μ*M* (Sakane *et al.*, 1989). Furthermore, sphingosine induced the functional conversion of the EGF receptor into an active state that expressed both higher affinity and an increased tyrosine kinase activity (Davis *et al.*, 1988).

These findings concerning the *in vitro* inhibition of PKC and other enzymes, coupled with the known ability of sphingosine to traverse cellular membranes (Merrill, 1991), led to testing of the effects of exogenous sphingosine on cells. Exogenous sphingosine has been found to evoke a variety of responses in cells (Hannun and Bell, 1989; Kolesnick, 1991; Merrill, 1991). The PKC-dependent phenomena are detailed in several excellent reviews from Bell and colleagues (1988; Hannun and Bell, 1987, 1989). Many examples of sphingosine-mediated cellular responses which are PKC-independent have also been reported and recently reviewed (Merrill, 1991).

Despite the wealth of phenomena associated with exogenous sphingosine, relatively little is known about whether endogenous sphingosine may have similar activities in cells. Recently, a spate of information has been reported concerning the levels of free sphingosine in a wide variety of cells (Merrill, 1991). This is largely due to the development of a sensitive HPLC method for the detection of free sphingosine in cells (Merrill *et al.*, 1988). Sphingosine levels have been found to be quite low in cells and, as demonstrated in a few cases thus far, may be subject to regulation (Merrill, 1991). Treatment of 3T3–L1 fibroblasts with dexamethasone has been shown to induce a modest increase in free sphingosine (Ramachandran *et al.*, 1990). An interesting modulation of sphingosine levels has been found in human neutrophils where the phorbol ester 12-*O*-tetradecanoylphorbol 13-acetate (TPA) decreased the level of free sphingosine, whereas serum, plasma, or serum lipoproteins induced increased levels (Wilson *et al.*, 1988).

II. Sphingosine in Cellular Proliferation

We have found that sphingosine may play an important role as a positive regulator of cell growth. Previously, we have extensively investigated the role of endogenous ganglioside GM1 in growth regulation of a variety of cell types (Spiegel *et al.*, 1985; Spiegel and Fishman, 1987; Spiegel, 1988a,b; Spiegel and Panagiotopoulos, 1988; Buckley *et al.*, 1990; Masco *et al.*, 1991). As an extension of this work, and as a continuing effort to understand the role of sphingolipids in the regulation of cell growth, we tested the effects of exogenous sphingosine on growth of Swiss 3T3 cells, a convenient model system for the study of cell activation and growth. Confluent cultures of Swiss 3T3 fibroblasts become quiescent in the G_1 to G_0 phase of the cell cycle when deprived of serum (Rozengurt, 1986). However, a variety of growth factors can re-initiate active proliferation in these quiescent cultures (Rozengurt, 1986). At low concentrations, sphingosine stimulated an increase in DNA synthesis and in cell number of quiescent cultures of Swiss 3T3 fibroblasts (Zhang *et al.*, 1990a). Furthermore, sphingosine potentiated the mitogenic action of maximal concentrations of many known growth factors (Table I). In contrast, analogs of sphingosine, including *N*-stearoylsphingosine and the aliphatic amines stearylamine and oleoylamine, had no effect on mitogenesis alone or in combination with other growth factors (Zhang *et al.*, 1990a). Furthermore, we have also found that only the naturally occurring analog of sphingosine (D-*erythro*-sphingosine) has this mitogenic activity (Carlson *et al.*, unpublished).

In contrast to its positive effects on cell growth, sphingosine was characterized originally as a negative regulator of cell growth and viability. Stoffel and co-workers were the first to describe the high toxicity of sphingoid bases in

Table I
EFFECTS OF SPHINGOSINE AND SPP ON DNA SYNTHESIS IN SWISS 3T3 FIBROBLASTS[a]

	[^{3}H]Thymidine incorporation (-fold increase)			
Stimulants	None	SPH	SPP	SPH + SPP
None	1.0	1.8	2.9	1.5
Insulin	1.7	7.4	8.0	8.1
EGF	4.1	6.1	7.2	7.8
Insulin + EGF	9.4	15.7	17.3	17.8

[a]Quiescent cultures of Swiss 3T3 cells were exposed to the indicated mitogens in the absence or presence of sphingosine (SPH, 20 μ*M*) and/or sphingosine-1-phosphate (SPP, 5 μ*M*) and [^{3}H]thymidine incorporation was measured (Zhang *et al.*, 1990a,b). The data are expressed as -fold increase compared to unstimulated controls. The standard deviations were less than 10% of the mean. The concentrations of the mitogenic agents were as follows: insulin, 4 μg/ml; epidermal growth factor (EGF), 10 ng/ml.

animal cells (Stoffel and Bister, 1973), observations which were later substantiated by many others (Campbell, 1983; Merrill, 1983; Merrill and Stevens, 1989; Stevens *et al.*, 1990; Zhang *et al.*, 1990a). The cytotoxic effects of sphingoid bases were interpreted initially as a consequence of their detergent properties (Merrill and Stevens, 1989), although it has since been proposed that inhibition of PKC activity may mediate the effect (Merrill and Stevens, 1989; Stevens *et al.*, 1990). In addition to the decrease in cell viability at high concentrations, sphingosine has been shown to inhibit cell growth in chinese hamster ovary cells (Merrill, 1983) and in human promyelocytic leukemia (HL-60) cells (Stevens *et al.*, 1990), effects which also may be due to PKC inhibition (Merrill and Stevens, 1989; Stevens *et al.*, 1990). In those studies, however, low concentrations of sphingosine were found to induce a modest increase in cell number (Merrill, 1983; Stevens *et al.*, 1990). Also, sphingosine has been reported to restore proliferation in HL-60 cells after induction of cell growth arrest by phorbol ester (Merrill *et al.*, 1986; Kolesnick, 1989). In other cell types, including A431 human epidermoid carcinoma cells (Faucher *et al.*, 1988) and WI38 human fetal lung fibroblasts (Davis *et al.*, 1988), exogenous sphingosine had no effect on proliferation, in spite of sphingosine-induced changes in EGF receptor activity. Further work is needed to determine the exact role of sphingosine on cell growth.

III. Involvement of Sphingosine in Signal Transduction

The biochemical mechanisms by which eukaryotic cells regulate their proliferation are not well understood. In Swiss 3T3 cells, which are sensitive to a wide range of mitogenic agents, some growth factors appear to function through intracellular messengers such as cyclic AMP, whereas others utilize the signal pathways associated with increased degradation of polyphosphoinositides leading to the generation of the intracellular messengers DAG and IP_3. DAG is an endogenous activator of PKC (Bell, 1986; Nishizuka, 1986, 1989) and IP_3 causes a release of calcium from intracellular stores (Berridge, 1984; Berridge and Irvine, 1989). Attention has been focused on other metabolites of membrane lipids, including PA (Moolenaar *et al.*, 1986; Yu *et al.*, 1988; Exton, 1990a,b), and AA and its metabolites (Millar and Rozengurt, 1990; Takuwa *et al.*, 1991), as potential mediators of growth factor-induced mitogenesis. Although, in general, the mitogenic roles of these intracellular messengers have been well characterized, a complete elucidation of the intracellular messengers involved in mitogenesis has not been accomplished (Berridge, 1985; Besterman *et al.*, 1986; Hesketh *et al.*, 1988; Spiegel and Panagiotopoulos, 1988). In particular, it is clear that the early responses of quiescent cells to a variety of growth factors, such as changes in intracellular Ca^{2+} and pH, and activation of phospholipase C and PKC, are insufficient by themselves to cause the cells to progress to DNA synthesis (Hesketh *et al.*, 1988).

In recent years, we have undertaken an extensive effort to identify the biochemical pathways involved in sphingosine-induced mitogenesis. This search, detailed below, has led us to explore the metabolism of sphingosine in 3T3 cells and to investigate the effects of exogenous sphingosine on many of the intracellular signaling molecules thought to play a role in proliferation, including PKC, calcium, PA, IP_3, and cAMP.

A. Protein Kinase C

Protein Kinase C is a lipid-activated kinase which plays a prominent role in the regulation of many fundamental cellular processes (Bell, 1986; Nishizuka, 1989). As discussed previously, Bell and co-workers first discovered the potential role for sphingosine as a negative effector of PKC. This led us to determine whether PKC might be involved in the mitogenic effect of sphingosine on Swiss 3T3 cells. After extensive studies, we are now convinced that the effect of sphingosine on cellular proliferation in Swiss 3T3 cells is clearly independent of PKC: First, sphingosine potentiated, rather than inhibited, the mitogenic effect of the phorbol ester TPA, whose action is known to be mediated through activation of PKC. The synergism that we found between optimal concentrations of sphingosine and TPA suggests that these compounds do not share a common pathway for induction of mitogenesis. Second, H7, a compound which is known to inhibit PKC (Hidaka *et al.*, 1984; Kawamoto and Hidaka, 1984), inhibited the mitogenic response to TPA. Furthermore, whereas sphingosine had no effect, H7 inhibited the stimulation of TPA-induced phosphorylation of an acidic cellular 80-kDa protein, which is a specific marker for the activation of PKC in Swiss

3T3 fibroblasts (Rozengurt, 1986). Third, stearylamine or *N*-stearoylsphingosine, sphingosine analogs which have been shown to inhibit PKC *in vitro* and *in vivo* with a potency equivalent to sphingosine (Merrill *et al.*, 1989), did not mimic the mitogenic effect of sphingosine nor did either potentiate the effect of other growth factors. Rather, similar to the effect of H7, either analog inhibited TPA-induced mitogenesis. Fourth, mitogenic concentrations of sphingosine do not interfere with phorbol dibutyrate binding in intact cells. Only at higher concentrations of sphingosine, in the cytotoxic range, was there a displacement of phorbol dibutyrate from its cellular binding sites (Fig. 2 in Zhang *et al.*, 1990a). Perhaps the best evidence that sphingosine is not likely to be acting through PKC was the finding that downregulation of PKC did not affect the magnitude of sphingosine-induced mitogenesis in 3T3 cells (Zhang *et al.*, 1990a). Therefore, at concentrations known to induce mitogenesis, sphingosine is not likely to be acting through PKC. Only high, cytotoxic concentrations of sphingosine have any effect on PKC in our system, indicating that sphingosine may not be a physiological modulator of PKC in Swiss 3T3 cells and suggesting the existence of other cellular targets for sphingosine action.

However, there are examples of sphingosine-mediated phenomena in cells which also occur at low concentrations of sphingosine and are apparently due to inhibition of PKC. Sphingosine was found to inhibit nerve growth factor-induced neurite outgrowth in PC12 cells with half-maximal activity at 2.5–5 μM, an effect which was antagonized by TPA (Hall *et al.*, 1988). Sphingosine (0.5–10 μM) inhibited attachment of Lewis lung carcinoma cells to an extracellular matrix, an effect which was also antagonized by TPA (Inokuchi *et al.*, 1991). Similar to the effects of other inhibitors of PKC, sphingosine inhibited angiotensin-stimulated aldosterone synthesis in adrenal glomerulosa cells with an IC_{50} of 5 μM (Elliott *et al.*, 1991).

One likely explanation for the differences between the concentration of sphingosine required for inhibition of PKC *in vitro* and *in vivo* in some cases is that the apparently high concentration of sphingosine required for inhibition in mixed micellar assays may be similar to the effective concentration of sphingosine in cells due to sequestration of sphingosine in membranes. Alternatively, specific isozymes of PKC may be much more sensitive to inhibition by sphingosine than the isozyme mixture previously studied *in vitro*. To date, eight distinct isozymes of PKC have been identified and divided into two subgroups, primarily based on the presence (α, βI, βII, and γ) or absence (δ, ε, ζ, and L) of the C2 region in the regulatory domain (Nishizuka, 1989). Only the PKC-α isozyme has been detected in Swiss 3T3 fibroblasts (McCaffrey and Rosner, 1987; Adams and Gullick, 1989). If PKC-α proves to be insensitive to inhibition by sphingosine relative to other PKC isozymes, this may explain why low concentrations of sphingosine do not significantly inhibit PKC in Swiss 3T3 fibroblasts but effectively inhibit PKC in other cells with a different isozyme content.

B. Sphingosine-1-Phosphate

1. *Exogenous Sphingosine is Converted to Sphingosine-1-Phosphate*

Cellular proliferation is closely correlated with active metabolism of phospholipids (Hasegawa-Sasaki, 1985). Two-dimensional thin-layer chromatographic (TLC) analysis of lipid extracts from 3T3 cells prelabeled with $^{32}P_i$ revealed that mitogenic concentrations of sphingosine significantly stimulated the incorporation of $^{32}P_i$ into a lipid which did not comigrate with any of the major known glycerophospholipids of Swiss 3T3 fibroblasts [phosphatidylcholine (PC), phosphatidylethanolamine (PE), phosphatidylinositol (PI), phosphatidylserine (PS), or PA] (Fig. 1). Also, in contrast to these phospholipids, this unknown lipid did not incorporate labeled [^{3}H]glycerol. Furthermore, this lipid was unaffected by a mild alkali treatment which degrades glycerophospholipids (Zhang *et al.*, 1991). Having demonstrated that the unidentified lipid was not a glycerophospholipid, we explored the possibility that this lipid was a phosphorylated derivative of exogenous sphingosine. It is known that cells contain a sphingosine kinase which catalyzes the phosphorylation of free sphingosine at the 1-OH position to produce SPP (Hirschberg *et al.*, 1970; Stoffel *et al.*, 1970, 1973a; Louie *et al.*, 1976; Buehrer and Bell, 1992). To identify this lipid further,

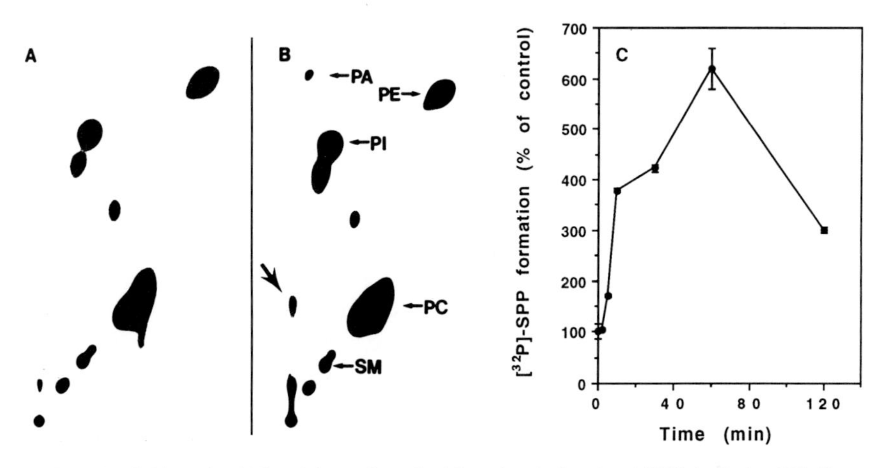

Fig. 1. Sphingosine-induced formation of sphingosine-1-phosphate (SPP) in Swiss 3T3 fibroblasts. Confluent and quiescent cultures of Swiss 3T3 fibroblasts were prelabeled with $^{32}P_i$ for 24 hours and stimulated with vehicle (A) or with sphingosine–BSA complex (10 μM) for 1 hour (B) or for the indicated time periods (C). The lipids were extracted and separated by two-dimensional TLC as described (Zhang *et al.*, 1991). In C, the results are expressed as percentage increase relative to untreated controls. The location of the standard phospholipids is indicated: phosphatidic acid (PA), phosphatidylethanolamine (PE), phosphatidylinositol (PI), phosphatidylcholine (PC), sphingomyelin (SM). The location of sphingosine-1-phosphate is indicated by the bold arrow. (Modified from Zhang *et al.*, 1991.)

we prepared standard SPP by a facile enzymatic synthesis from sphingosinephosphorylcholine using phospholipase D. This elegant procedure, described by Bell's group (Van Veldhoven *et al.*, 1989), enabled us to prepare milligram quantities of pure SPP, which was found to comigrate in TLC with the unidentified ^{32}P-labeled lipid using a variety of solvent systems (Zhang *et al.*, 1991). Furthermore, the putative SPP from cells and the authentic SPP standard were subjected to periodate oxidation followed by borohydride reduction (Hirschberg *et al.*, 1970). This procedure converted both the putative and the authentic SPP to an aqueous-extractable material which comigrated with the expected product, ethylene glycol monophosphate. Sphingosine-3-phosphate, in contrast to SPP, would be converted to a chloroform-soluble, long-chain alcohol by this procedure. Since no chloroform-soluble radioactivity was detected, and based on the other results discussed above, we suggested that the unidentified lipid was most likely SPP (Zhang *et al.*, 1991).

We subsequently showed that mitogenic concentrations of sphingosine induced a rapid rise in the levels of ^{32}P-labeled SPP. A significant increase in ^{32}P-labeled SPP was detected within 5 minutes after exposure of the cells to sphingosine, which reached a peak within 60 minutes, and then gradually decreased over a period of 8 hours. The accumulation of ^{32}P-labeled SPP in response to sphingosine was also concentration dependent, with a maximal accumulation at 20 μM, which correlated closely with the dose response for sphingosine-induced stimulation of DNA synthesis (Zhang *et al.*, 1990b). In PKC downregulated cells, the accumulation of SPP in response to sphingosine was unaltered, suggesting that PKC does not play a role in this pathway (Zhang *et al.*, 1991).

The phosphorylated forms of sphingoid bases (e.g., SPP, dihydroSPP, and phytoSPP) were described initially as intermediates in the degradation of long-chain bases (Stoffel *et al.*, 1968a, 1970; Stoffel and Bister, 1973). Extensive studies more than 20 years ago showed that sphingoid bases can be degraded in an ATP-dependent manner (Stoffel *et al.*, 1968a), yielding ethanolamine phosphate and a fatty aldehyde which is two carbon atoms shorter than the precursor sphingoid bases (Stoffel *et al.*, 1968b). Subsequently, it was established *in vivo* and *in vitro* that the breakdown of sphingoid bases required two distinct enzymes: a kinase which catalyzes phosphorylation of the sphingoid base (Stoffel *et al.*, 1968a, 1973a,b; Keenan and Haeglin, 1969; Hirschberg *et al.*, 1970; Keenan, 1972; Louie *et al.*, 1976), and a lyase which catalyzes cleavage of the phosphoryl derivative (Stoffel *et al.*, 1968b, 1969; Keenan and Maxam, 1969; Stoffel and Bister, 1973), activities which have been found to be ubiquitous in cells (Stoffel *et al.*, 1968a, 1969, 1970, 1973a,b; Keenan and Haeglin, 1969; Hirschberg *et al.*, 1970; Stoffel and Assmann, 1970; Keenan, 1972; Stoffel and Bister, 1973; Louie *et al.*, 1976; Van Veldhoven and Mannaerts, 1991; Buehrer and Bell, 1992).

Sphingosine kinase has been isolated from pig platelet cytosol (Stoffel *et al.*, 1973b) and partially purified from soluble fractions of rat liver (Hirschberg *et al.*, 1970) and of bovine brain (Louie *et al.*, 1976). This kinase activity has also been detected in microsomal fractions from *Tetrahymena pyriformis* (Keenan, 1972) and, more recently, in rat brain and platelets (Buehrer and Bell, 1992). Studies on kinase activity *in vitro* have demonstrated a high degree of stereospecificity. The naturally occurring D(+)-*erythro* isomer of sphingosine is the most favored substrate (Stoffel *et al.*, 1973b; Buehrer and Bell, 1992). The D(+)-*threo* (Stoffel *et al.*, 1973b; Buehrer and Bell, 1992) and L(–)-*threo* forms (Buehrer and Bell, 1992) have been found to inhibit sphingosine kinase activity, whereas, the L(–)-*erythro* isomer has been reported to be either a weaker substrate (Buehrer and Bell, 1992) or an inhibitor of the enzyme (Stoffel *et al.*, 1973b). Surprisingly, studies *in vivo* showed that the four stereoisomers were equally phosphorylated in the rat liver (Stoffel and Bister, 1973).

Sphingosine lyase is a pyridoxal phosphate-dependent enzyme (Keenan and Haeglin, 1969; Stoffel *et al.*, 1969; Van Veldhoven and Mannaerts, 1991) that has been localized in the microsomal fraction of several tissues (Stoffel *et al.*, 1968b, 1969; Keenan and Haeglin, 1969; Van Veldhoven and Mannaerts, 1991). It has been characterized as an integral membrane protein, with the catalytic domain probably facing the cytosol (Van Veldhoven and Mannaerts, 1991). Unlike the phosphorylation of sphingosine, studies *in vivo* showed that the subsequent degradation of phosphorylated sphingoid bases is completely stereospecific, as only D(+)-*erythro* sphingosine could be degraded in rat liver to the expected products of lyase activity (Stoffel and Bister, 1973). The lyase activity has been postulated to provide ethanolamine phosphate for the synthesis of PE and PC, and fatty aldehyde for the synthesis of plasmalogens or fatty acids (Keenan, 1972; Stoffel *et al.*, 1968b; Van Veldhoven and Mannaerts, 1991).

To date, there are only a few reports on the detection of SPP in cells (Stoffel and Assmann, 1970; Stoffel *et al.*, 1973a; Zhang *et al.*, 1991), likely due to high lyase activity in most tissues (Van Veldhoven and Mannaerts, 1991). Recently, we have found that SPP has novel signaling properties in Swiss 3T3 fibroblasts, which may provide new interest and promising insights into the physiological role of SPP (Zhang *et al.*, 1990a,b, 1991).

2. *Effects of Sphingosine-1-Phosphate on Cellular Proliferation*

Similar to sphingosine, albeit at lower concentrations, SPP was found to induce increased DNA synthesis and cell division in quiescent cultures of Swiss 3T3 fibroblasts. SPP alone at optimal concentrations was more mitogenic than sphingosine or insulin and was equally as effective as epidermal growth factor (EGF), a potent growth stimulator for Swiss 3T3 cells. Also similar to sphingosine, SPP acted synergistically with a wide variety of growth factors, such as insulin, EGF, TPA, fibroblast growth factor, and unfractionated fetal calf serum

(FCS). However, in the presence of an optimal dose of sphingosine, SPP had no additional effect on DNA synthesis (Table I). Furthermore, SPP not only stimulated [^{3}H]thymidine incorporation with similar efficiency and kinetics as sphingosine, it also induced a similar morphological transformation (Zhang *et al.*, 1990a, 1991). It is interesting to note that SPP, in contrast to sphingosine, has not been reported to have any effects on PKC *in vitro* (Van Veldhoven *et al.*, 1989). These results, together with those regarding sphingosine conversion to SPP in 3T3 cells, support the notion that SPP mediates the mitogenic effects of sphingosine.

In preliminary experiments, we found that ^{32}P-labeled SPP was rapidly taken up by cells and, after 1 hour of incubation, [^{32}P]-SPP was the only labeled lipid detected in cells. Therefore, at least initially, the exogenous SPP taken up was not significantly degraded, suggesting that the active species was SPP itself and was not sphingosine (formed by the action of cellular phosphatases). We would like to point out that the amount of SPP formed intracellularly in response to a mitogenic concentration of sphingosine was approximately the same order of magnitude as that taken up by the cells after treatment with SPP (A. Olivera *et al.*, unpublished). Unfortunately, the subcellular localizations of sphingosine and SPP are still unknown, which makes the comparison of extracellular to intracellular concentrations ambiguous.

C. Calcium

1. Sphingosine or Sphingosine-1-Phosphate Increases Intracellular Free Calcium

Recently, sphingosine or sphingosylphophorylcholine (SPC), a derivative of sphingosine, was found to induce the rapid release of calcium from permeabilized smooth muscle cells loaded with ^{45}Ca (Ghosh *et al.*, 1990). This calcium release arose from both IP_3-sensitive and -insensitive pools of intracellular calcium. Whereas the response to SPC was almost instantaneous, the calcium release occurred after a significant lag following addition of sphingosine. Furthermore, decreasing the temperature from 37 to 4°C completely blocked the response to sphingosine but had no effect on the response to SPC. The time lag and the temperature dependence of the response to sphingosine suggested that there was an intervening step, likely involving enzymatic conversion of sphingosine to SPP, which then induces calcium mobilization (Ghosh *et al.*, 1990). However, the effect of SPP itself was not tested.

We have recently presented direct evidence that exogenous SPP potently stimulates release of intracellular calcium in viable Swiss 3T3 fibroblasts. Using dual-imaging fluorescence microscopy, we observed that exogenous sphingosine or SPP stimulated about a fourfold increase in the intracellular calcium concentration ($[Ca^{2+}]_i$) in individual, quiescent Swiss 3T3 fibroblasts (Fig. 2). The

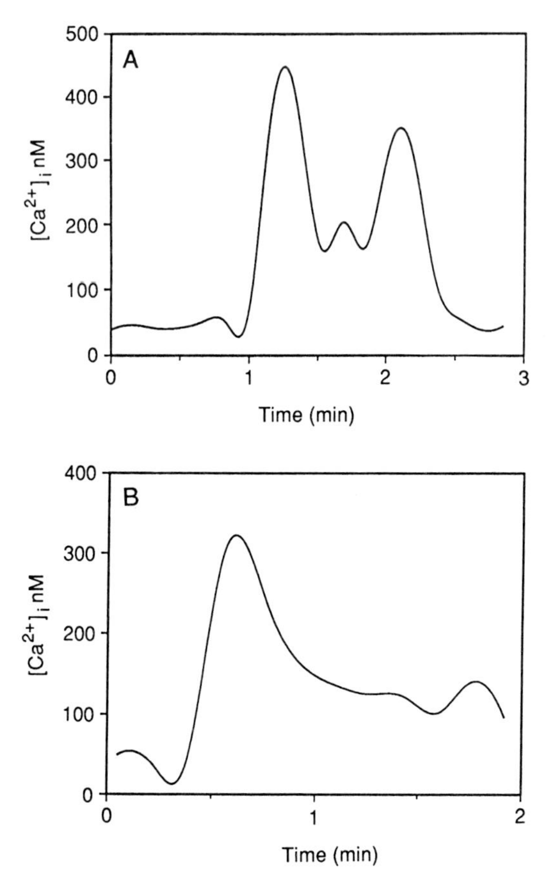

FIG. 2. Changes in cytoplasmic free calcium following addition of sphingosine or sphingosine-1-phosphate at 37°C. Dual-imaging fluorescence microscopy was used to monitor $[Ca^{2+}]_i$ in quiescent 3T3 fibroblasts as described (Zhang *et al.*, 1991). At time 0, 50 μ*M* sphingosine (A) or 2 μ*M* SPP (B) was added. In each case, the calcium changes represents the average of several responsive cells. (Modified from Zhang *et al.*, 1991.)

effect of sphingosine on $[Ca^{2+}]_i$ was highly temperature-dependent, in that virtually no effect was observed at room temperature, whereas at 37°C sphingosine induced the full increase in $[Ca^{2+}]_i$ within less than 1 minute. The typical response to sphingosine involved an apparent oscillation of $[Ca^{2+}]_i$. Conversely, the response to SPP was not altered upon lowering the temperature from 37°C to room temperature. Furthermore, the SPP-induced changes in $[Ca^{2+}]_i$ did not resemble the oscillatory response to sphingosine, but rather $[Ca^{2+}]_i$ rose to a peak within 1 to 2 minutes which returned to resting levels after several minutes

(Fig. 2). Neither the response to sphingosine nor to SPP was notably altered in the absence of extracellular free calcium, suggesting mobilization from intracellular stores as the primary source of increased $[Ca^{2+}]_i$ (Zhang *et al.*, 1991). Preliminary results from our laboratory suggest that the intracellular source of calcium is mainly derived from IP_3-insensitive pools in Swiss 3T3 cells (M. Mattie *et al.*, unpublished).

In addition to measuring the magnitude of changes in $[Ca^{2+}]_i$ in individual intact cells, the digital imaging system has the added advantage of temporal and spatial monitoring of changes in $[Ca^{2+}]_i$ continuously with great accuracy, sensitivity, and resolution in single living cells. Both sphingosine and SPP induced a wave of increased $[Ca^{2+}]_i$ which spread to involve almost the entire fibroblast. The rate of this spread varied slightly among cells. However, there was a population of cells unresponsive to sphingosine or SPP which did not have any obvious distinguishing morphological features (Zhang *et al.*, 1991). These results are similar to recent observations on heterogeneous responses to other growth factors, such as bombesin, EGF, and vasopressin (Hesketh *et al.*, 1988), where 40% of cells were unresponsive to any mitogens.

The effective concentrations of either sphingosine or SPP for induction of increased $[Ca^{2+}]_i$ correlate with their dose dependence for mitogenesis in 3T3 cells, which suggests a possible role for changes in intracellular calcium in these mitogenic effects. The temperature dependency of the sphingosine-induced (but not SPP-induced) rise in intracellular calcium supports the suggestion that sphingosine must require a temperature-dependent enzymatic conversion to SPP for its function (Ghosh *et al.*, 1990). Also, SPP is much more potent than sphingosine at stimulating release of intracellular calcium, consistent with the fact that only a small fraction of exogenous sphingosine is converted to SPP intracellularly (H. Zhang *et al.*, unpublished). Thus, SPP probably mediates the effects of sphingosine on $[Ca^{2+}]_i$ and may play a crucial role in the regulation of calcium homeostasis in cells as an intermediate step in sphingosine-induced mitogenesis.

More recently, it was observed that sphingosine also stimulated release of intracellular calcium in a dose-dependent manner in rat parotid acinar cells (Sugiya and Furuyama, 1991). However, in contrast to its effects on smooth muscle cells and on fibroblasts, the response to sphingosine in parotid acinar cells showed a virtually complete dependence on extracellular calcium, in that the rise in intracellular calcium was blocked by addition of $LaCl_3$ (a voltage-dependent calcium channel blocker) or by the removal of extracellular calcium.

2. *Calcium and Mitogenesis*

The importance of calcium in mitogenesis is well established (Metcalfe *et al.*, 1986; Rozengurt, 1986). A rise in intracellular free calcium appears essential for induction of the early events of mitogenesis (Tombes and Borisy, 1989; Seuwen *et al.*, 1990); however, increases in $[Ca^{2+}]_i$ alone are not sufficient for induction of a complete cycle of cell division. In fibroblasts, the exact role of calcium remains controversial. Changes in $[Ca^{2+}]_i$ have been shown to be involved in nuclear envelope breakdown, but are not necessary for the anaphase to metaphase transition (Kao *et al.*, 1990). Although many mitogens, including EGF, PDGF, vasopressin, $PGF_{2\alpha}$, and bombesin, stimulate increases in $[Ca^{2+}]_i$ in quiescent 3T3 fibroblasts, others, such as insulin and TPA, have no detectable effects (Hesketh *et al.*, 1988). In an extensive study designed to compare calcium mobilization and DNA synthesis induced by pairs of mitogens, no correlation between increased $[Ca^{2+}]_i$ and mitogenesis was discernible (Hesketh *et al.*, 1988). Furthermore, pretreatment of Swiss 3T3 fibroblasts with vasopressin blocks subsequent bombesin-induced mitogenesis without altering the bombesin-induced mobilization of intracellular calcium (Millar and Rozengurt, 1990). Conversely, Takuwa *et al.* (1991) showed that the bombesin-induced mitogenesis was dependent on bombesin-stimulated calcium influx.

In summary, sphingosine and, in particular, SPP are potent stimulators of intracellular calcium mobilization, acting through a previously undescribed mechanism. Further characterization of this pathway may provide new insights into the regulation of mitogenesis. However, until the precise role of calcium in mitogenesis is understood, we cannot fully appreciate the significance of sphingosine- and SPP-induced changes in intracellular calcium.

D. Phosphatidic Acid

1. Sphingosine or SPP Stimulates Accumulation of Phosphatidic Acid

Phosphatidic acid has been implicated in the regulation of mitogenesis as a possible intracellular signaling molecule. Exogenous PA is a potent mitogen for a variety of cell types and has been shown to induce changes in signaling pathways similar to the effects of some growth factors (Moolenaar *et al.*, 1986; Murayama and Ui, 1987; Siegmann, 1987; Yu *et al.*, 1988; Zhang *et al.*, 1990b). Also, PA is produced in cells in response to several growth factors (Ben-Av and Liscovitch, 1989; Exton, 1990b; Huang and Cabot, 1990). We observed that, in quiescent Swiss 3T3 fibroblasts, mitogenic concentrations of sphingosine induced a significant increase in the levels of PA (Zhang *et al.*, 1990b). In contrast, structurally related analogs of sphingosine, such as *N*-stearoylsphingosine and other long-chain aliphatic amines, which lacked mitogenic activity, did not induce elevated PA levels. The accumulation of PA in response to mitogenic concentrations of sphingosine was rapid and transient. Therefore, the formation of PA should be considered to be an early event preceding entry into the S phase of the cell cycle in Swiss 3T3 fibroblasts.

There is a striking similarity in the mitogenic properties of PA and those of sphingosine. First, the time course of increased DNA synthesis in quiescent cultures of Swiss 3T3 cells was essentially identical. Second, not only did sphingosine stimulate [^{3}H]thymidine incorporation with similar efficiency and kinetics as PA, it also induced similar changes in morphology (Zhang *et al.*, 1990a,b). Third, sphingosine and PA potentiated the effects of a wide variety of growth factors to the same extent. Most importantly, PA had no additional effect on DNA synthesis in the presence of an optimal dose of sphingosine, in the presence or absence of other growth factors. Furthermore, sphingosine and PA stimulated DNA synthesis in cells made PKC-deficient by prolonged treatment with phorbol ester and sphingosine stimulated similar increases in PA in these cells. Moreover, TPA and sphingosine have been shown to synergize both in inducing mitogenesis (Zhang *et al.*, 1990b) and in increasing PA levels (Kiss, 1990). We conclude that sphingosine and PA control cellular responses in Swiss 3T3 cells by a common mechanism that is unrelated to PKC, and further, that the mitogenic effects of sphingosine are mediated, at least in part, through PA generation.

Low concentrations of SPP were also found to induce an elevation of PA levels in Swiss 3T3 fibroblasts (Desai *et al.*, 1992). Consistent with its effect on intracellular calcium levels and mitogenesis, SPP was also more potent than sphingosine in increasing PA levels. The dose responses for accumulation of PA and for mitogenesis in response to SPP correlated precisely. Kinetic studies revealed that the production of SPP preceded the increase in PA levels and that the subsequent decrease in SPP to basal levels was followed by a similar decrease in PA levels (Desai *et al.*, 1992; Zhang *et al.*, 1991). These results imply that SPP may mediate the effects of sphingosine on the rapid increases in levels of PA.

2. *Regulation of Phosphatidic Acid Metabolism*

Rapid increases in PA occur mainly as a consequence of the activation of certain phospholipases (Fig. 3). Phospholipase D (PLD)-catalyzed hydrolysis of the phosphomonoester bond of phospholipids, primarily PC or PE, is a direct pathway for production of PA (Dennis *et al.*, 1991). Phospholipase C (PLC) activity provides an indirect pathway in catalyzing the breakdown of phospholipids to phosphorylated polar head groups and DAG, the latter of which can be phosphorylated by DAG kinase to form PA. PA is also produced *de novo* via either glycerophosphate acyltransferase or dihydroxyacetonephosphate acyltransferase activity (Manning and Brindley, 1972; Mok and McMurray, 1990). *De novo* synthesis can be stimulated in some cells by extracellular stimuli such as ionophores (Reinhold *et al.*, 1989) or insulin (Hoffman *et al.*, 1991), but such activation is most likely associated with cellular phospholipid remodeling rather than signal transduction.

Phosphatidic acid phosphohydrolase (PAH) catalyzes the degradation of PA. This pathway may also be a major source of DAG in cells during signal transduction, as PLD activation has been shown to be followed by a sustained

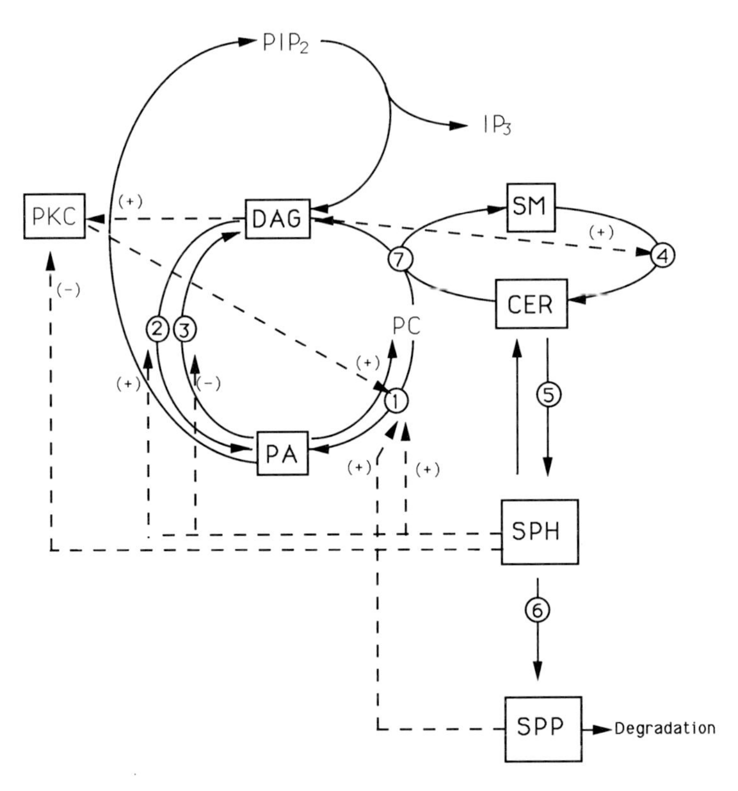

FIG. 3. Interactions between phospholipid cycles and effects of sphingosine and sphingosine-1-phosphate. A simplified schematic representation of phospholipid and sphingomyelin metabolic cycles which illustrates interactions between these cycles and potential target sites for sphingosine and SPP is shown. Numbers indicate enzyme activities involved: (1) PLD, (2) DAG kinase, (3) PAH, (4) sphingomyelinase, (5) ceramidase, (6) sphingosine kinase, (7) sphingomyelin synthase. For detailed information about these enzymes, see text. Dashed arrows indicate stimulatory (+) or inhibitory (–) effects on pathways: DAG stimulates PKC activity, and thus can indirectly activate PLD (1). DAG levels increase during the synthesis of sphingomyelin (SM) from PC and ceramide (CER) (7). Conversely, DAG may induce SM breakdown by activating sphingomyelinase (4), giving rise to CER. SPH and SPP can stimulate PLD activity (1). Additionally, SPH can activate DAG kinase (2) and inhibit PAH (3), leading to an increase of PA.

increase in DAG in a wide variety of stimulated cells (Exton, 1990b; Dennis *et al.*, 1991; Divecha *et al.*, 1991; Plevin *et al.*, 1991).

There has been a recent surge of information regarding spingosine-mediated regulation of PA metabolism. Sphingosine has been shown to stimulate PLD activity in several cell types (Lavie and Liscovitch, 1990; Mullmann *et al.*, 1991). PC was the primary PLD substrate in sphingosine-treated NG108-15 neuroblastoma–glioma hybrid cells (Lavie and Liscovitch, 1990) and in Swiss 3T3 fibroblasts (Zhang *et al.*, 1990b), but a specific hydrolysis of PE induced by sphingosine has also been observed in NIH 3T3 fibroblasts (Kiss and Anderson, 1990). Sphingosine can also activate the 80-kDa form of DAG kinase *in vitro* while inhibiting the 150-kDa form (Sakane *et al.*, 1989). Thus, the effect of sphingosine on DAG kinase, and hence on PA production, could be different depending on the type of DAG kinase isozyme present in a particular cell type. Sphingosine may also increase PA levels by inhibition of PAH. Sphingosine has been shown to inhibit PAH in cell lysates from NG108-15 cells (Lavie *et al.*, 1990) and neutrophils (Mullmann *et al.*, 1991) and in plasma membrane and endoplasmic reticulum fractions from rat liver (Jamal *et al.*, 1991).

In a recent study, we observed that SPP stimulated PLD activity in Swiss 3T3 fibroblasts (Desai *et al.*, 1992). PLD has transphosphatidylase activity with a large variety of alcohol acceptors, producing the corresponding phosphatidyl alkyl ester (Pai *et al.*, 1991). We utilized the formation of phosphatidylpropanol, produced by PLD when 1-propanol acts as the phosphatidyl group acceptor, as a measure of PLD activity. SPP induced a rapid stimulation of phosphatidylpropanol formation with a dose response similar to that for stimulation of PA accumulation and of cell division. Conversely, we were unable to detect an effect of SPP on PAH in Swiss 3T3 cells. Therefore, the most likely mechanism for SPP-induced PA accumulation is the activation of PLD.

The molecular mechanism by which sphingoid bases modify the activity of PLD, PAH, or DAG kinase is still obscure. However, it is interesting to note that in a wide variety of cells with different pathways for formation of PA, sphingosine uniformly has a stimulatory effect on PA levels. The molecular features of sphingosine necessary for action on PLD (Lavie and Liscovitch, 1990) and PAH (Mullmann *et al.*, 1991) include a long-chain base, a free amino group, and the absence of a bulky substituent in the hydroxyl position. Interestingly, these structural features are quite similar to those required for sphingosine-induced inhibition of PKC (Merrill *et al.*, 1989; Lavie and Liscovitch, 1990).

The metabolic cycles which regulate levels of PA and other phospholipids interact with sphingolipid metabolism mainly through DAG (Koval and Pagano, 1991). The cross talk between these lipid metabolites, which act as second messengers, and the function of these interlocking pathways remain unclear (Fig. 3). Sphingosine- and SPP-mediated regulation of PA levels may provide another pathway for cross talk between these important lipid metabolic and (putative) signal transducing pathways.

3. *Phosphatidic Acid and Mitogenesis*

The mitogenicity of exogenous PA is well documented (Moolenaar *et al.*, 1986; Siegmann, 1987; Yu *et al.*, Zhang *et al.*, 1990b). It has been shown that PA can be produced endogenously via receptor-coupled activation of PLD (Ben-Av and Liscovitch, 1989; Cook and Wakelam, 1991; Divecha *et al.*, 1991; Kiss *et al.*, 1991; Olson *et al.*, 1991; Plevin *et al.*, 1991). This fact highlights the potential participation of PA in cellular proliferation as an additional signaling mechanism. PA can affect various biochemical signaling pathways involving the classical intracellular second messengers (Moolenaar *et al.*, 1986; Murayama and Ui, 1987) and also can induce protein phosphorylation, a versatile transduction mechanism (Negami *et al.*, 1985; Bocckino *et al.*, 1991). However, the exact mechanism by which PA mediates its proliferative effects still remains unclear (Yu *et al.*, 1988).

PA is known to alter the expression or the activity of certain protooncogenes which may play roles in cellular proliferation. Similar to other growth factors, PA induces the expression of c-*fos* and c-*myc* in association with increased mitogenesis (Moolenaar *et al.*, 1986). Perhaps most importantly, PA has been shown to regulate the biological action of cellular *ras* activity (Hasegawa-Sasaki, 1985; Yu *et al.*, 1988; Tsai *et al.*, 1989a,b, 1990, 1991). Cellular *ras* protein participates as a molecular switch in the early steps of the signal transduction pathway associated with cell growth (Barbacid, 1987). When the protein is in the GTP-complexed form it is active in signal transduction pathways, whereas it is inactive in its GDP-complexed form. A cytoplasmic GTPase-activating protein (GAP) stimulates the conversion to the inactive form (Trahey and McCormick, 1987). It has been shown that GAP activity is inhibited by PA (Tsai *et al.*, 1989a,b, 1991). Moreover, PA also binds to a GTPase inhibitory protein, increasing its potency (Tsai *et al.*, 1990). The combined effect is an inhibition in the intrinsic GTPase activity of *ras,* stabilizing its active form (Tsai *et al.*, 1990). This suggests that PA might be involved in positively modulating *ras* activity during mitogenic stimulation (Tsai *et al.*, 1989a,b).

E. Other Intracellular Signaling Pathways

1. Inositol Trisphosphate

Activation of PLC, leading to generation of IP_3 and DAG, has been implicated in the mechanism of action of a variety of growth factors (Hesketh *et al.*, 1988). Recently, the importance of this pathway for mitogenesis has become controversial (Hesketh *et al.*, 1988; Millar and Rozengurt, 1990; Takuwa *et al.*, 1991). We found that sphingosine stimulated the accumulation of inositol phosphates in Swiss 3T3 fibroblasts. Although a twofold increase was observed after 1 hour of incubation with sphingosine, no change were detectable during the first 5 minutes (Zhang *et al.*, 1990b). Since the sphingosine-induced change in $[Ca^{2+}]_i$ occurs within 2 minutes, it is unlikely that IP_3 mediates the sphingosine-induced intracellular calcium mobilization. Murayama and Ui (1987) showed that exogenous PA induced production of IP_3 in 3T3 fibroblasts; therefore, the sphingosine-induced increase in inositol phosphates may be mediated via PA. Sphingosine has been shown to potentiate the EGF-induced increase in inositol phosphate levels in A431 cells without altering the rate of proliferation (Wahl and Carpenter, 1988). The exact role, if any, of the sphingosine-induced increase in inositol phosphates in Swiss 3T3 cells remains unclear.

2. cAMP

Similar to previous reports that PA induced a dramatic decrease in cAMP levels in isoproterenol-treated 3T3 fibroblasts (Murayama and Ui, 1987), we observed that sphingosine also caused a drastic decrease in the cellular cAMP levels (Zhang *et al.*, 1990b). Similar effects were reported in S49 lymphoma cells in which sphingosine inhibited adrenaline-, propranolol-, or forskolin-stimulated increases in cAMP (Johnson and Clark, 1990), which suggests that sphingosine is involved in either inhibition of adenylate cyclase or activation of phosphodiesterase (PDE) which is independent of receptor function. In Swiss 3T3 fibroblasts, activation of PDE appears an unlikely mechanism since sphingosine inhibited cAMP accumulation in the presence of the PDE inhibitor isobutylmethylxanthine.

In Swiss 3T3 cells, cAMP has been shown unequivocally to act as a positive effector of proliferation (Rozengurt *et al.*, 1981); thus, it is unlikely that the sphingosine-induced decrease in cAMP plays any significant role in the induction of proliferation. However, in a myriad of other cell types, cAMP is most likely a negative effector of mitogenesis (Leitman *et al.*, 1986; Pines *et al.*, 1988; Lingk *et al.*, 1990; Moodie and Martin, 1991; Kano *et al.*, 1991; Reuse *et al.*, 1991; Vazquez *et al.*, 1991; Vittet *et al.*, 1992), and therefore, the relevance of changes in cAMP levels to the mechanism of action of sphingosine in other cell types is presently unclear.

3. Kinases Other Than Protein Kinase C

As discussed above, relatively low concentrations of sphingosine and some analogs have been shown to regulate the activities of a variety of kinases *in vitro* and *in vivo,* including CAM kinase (Jefferson and Schulman, 1988), myosin light chain kinase (Jefferson and Schulman, 1988), DAG kinases (Sakane *et al.*, 1989), insulin receptor kinase (Arnold and Newton, 1991), EGF-receptor tyrosine kinase (Davis *et al.*, 1988), and casein kinase (McDonald *et al.*, 1991). Effects of sphingosine on these kinases are within the dose range for induction of mitogenesis in 3T3 cells.

The reported effects of sphingosine on growth factor receptors may also prove relevant to its effects on mitogenesis. Sphingosine, or its derivative *N,N*-dimethyl-sphingosine, has been reported to increase the number of EGF-binding sites and affinity and phosphorylation state of the EGF receptor in A431 human epidermoid carcinoma cells (Faucher *et al.*, 1988; Igarashi *et al.*, 1990) and in WI 38 human fetal lung fibroblasts (Davis *et al.*, 1988). However, as mentioned above, the sphingosine-induced changes were not associated with altered growth in either of these cell types, bringing into question the relevance of these sphingosine-induced effects for proliferation. Another action of sphingosine on growth factor receptors is the inhibition of insulin receptor autophosphorylation in liver preparations, NIH 3T3 cell membranes, and intact NIH 3T3 cells (Arnold and Newton, 1991). In spite of this inhibition of receptor tyrosine kinase activity, sphingosine did not alter insulin binding (Arnold and Newton, 1991; Robertson *et al.*, 1989; Smal and De Meyts, 1989).

In recent years, tyrosine kinases have been shown to play potentially important roles in the process of cellular proliferation (Comoglio *et al.*, 1990). Various inhibitors of tyrosine kinases have been reported to inhibit TPA-dependent proliferation of T cells (Munoz *et al.*, 1991) and EGF-induced mitogenesis of A431 cells (Yaish *et al.*, 1988) and NIH 3T3 cells (Lyall *et al.*, 1989). Such inhibitors could be used to probe the involvement of tyrosine kinase activity in sphingosine-stimulated mitogenesis. Although we have ruled out a major role for PKC with regard to sphingosine-induced mitogenesis in 3T3 fibroblasts, inhibition or activation of other kinases may still prove to be important.

IV. Concluding Remarks

We have described here our recent efforts to understand the mitogenic action of exogenous sphingosine in Swiss 3T3 fibroblasts in which we have identified a number of novel intracellular signaling pathways as potential mediators of sphingosine-induced mitogenesis. This includes elucidation of a potential dual-action intracellular messenger role for SPP, involving pathways both for rapid mobilization of intracellular free calcium and for stimulation of PA levels (Fig. 4). Both of these events may lead directly to increased cell division. A crucial area for future research will be to determine whether this previously unknown potential modulator of cell growth has endogenous relevance. Certain growth factors may stimulate the degradation of complex glycosphingolipids or sphingomyelin, leading to increased levels of sphingosine and/or SPP. Whether such pathways exist and to what extent these pathways are essential for mitogenesis await further study.

Although the existence of a sphingolipid cycle which can act as a signal transducing element, similar to phospholipid cycles, is an intriguing hypothesis, many questions still remain unanswered. Is sphingosine and/or SPP the actual bioactive molecule for this cycle, or are other sphingolipid metabolites involved? Is this cycle a universal mechanism of signal transduction in cells? Are all the characteristics of the cycle conserved in different cell types? Attempts at answering these questions are sure to broaden our understanding of mitogenesis and likely of other fundamental cellular events.

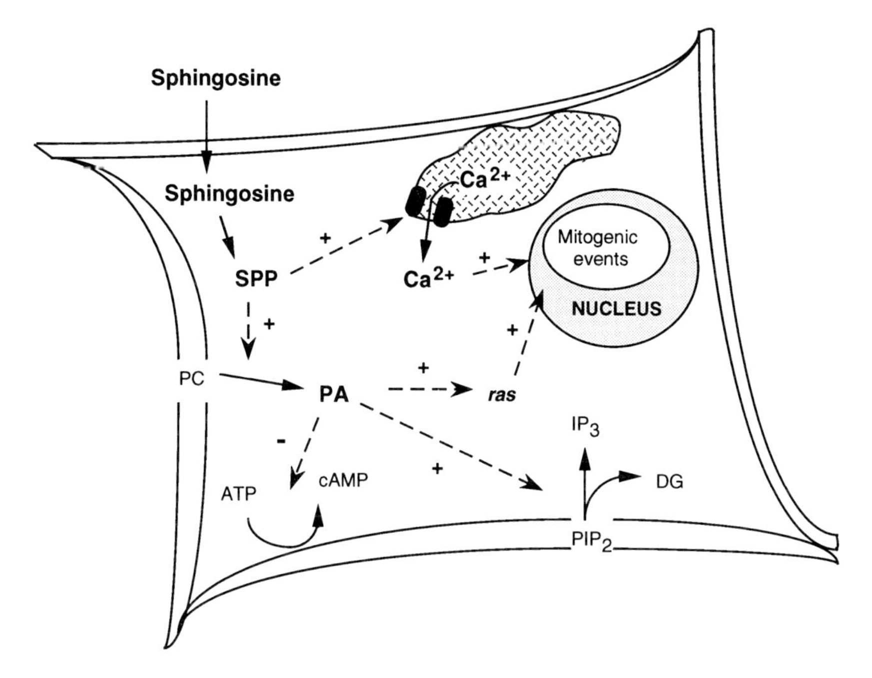

FIG. 4. Model of the possible mechanism of action of sphingosine in the induction of mitogenesis. For more detailed information, see text. Dashed arrows indicate influences on pathways. Stimulatory or inhibitory effects are indicated by + or –, respectively.

Acknowledgments

Supported by Research Grants 1RO1 GM43880 and 1R29 GM39718 from the National Institutes of Health and 3018M from The Council For Tobacco Research. Dr. Robert O. Carlson wishes to acknowledge support from The Cancer Research Foundation of America. Dr. Ana Olivera was supported by a fellowship from The Ministry of Education and Science (Spain).

References

Adams, J. C., and Gullick, W. J. (1989). *Biochem. J.* **257,** 905–911.

Al, B. J. M., Tiffany, C. W., Gomes de Mesquita, D. S., Moser, H. W., Tager, J. M., and Schram, A. W. (1989). *Biochim. Biophys. Acta* **1004,** 245–251.

Arnold, R. S., and Newton, A. C. (1991). *Biochemistry* **30,** 7747–7754.

Barbacid, M. (1987). *Annu. Rev. Biochem.* **56,** 779–827.

Bell, R. M. (1986). *Cell (Cambridge, Mass.)* **45,** 631–632.

Bell, R. M., Loomis, C. R., and Hannun, Y. A. (1988). *Cold Spring Harbor Symp. Quant. Biol.* **1,** 103–110.

Ben-Av, P., and Liscovitch, M. (1989). *FEBS Lett.* **259,** 64–66.

Berridge, M. J., (1984). *Biochem. J.* **220,** 345–360.

Berridge, M. J. (1985). *Sci. Am.* **253,** 95–106.

Berridge, M. J., and Irvine, R. F. (1989). *Nature (London)* **341,** 197–205.

Besterman, J. M., Watson, S. P., and Cuatrecasas, P. (1986). *J. Biol. Chem.* **261,** 723–727.

Bocckino, S. B., Wilson, P. B., and Exton, J. H. (1991). *Proc. Natl. Acad. Sci. U.S.A.* **88,** 6210–6213.

Buckley, N. E., Matyas, G. R., and Spiegel, S. (1990). *Exp. Cell Res.* **189,** 13–21.

Buehrer, B. M., and Bell, R. M. (1992). *J. Biol. Chem.* **267,** 3154–3159.

Campbell, P. I. (1983). *Cytobios* **37,** 21–26.

Comoglio, P., Di Renzo, M., Gaudino, G., Ponzetto, C., and Prat, M. (1990). *Am. Rev. Respir. Dis.* **142,** S16–S19.

Cook, S. J., and Wakelam, M. J. O. (1991). *Biochim. Biophys. Acta* **1092,** 265–272.

Davis, R. J., Girones, N., and Faucher, M. (1988). *J. Biol. Chem.* **263,** 5377–5379.

Dennis, E. A., Rhee, S. G., Billah, M. M., and Hannun, Y. A. (1991). *FASEB J.* **5,** 2068–2077.

Desai, N. N., Zhang, H., Olivera, A., Mattie, M. E., and Spiegel, S. (1992). *J. Biol. Chem.* **267,** 23122–23128.

Divecha, N., Lander, D. J., Scott, T. W., and Irvine, R. F. (1991). *Biochim. Biophys. Acta* **1093,** 184–188.

Elliott, M. E., Jones, H. M., Tomasko, S., and Goodfriend, T. L. (1991). *Biochim. Biophys. Acta* **1083,** 179–186.

Exton, J. H. (1990a). *Adv. Second Messenger Phosphoprotein Res.* **24,** 152–157.

Exton, J. H. (1990b). *J. Biol. Chem.* **265,** 1–4.

Faucher, M., Girones, N., Hannun, Y. A., Bell, R. M., and Davis, R. J. (1988). *J. Biol. Chem.* **263,** 5319–5327.

Ghosh, T. K., Bian, J., and Gill, D. L. (1990). *Science* **248,** 1653–1656.

Hakomori, S. (1990). *J. Biol. Chem.* **265,** 18713–18716.

Hall, F. L., Fernyhough, P., Ishii, D. N., and Vulliet, P. R. (1988). *J. Biol. Chem.* **263,** 4460–4466.

Hannun, Y. A., and Bell, R. M. (1987). *J. Biol. Chem.* **235,** 670–674.

Hannun, Y. A., and Bell, R. M. (1989). *Science* **243,** 500–507.

Hannun, Y. A., Loomis, C. R., Merrill, A. H., and Bell, R. M. (1986). *J. Biol. Chem.* **261,** 12604–12609.

Hasegawa-Sasaki, H. (1985). *Biochem. J.* **232,** 99–109.

Hesketh, T. R., Morris, J. D. H., Moore, J. P., and Metcalfe, J. C. (1988). *J. Biol. Chem.* **263,** 11879–11886.

Hidaka, H., Inagaki, M., Kawamoto, S., and Sasaki, Y. (1984). *Biochemistry* **23,** 5036–5041.

Hirschberg, C. B., Kisič, A., and Schroepfer, G. J. (1970). *J. Biol. Chem.* **245,** 3084–3090.

Hoffman, J. M., Standaert, M. L., Nair, G. P., and Farese, R. V. (1991). *Biochemistry* **30,** 3315–3322.

Huang, C., and Cabot, M. C. (1990). *J. Biol. Chem.* **265,** 17468–17473.

Igarashi, Y., Kitamura, K., Toyokuni, T., Dean, B., Fenderson, B., Ogawass, T., and Hakomori, S. (1990). *J. Biol. Chem.* **265,** 5385–5389.

Inokuchi, J., Kumamoto, Y., Jimbo, M., Shimeno, H., and Nagamatsu, A. (1991). *FEBS Lett.* **286,** 39–43.

Jamal, Z., Martin, A., Munoz, A. G., and Brindley, D. N. (1991). *J. Biol. Chem.* **266,** 2988–2996.

Jefferson, A., and Schulman, H. (1988). *J. Biol. Chem.* **263,** 15241–15244.

Johnson, J. A., and Clark, R. B. (1990). *J. Biol. Chem.* **265,** 9333–9339.

Kano, J., Sugimoto, T., Fukase, M., and Fujita, T. (1991). *Biochem. Biophys. Res. Commun.* **177,** 365–369.

Kao, J. P., Alderton, J. M., Tsien, R. Y., and Steinhardt, R. A. (1990). *J. Cell Biol.* **111,** 183–196.

Kawamoto, S., and Hidaka, H. (1984). *Biochem. Biophys. Res. Commun.* **125,** 258–264.

Keenan, R. W. (1972). *Biochim. Biophys. Acta* **270,** 383–396.

Keenan, R. W., and Haeglin, B. (1969). *Biochem. Biophys. Res. Commun.* **37,** 888–894.
Keenan, R. W., and Maxam, A. (1969). *Biochim. Biophys. Acta* **176,** 348–356.
Kim, M. Y., Linardič, C., Obeid, L., and Hannun, Y. (1991). *J. Biol. Chem.* **266,** 484–489.
Kiss, Z. (1990). *Prog. Lipid Res.* **29,** 141–166.
Kiss, Z., and Anderson, W. B. (1990). *J. Biol. Chem.* **265,** 7188–7194.
Kiss, Z., Chattopadhyay, J., and Pettit, G. R. (1991). *Biochem. J.* **273,** 189–194.
Kolesnick, R. N. (1989). *J. Biol. Chem.* **264,** 7617–7623.
Kolesnick, R. N. (1991). *Prog. Lipid Res.* **30,** 1–38.
Koval, M., and Pagano, R. E. (1991). *Biochim. Biophys. Acta* **1082,** 113–125.
Lavie, Y., and Liscovitch, M. (1990). *Biochem. Biophys. Res. Commun.* **167,** 607–613.
Lavie, Y., Piterman, O., and Liscovitch, M. (1990). *FEBS Lett.* **277,** 7–10.
Leitman, D., Fiscus, R., and Murad, F. (1986). *J. Cell. Physiol.* **127,** 237–243.
Lingk, D., Chan, M., and Gelfand, E. (1990). *J. Immunol.* **145,** 449–455.
Louie, D. D., Kisič, A. K., and Schroepfer, G. J. J. (1976). *J. Biol. Chem.* **52,** 4557–4564.
Lyall, R., Zilberstein, A., Gazit, A., Gilon, C., Levitzki, A., and Schlessinger, J. (1989). *J. Biol. Chem.* **264,** 14503–14509.
Manning, R., and Brindley, D. N. (1972). *Biochem. J.* **130,** 1003–1012.
Masco, D., Van de Walle, M., and Spiegel, S. (1991). *J. Neurosci.* **11,** 2443–2452.
McCaffrey, P. G., and Rosner, M. R. (1987). *Biochem. Biophys. Res. Commun.* **146,** 140–146.
McDonald, O., Hannun, Y., Reynolds, C., and Sahyoun, N. (1991). *J. Biol. Chem.* **266,** 21773–21776.
Medlock, K. A., and Merrill, A. H., Jr. (1988). *Biochem. Biophys. Res. Commun.* **157,** 232–237.
Merrill, A. H. (1983). *Biochim. Biophys. Acta* **754,** 284–291.
Merrill. A. H. J. (1991). *J. Bioenerg. Biomembr.* **23,** 83–104.
Merrill, A. H., Sereni, A. M., Stevens, V. L., Hannun, Y. A., Bell, R. M., and Kinkade, J. M. (1986). *J. Biol. Chem.* **261,** 12610–12615.
Merrill, A. J., and Stevens, V. L. (1989). *Biochim. Biophys. Acta* **1010,** 131–139.
Merrill, A. J., Wang, E., Mullins, R. E., Jamison, W. C., Nimkar, S., and Liotta, D. C. (1988). *Anal. Biochem.* **171,** 373–381.
Merrill, A. J., Nimkar, S., Menaldino, D., Hannun, Y. A., Loomis, C., Bell, R. M., Tyagi, S. R., Lambeth, J. D., Stevens, V. L., and Hunter, R. (1989). *J. Biol. Chem.* **264,** 6773–6779.
Metcalfe, J. C., Moore, J. P., Smith, G. A., and Hesketh, T. R. (1986). *Br. Med. Bull.* **42,** 405–412.
Millar, J. B. A., and Rozengurt, E. (1990). *J. Biol. Chem.* **265,** 19973–19979.
Mok, A. P., and McMurray, W. C. (1990). *Biochem. Cell Biol.* **68,** 1380–1392.
Moodie, S., and Martin, W. (1991). *Br. J. Pharmacol.* **102,** 101–106.
Moolenaar, W. H., Krujer, W., Tilly, B. C., Verlaan, I., Bierman, A. J., and deLaat, S. W. (1986). *Nature (London)* **323,** 171–173.
Mullmann, T. J., Siegel, M. I., Egan, R. W., and Billah, M. M. (1991). *J. Biol. Chem.* **266,** 2013–2016.
Munoz, A. G., Zubiaga, A., and Huber, B. (1991). *FEBS Lett.* **279,** 319–322.
Murayama, T., and Ui, M. (1987). *J. Biol. Chem.* **262,** 5522–5529.
Negami, A. I., Sasaki, H., and Yamamura, H. (1985). *Biochem. Biophys. Res. Commun.* **131,** 712–719.
Nishizuka, Y. (1986). *Science* **233,** 305–312.
Nishizuka, Y. (1989). *Cancer (Philadelphia)* **63,** 1892–1903.
Oishi, K., Zheng, B., and Kuo, J. F. (1990). *Carbohydr. Res.* **195,** 199–224.
Okazaki, T., Bell, R. M., and Hannun, Y. A. (1989). *J. Biol. Chem.* **264,** 19076–19080.
Okazaki, T., Bielawska, A., Bell, R. M., and Hannun, Y. A. (1990). *J. Biol. Chem.* **265,** 15823–15831.
Olson, S. C., Bowman, E. P., and Lambeth, J. D. (1991). *J. Biol. Chem.* **266,** 17236–17242.
Pagano, R. E. (1988). *Trends Biochem. Sci.* **13,** 202–205.
Pai, J. K., Pachter, J. A., Weinstein, I. B., and Bishop, W. R. (1991). *Proc. Natl. Acad. Sci. U.S.A.* **88,** 598–602.

Pines, M., Ashkenazi, A., Cohen-Chapnik, N., Binder, L., and Gertler, A. (1988). *J. Cell. Biochem.* **37,** 119–129.
Plevin, R., Cook, S. J., Palmer, S., and Wakelam, S. J. (1991). *Biochem. J.* **279,** 559–565.
Ramachandran, C. K., Murray, D. K., and Nelson, D. H. (1990). *Biochem. Biophys. Res. Commun.* **167,** 607–613.
Reinhold, S. L., Zimmerman, G. A., Prescott, S. M., and McIntyre, T. M. (1989). *J. Biol. Chem.* **264,** 21652–21659.
Reuse, S., Pirson, I., and Dumont, J. (1991). *Exp. Cell Res.* **196,** 210–215.
Robertson, D. G., DiGirolamo, M., Merrill, A. J., and Lambeth, J. D. (1989). *Biochim. Biophys. Acta* **980,** 319–325.
Rozengurt, E. (1986). *Science* **234,** 161–166.
Rozengurt, E., Legg, A., Strang, G., and Courtenay-Luck, N. (1981). *Proc. Natl. Acad. Sci. U.S.A.* **78,** 4392–4396.
Sakane, F., Yamada, K., and Kanoh, H. (1989). *Eur. J. Pediatr.* **149,** 31–39.
Schwarzmann, G., and Sandhoff, K. (1990). *Biochemistry* **29,** 10865–10871.
Seuwen, K., Kahan, C., Hartmann, T., and Pouyssegur, J. (1990). *J. Biol. Chem.* **265,** 22292–22299.
Siegmann, D. W. (1987). *Biochem. Biophys. Res. Commun.* **145,** 228–233.
Smal, J., and De Meyts, P. (1989). *Proc. Natl. Acad. Sci. U.S.A.* **86,** 4705–4709.
Sohal, P. S., and Cornell, R. B. (1990). *Biochem. Biophys. Res. Commun.* **170,** 162–168.
Spiegel, S. (1988a). *In* "New Trends in Ganglioside Research: Neurochemical and Neuroregenerative Aspects" (R. W. Ledeen, E. L. Hogan, G. Tettamanti, A. J. Yates, and R. K. Yu, eds.), Fidia Res. Ser., Vol. 14, pp. 405–421. Liviana Press, Springer-Verlag, Padova/Berlin.
Spiegel, S. (1988b). *Biochim. Biophys. Acta* **969,** 249–256.
Spiegel, S., and Fishman, P. H. (1987). *Proc. Natl. Acad. Sci. U.S.A.* **84,** 141–147.
Spiegel, S., and Panagiotopoulos, C. (1988). *Exp. Cell Res.* **177,** 414–427.
Spiegel, S., Fishman, P. H., and Weber, R. J. (1985). *Science* **230,** 1283–1287.
Stevens, V. L., Nimkar, S., Jamison, W. C., Liotta, D. C., and Merrill, A. J. (1990). *Biochim. Biophys. Acta* **1051,** 37–45.
Stoffel, W., and Assmann, G. (1970). *Hoppe-Seyler's Z. Physiol. Chem.* **351,** 1041–1049.
Stoffel, W., and Bister, K. (1973). *Hoppe-Seyler's Z. Physiol. Chem.* **354,** 169–181.
Stoffel, W., Sticht, G., and LeKim, D. (1968a). *Hoppe-Seyler's Z. Physiol. Chem.* **349,** 1149–1156.
Stoffel, W., Sticht, G., and LeKim, D. (1968b). *Hoppe-Seyler's Z. Physiol. Chem.* **349,** 1745–1748.
Stoffel, W., Lekim, D., and Sticht, G. (1969). *Hoppe-Seyler's Z. Physiol. Chem.* **350,** 1233–1241.
Stoffel, W., Assmann, G., and Binczek, E. (1970). *Hoppe-Seyler's Z. Physiol. Chem.* **351,** 635–642.
Stoffel, W., Heimann, G., and Hellenbroich, B. (1973a). *Hoppe-Seyler's Z. Physiol. Chem.* **354,** 562–566.
Stoffel, W., Hellenbroich, B., and Heimann, G. (1973b). *Hoppe-Seyler's Z. Physiol. Chem.* **354,** 1311–1316.
Sugiya, H., and Furuyama, S. (1991). *FEBS Lett.* **286,** 113–116.
Sweeley, C. C. (1985). *In* "Biochemistry of Lipids and Membranes" (J. E. Vance and M. E. Vance, eds.), pp. 361–403. Benjamin/Cummings, Menlo Park, CA.
Takuwa, N., Iwamoto, A., Kumada, M., Yamashita, K., and Takuwa, Y. (1991). *J. Biol. Chem.* **266,** 1403–1409.
Tombes, R. M., and Borisy, G. G. (1989). *J. Cell. Biol.* **109,** 627–636.
Trahey, M., and McCormick, F. (1987). *Science* **238,** 524–525.
Tsai, M.-H., Hall, A., and Stacey, D. W. (1989a). *Mol. Cell Biol.* **9,** 5260–5264.
Tsai, M.-H., Yu, C.-L., Wei, F. S., and Stacey, D. W. (1989b). *Science* **243,** 522–526.
Tsai, M.-H., Yu, C.-L., and Stacey, D. W. (1990). *Science* **250,** 982–985.
Tsai, M.-H., Roudebush, M., Dobrowolski, Yu, C.-L., Gibbs, J. B., and Stacey, D. W. (1991). *Mol. Cell. Biol.* **11,** 2785–2793.

Van Veldhoven, P. P. V., and Mannaerts, G. P. (1991). *J. Biol. Chem.* **266,** 12502–12507.
Van Veldhoven, P. P., Foglesong, R. J., and Bell, R. M. (1989). *J. Lipid Res.* **30,** 611–616.
Vazquez, A., Auffredou, M., Galanaud, P., Leca, G. (1991). *J. Immunol.* **146,** 4222–4227.
Vittet, D., Mathieu, M., Launay, J., and Chevillard, C. (1992). *J. Cell. Physiol.* **150,** 65–75.
Wahl, M., and Carpenter, G. (1988). *J. Biol. Chem.* **263,** 7581–7590.
Weis, F., and Davis, R. J. (1990). *J. Biol. Chem.* **265,** 12059–12066.
Wilson, E., Wang, E., Mullins, R. E., Uhlinger, D. J., Liotta, D. C., Lambeth, J. D., and Merrill, A. H. J. (1988). *J. Biol. Chem.* **263,** 9304–9309.
Yaish, P., Gazit, A., Gilon, C., and Levitzki, A. (1988). *Science* **242,** 933–935.
Yu, C.-L., Tsai, M. H., and Stacey, D. W. (1988). *Cell (Cambridge, Mass.)* **52,** 63–71.
Zhang, H., Buckley, N. E., Gibson, K., and Spiegel, S. (1990a). *J. Biol. Chem.* **265,** 76–81.
Zhang, H., Desai, N. N., Murphey, J. M., and Spiegel, S. (1990b). *J. Biol. Chem.* **265,** 21309–21316.
Zhang, H., Desai, N. N., Olivera, A., Seki, T., Brooker, G., and Spiegel, S. (1991). *J. Cell Biol.* **114,** 155–167.

ADVANCES IN LIPID RESEARCH, VOL. 25

Sphingolipid Regulation of the Epidermal Growth Factor Receptor

ROGER J. DAVIS

Howard Hughes Medical Institute
Program in Molecular Medicine
Department of Biochemistry and Molecular Biology
University of Massachusetts Medical School
Worcester, Massachusetts 01605

It has been established that sphingolipids represent a major class of tumor-specific cell surface antigens (Hakomori, 1981, 1984) and that the expression of sphingolipids is regulated during the cell cycle, density-dependent growth inhibition, and oncogenic transformation (Hakomori, 1981, 1984). However, the role of these molecules in the control of cellular proliferation is not understood. Recently, evidence has been presented demonstrating that sphingolipids are bimodal regulators of the epidermal growth factor (EGF) receptor tyrosine kinase activity (Hakomori, 1990). It is therefore possible that the *in vivo* function of the EGF receptor is modulated by the level of endogenous sphingolipids. The purpose of this review is to summarize the results of recent studies that have been designed to test this hypothesis.

I. The EGF Receptor: An Allosteric Enzyme Regulated by Multisite Phosphorylation

The EGF receptor is a 170-kDa transmembrane glycoprotein (Ullrich and Schlessinger, 1990). The receptor is composed of an extracellular domain that binds EGF, a single transmembrane spanning domain, and a cytoplasmic domain with intrinsic tyrosine kinase activity (Fig. 1). Mutational analysis of the EGF receptor has demonstrated that the tyrosine kinase activity is required for signal transduction (Ullrich and Schlessinger, 1990). The binding of EGF to the

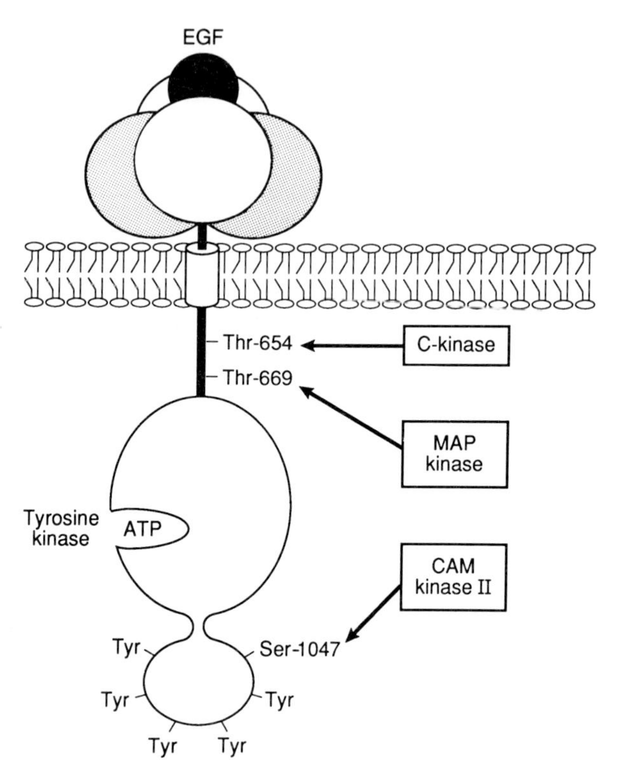

FIG. 1. Schematic diagram illustrating the phosphorylation of the EGF receptor by three protein kinases. The EGF receptor is a transmembrane protein consisting of an extracellular EGF-binding domain, a single membrane-spanning domain, and a cytoplasmic domain with intrinsic tyrosine kinase activity. The COOH-terminal region of the cytoplasmic domain contains all the sites of autophosphorylation of the receptor on tyrosine residues. The EGF receptor cytoplasmic domain is also a substrate for phosphorylation by C-kinase at Thr^{654} (Davis and Czech, 1985a; Downward *et al.*, 1985; Hunter *et al.*, 1984), MAP kinases at Thr^{669} (Northwood *et al.*, 1991; Takishima *et al.*, 1991), and Ca^{2+}/calmodulin-dependent protein kinase II at $Ser^{1046/7}$ (Countaway *et al.*, 1990, 1992).

extracellular domain of the receptor results in the stimulation of the cytoplasmic domain tyrosine kinase activity (Ullrich and Schlessinger, 1990). This process has been proposed to be mediated by an allosteric mechanism that involves receptor dimerization (Schlessinger, 1988). Activation of the EGF receptor causes the physical association of the receptor with several cytoplasmic substrates for tyrosine phosphorylation, including phospholipase C-γ (PLC-γ1) and the GTPase activating protein (Koch *et al.*, 1991; Ullrich and Schlessinger, 1990). The binding of these substrates can be accounted for by the presence of conserved *Src* homology domains (SH2) in these proteins (Koch *et al.*, 1991). The binding site on the EGF receptor has been identified as the tyrosine phosphorylated COOH-terminal domain of the receptor (Koch *et al.,* 1991; Ullrich and Schlessinger, 1990). This COOH-terminal region therefore represents an important effector domain for EGF receptor function that interacts with this class of signaling molecules (Fig. 1). The tyrosine phosphorylation of substrate proteins by the EGF receptor is thought to be an important initial step in signal transduction that leads to cellular proliferation (Koch *et al.,* 1991; Ullrich and Schlessinger, 1990).

Treatment of cultured cells with EGF causes the initiation of signal transduction by the EGF receptor followed by the rapid desensitization of receptor function (Cunningham *et al.,* 1989; Davis and Czech, 1986). This process of desensitization is a property that is common to many cell surface receptors (Hausdorff *et al.,* 1990). One mechanism of long-term (hours) desensitization of the EGF receptor is mediated by the internalization and degradation of the receptor (Ullrich and Schlessinger, 1990). In addition, there exists a rapid (minutes) mechanism of EGF-induced EGF receptor desensitization (Chinkers and Garbers, 1986; Countaway *et al.,* 1992; Lai *et al.,* 1989; McCune *et al.,* 1990) that is mechanistically distinct from the desensitization caused by the treatment of cells with phorbol ester (Countaway *et al.,* 1990, 1992; Davis, 1988; Decker *et al.,* 1990; Livneh *et al.,* 1988; Lund *et al.,* 1990). The rapid desensitization of the EGF receptor is associated with (1) a decrease in the high-affinity binding of EGF to the receptor (Countaway *et al.,* 1992), and (2) an inhibition of the receptor tyrosine kinase activity (Chinkers and Garbers, 1986; Countaway *et al.,* 1992; Lai *et al.,* 1989; McCune *et al.,* 1990). The mechanism that accounts for the loss of high-affinity binding of EGF is not understood (Ullrich and Schlessinger, 1990). However, the decreased EGF receptor tyrosine kinase activity has been proposed to be caused by an increase in the phosphorylation state of the receptor (Fig. 2).

A. Protein Kinase C Phosphorylation at Thr^{654}

The EGF receptor is phosphorylated at Thr^{654} by C-kinase (Davis and Czech, 1985a; Downward *et al.,* 1985; Hunter *et al.,* 1984). Treatment of intact cells with phorbol ester or growth factors (e.g., platelet-derived growth factor) causes (1) an increase in Thr^{654} phosphorylation, and (2) an inhibition of EGF-stimulated tyrosine phosphorylation (Cochet *et al.,* 1984; Countaway *et al.,* 1989a,b, 1990, 1992; Davis, 1988; Davis and Czech, 1984, 1985a,b, 1986, 1987; Davis *et al.,* 1985a,b, 1988; Decker *et al.,* 1990; Downward *et al.,* 1985; Friedman *et al.,* 1984; Livneh *et al.,* 1988; Lund *et al.,* 1990; Northwood and Davis, 1989, 1990). A causal relationship between phosphorylation at Thr^{654} and decreased EGF-stimulated tyrosine kinase activity has been established using *in vitro* experiments with purified protein kinase C and the EGF receptor (Cochet

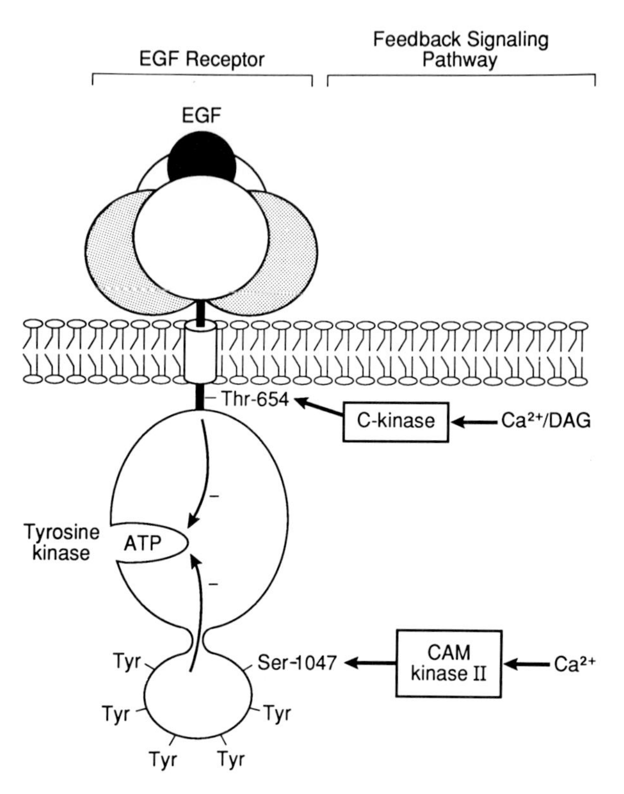

FIG. 2. Regulation of the EGF receptor protein-tyrosine kinase activity by multisite phosphorylation. Thr^{654}: A substrate for C-kinase. Phosphorylation of the EGF receptor at this site is the major mechanism that accounts for the inhibition of the receptor protein-tyrosine kinase activity observed in phorbol ester-treated cells (Davis, 1988). $Ser^{1046/7}$: Phosphorylation of the EGF receptor at $Ser^{1046/7}$ is the major mechanism that accounts for the desensitization of the receptor protein-tyrosine kinase activity observed in EGF-treated cells (Countaway *et al.*, 1992). The identity of the protein kinase that accounts for the *in vivo* phosphorylation of the EGF receptor at $Ser^{1046/7}$ is unknown. However, Ser^{1047} is an *in vitro* substrate for CAM kinase II (Countaway *et al.*, 1992).

et al., 1984; Davis, 1988; Downward *et al.,* 1985). This conclusion is also supported by the results of a mutational analysis. Replacement of Thr^{654} with Ala blocks the effect of protein kinase C phosphorylation to inhibit the EGF receptor tyrosine kinase activity *in vitro* (Davis, 1988). This mutation also blocks the effect of phorbol ester to inhibit EGF-stimulated tyrosine phosphorylation (Countaway *et al.,* 1989b, 1990, 1992; Davis, 1988; Decker *et al.,* 1990; Livneh *et al.,* 1988) and receptor internalization *in vivo* (Lund *et al.,* 1990). Furthermore, mutations that cause constitutive phosphorylation of the EGF receptor at this site are defective in EGF-stimulated mitogenic signal transduction (Bowen *et al.,* 1991). It is therefore established that the phosphorylation of the EGF receptor at Thr^{654} causes an inhibition of the EGF receptor tyrosine kinase activity (Fig. 2).

The functional consequence of the phosphorylation of the EGF receptor at Thr^{654} is an inhibition of signal transduction (Bowen *et al.,* 1991; Davis, 1988). However, the molecular mechanism that accounts for the effect of phosphorylation at Thr^{654} to regulate the EGF receptor tyrosine kinase activity is not understood. The location of Thr^{654} within the juxtamembrane region of the cytoplasmic domain of the EGF receptor suggests that it may disrupt the process of signal transmission between EGF binding to the extracellular domain and the cytoplasmic tyrosine kinase domain (Fig. 2). It has been proposed that the EGF receptor tyrosine kinase activity is stimulated by EGF by an allosteric mechanism that involves receptor dimerization (Schlessinger, 1988). This mechanism of activation suggests that Thr^{654} phosphorylation may inhibit EGF receptor dimerization. However, it has been found that the inhibition of tyrosine kinase activity caused by Thr^{654} phosphorylation is independent of the oligomeric state of the EGF receptor (Northwood and Davis, 1989). Further studies are therefore required to define the structural basis for the regulation of the EGF receptor by phosphorylation at Thr^{654}.

B. MAP Kinase Phosphorylation at Thr^{669}

The major site of growth factor-stimulated phosphorylation of the EGF receptor is Thr^{669}. This site of phosphorylation is located within the juxtamembrane region of the cytoplasmic domain of the EGF receptor (Fig. 1). The sequence surrounding Thr^{669} is unusual because it is rich in Pro residues (Countaway *et al.,* 1989). A growth factor-stimulated protein kinase activity that phosphorylates Thr^{669} was detected in cell extracts (Countaway *et al.,* 1989). Purification and characterization of this activity demonstrated that it was accounted for by a mixture of mitogen-activated protein (MAP) kinase isoforms (Northwood *et al.,* 1991; Takishima *et al.,* 1991). It has been demonstrated that the phosphorylation of both Thr and Tyr residues is required for growth factor activation of MAP kinase activity (Cobb *et al.,* 1991). Recently, molecular clones for several members of the MAP kinase family (also referred to as ERKs) have been isolated (Cobb *et al.,* 1991). Investigation of the effect of point mutations within the EGF receptor has demonstrated that the consensus sequence (Alvarez *et al.,* 1991; Clark-Lewis *et al.,* 1991; Gonzalez *et al.,* 1991) for substrate phosphorylation by MAP kinases is Pro-Xaa_n-Ser/Thr-Pro (where Xaa is a neutral or basic amino acid and $n = 1$ or 2).

The role of phosphorylation of the EGF receptor at Thr^{669} has been investigated by studying the effect of substitution of this residue with Ala (Countaway *et al.*, 1989a, 1990; Heisermann *et al.*, 1990). This mutation was found to cause a defect in EGF-stimulated endocytosis of the EGF receptor and in the EGF-stimulated tyrosine phosphorylation of an 85-kDa substrate (Heisermann *et al.*, 1990). It has therefore been proposed that the function of phosphorylation at Thr^{669} may be to regulate the interaction of substrates with the EGF receptor (Heisermann *et al.*, 1990). A direct experimental test of this hypothesis is required.

C. Ca^{2+}/Calmodulin-Dependent Protein Kinase II Phosphorylation at $Ser^{1046/7}$

The major site of growth factor-stimulated Ser phosphorylation of the EGF receptor is located within the COOH-terminal domain at $Ser^{1046/7}$ (Fig. 1). This site is an *in vitro* substrate for Ca^{2+}/calmodulin-dependent protein kinase II [CAM kinase II (Countaway *et al.*, 1990, 1992)], but it is likely that other protein kinases [such as S6 kinases (Erikson, 1991)] may also contribute to the phosphorylation of the EGF receptor at $Ser^{1046/7}$ in intact cells (Countaway *et al.*, 1990, 1992; Heisermann and Gill, 1988).

The EGF receptor tyrosine kinase is desensitized during incubation of tissue culture cells with EGF (Chinkers and Garbers, 1986; Countaway *et al.*, 1992; Lai *et al.*, 1989; McCune *et al.*, 1990). The EGF-stimulated phosphorylation of the EGF receptor at $Ser^{1046/7}$ has been proposed to account for this desensitization (Countaway *et al.*, 1992). Evidence to support this hypothesis is derived from the following observations: (1) phosphorylation at $Ser^{1046/7}$ *in vitro* with CAM kinase II causes an inhibition of EGF receptor tyrosine kinase activity (Countaway *et al.*, 1992); and (2) substitution of $Ser^{1046/7}$ (but not Thr^{654} or Thr^{669}) with Ala residues blocks the EGF-stimulated desensitization of the receptor tyrosine kinase (Countaway *et al.*, 1992) and EGF-stimulated receptor internalization (Countaway *et al.*, 1992). $Ser^{1046/7}$ is therefore a negative regulatory site of phosphorylation of the EGF receptor (Fig. 2). Deletion of this phosphorylation site occurs in oncogenic forms of the EGF receptor, and the loss of $Ser^{1046/7}$ is associated with a marked increase in disease potential (Théroux *et al.*, 1992).

II. Effect of Sphingolipids on EGF Receptor Phosphorylation

Long-chain sphingoid bases (e.g., sphingosine) have been characterized as inhibitors of C-kinase (Hannun and Bell, 1987, 1989; Hannun *et al.*, 1986; Merrill, 1991). The activation of C-kinase by growth factors and phorbol ester can be inhibited by treatment of cells in tissue culture (Faucher *et al.*, 1988; Igarashi *et al.*, 1989; Kah *et al.*, 1990, 1991; Lambeth *et al.*, 1988; Merrill *et al.*, 1989; Stevens *et al.*, 1989, 1990) or animals (Borek *et al.*, 1991; Endo *et al.*, 1991;

Enkretchakul *et al.*, 1989) with long-chain sphingoid bases. As the EGF receptor is phosphorylated and regulated by C-kinase (Fig. 2), sphingoid bases can therefore be predicted to alter EGF receptor function by inhibiting C-kinase. Indeed, it was found that the treatment of cells with sphingosine caused an inhibition of phorbol ester-stimulated phosphorylation of the EGF receptor at the C-kinase site, Thr^{654} (Faucher *et al.*, 1988). Sphingosine therefore blocks phorbol ester action to inhibit EGF-stimulated tyrosine phosphorylation (Faucher *et al.*, 1988).

Although evidence for the inhibition of C-kinase activity was obtained (Faucher *et al.*, 1988), it became apparent that the effects of sphingosine on the EGF receptor could not be fully accounted for by changes in C-kinase activity (Davis *et al.*, 1988; Faucher *et al.*, 1988). This was directly demonstrated by the observation that sphingosine treatment markedly altered the properties of the mutated [Ala^{654}]EGF receptor which lacks the C-kinase phosphorylation site, Thr^{654} (Faucher *et al.*, 1988). Furthermore, sphingosine was found to increase the phosphorylation state of the EGF receptor in experiments using cells in which C-kinase was not activated (Faucher *et al.*, 1988). Phosphoamino acid analysis and phosphopeptide mapping demonstrated that the sphingosine-stimulated phosphorylation of the EGF receptor can be accounted for by an increase in phosphorylation at the MAP kinase site, Thr^{669} (Countaway *et al.*, 1989a, 1990, 1992; Esko *et al.*, 1988; Heisermann and Gill, 1988). The mechanism by which sphingosine increases the phosphorylation of the EGF receptor at Thr^{669} is not understood (Faucher *et al.*, 1988). Previous studies have indicated that MAP kinases can be activated by a very wide variety of extracellular stimuli (Cobb *et al.*, 1991). It is therefore possible that sphingosine is able to activate a MAP kinase isoform in tissue culture cells. However, *in vitro* experiments demonstrate that sphingosine does not directly activate MAP kinases (R. J. Davis, 1990, unpublished observations). Thus, any effect of sphingosine on MAP kinase activity in intact cells must be mediated by an indirect mechanism.

The effect of exogenous sphingosine on the phosphorylation state of the EGF receptor (Faucher *et al.*, 1988) suggests a role for endogenous sphingosine in the regulation of EGF receptor function. The rapid metabolic turnover of sphingosine is consistent with a possible regulatory role for this molecule (Medlock and Merrill, 1988b). Recently, methods have become available to measure the level of endogenous sphingosine (Igarashi *et al.*, 1990a; Merrill *et al.*, 1988; Van Veldhoven *et al.*, 1989), and enzyme activities that result in the production of sphingosine have been characterized (Slife *et al.*, 1989; Wilson *et al.*, 1988). Free sphingosine has been measured in some tissues (Van Veldhoven *et al.*, 1989; Wilson *et al.*, 1988) and has been detected in A431 epidermoid carcinoma cells (Goldkorn *et al.*, 1991). In addition, A431 cells were observed to contain *N,N*-dimethylsphingosine (Igarashi *et al.*, 1990a). Thus, either sphingosine or a methylated metabolite (Igarashi and Hakomori, 1989) may mediate the effect of exogenous sphingosine on EGF receptor phosphorylation in A431 cells. Alternatively, it is possible that other sphingolipids may have regulatory roles. For example, it has been proposed that a ceramide may mediate the effect of sphingosine to cause increased phosphorylation of the EGF receptor at Thr^{669} (Goldkorn *et al.*, 1991).

In several cell types a potential signaling pathway involving increased turnover of sphingomyelin and the generation of ceramide (Dressler *et al.*, 1991; Kim *et al.*, 1991; Kolesnick, 1987, 1989a,b; Kolesnick and Clegg, 1988; Okazaki *et al.*, 1989, 1990) and ceramide-1-phosphate (Dressler and Kolesnick, 1990) has been described. A signaling role for ceramide is indicated by observations that this module stimulates the activity of a protein phosphatase (Dobrowsky and Hannun, 1992) and a protein kinase (Dressler and Kolesnick, 1990; Friedman *et al.*, 1984; Mathias *et al.*, 1991). The ceramide-stimulated protein kinase has been shown to phosphorylate a synthetic peptide corresponding to the sequence surrounding Thr^{669} in the EGF receptor (Dressler and Kolesnick, 1990; Mathias *et al.*, 1991). This observation of *in vitro* ceramide-stimulated phosphorylation of Thr^{669} provides a mechanism that can account for the effects of sphingolipids on the phosphorylation state of the EGF receptor observed in intact cells. A test of this hypothesis will require the molecular characterization of the ceramide-stimulated protein kinase and the establishment of the relationship of this enzyme with the other Thr^{669} protein kinases (MAP kinases and cyclin-dependent protein kinases) that have been identified in cell extracts (Northwood *et al.*, 1991; Takishima *et al.*, 1991).

III. Sphingolipid Regulation of EGF Binding to Cell Surface Receptors

Characterization of cell surface EGF receptors by examination of the EGF binding isotherm demonstrates the presence of a heterogeneous population of EGF receptors (Ullrich and Schlessinger, 1990). A small number of receptors (1–5%) exhibit a high apparent affinity, while most of the receptors (95–99%) exhibit a low apparent affinity for EGF. Treatment of cultured cells with sphingosine causes the conversion of the low-affinity receptors into the high-affinity state (Davis *et al.*, 1988; Faucher *et al.*, 1988). This is an interesting observation because the molecular basis for the different affinity states of the EGF receptor is not understood. Furthermore, the expression of the low-affinity state of the EGF receptor is acutely regulated *in vivo* by treatment of cells with growth factors (such as platelet-derived growth factor) and with phorbol ester (Countaway *et al.*, 1989a,b, 1990, 1992; Davis, 1988; Davis and Czech, 1984, 1985a,b, 1986, 1987; Davis *et al.*, 1985a,b, 1988; Decker *et al.*, 1990; Downward *et al.*, 1985; Friedman *et al.*, 1984; Lund *et al.*, 1990; Northwood and Davis, 1988, 1989, 1990).

As the EGF receptor is regulated by multisite phosphorylation (Fig. 2), the initial hypothesis that we tested was that the effects of sphingosine on the affinity of the EGF receptor were caused by changes in the phosphorylation state of the receptor. However, a mutational analysis demonstrated that EGF receptor phosphorylation was not required for the effect of sphingosine on the affinity of the EGF receptor (Faucher *et al.*, 1988). We therefore investigated the effect of sphingosine on the EGF receptor using *in vitro* experiments (Davis *et al.*, 1988). It was observed that sphingosine increased the affinity of the isolated EGF receptor and that this effect of sphingosine did not require the addition of ATP to the incubation (Davis *et al.*, 1988). Control experiments demonstrated that *N*-acetylsphingosine, glucosylsphingosine, lactosylsphingosine, and gangliosides (G_{M1} and G_{M3}) had no effect on the affinity of the EGF receptor (Davis *et al.*, 1988). Together, these data indicate that sphingosine has a direct action to regulate the affinity of the EGF receptor that is independent of protein phosphorylation. The physiological significance of this observation is unclear. The effect of sphingosine may be specific, but it is also possible that nonspecific detergent-like properties of sphingosine may contribute to the effect of this molecule on the affinity of the EGF receptor. Progress in this research area will require a more detailed understanding of the molecular basis for the heterogeneous expression of the EGF receptor in high and low apparent affinity states.

IV. Sphingolipid Regulation of the EGF Receptor Tyrosine Kinase Activity

Signal transduction by the EGF receptor requires the cytoplasmic tyrosine kinase domain (Fig. 1). Consequently, it is significant that sphingolipids have been observed to regulate the EGF receptor tyrosine kinase activity (Hakomori, 1990). Sphingolipids may therefore represent physiological regulators of the EGF receptor tyrosine kinase activity.

Increased tyrosine kinase activity has been reported when the EGF receptor is incubated with sphingosine (Davis *et al.*, 1988; Faucher *et al.*, 1988; Northwood and Davis, 1988; Wedegaertner and Gil, 1989), *N,N*-dimethylsphingosine (Igarashi *et al.*, 1990b), and de-*N*-acetyl-G_{M3} (Hanai *et al.*, 1988a; Igarashi *et al.*, 1990b). In contrast, *N*-acetylsphingosine, glucosylsphingosine, and lactosylsphingosine were observed to have no effect on the EGF receptor tyrosine kinase activity (Davis *et al.*, 1988; Faucher *et al.*, 1988). Marked effects of non-ionic detergents (e.g., Triton X-100) have been observed in these *in vitro* experiments. For example, the stimulatory effect of de-*N*-acetyl-G_{M3} on the EGF receptor tyrosine kinase activity was found to require the presence of non-ionic detergent, suggesting that the effect of de-*N*-acetyl-G_{M3} may be nonphysiological (Song *et al.*, 1991). In the case of sphingosine, stimulated tyrosine kinase activity was observed when the EGF receptor was inserted in a lipid bilayer (Davis *et al.*, 1988; Faucher *et al.*, 1988; Northwood and Davis, 1988).

Non-ionic detergents blocked this action of sphingosine (Davis *et al.*, 1988; Faucher *et al.*, 1988; Northwood and Davis, 1988), perhaps by sequestering the sphingosine within micelles. Together, these data support the hypothesis that sphingosine and *N,N*-dimethylsphingosine are candidate molecules that activate the EGF receptor tyrosine kinase within intact cells. Other sphingolipids may also have a significant role in this process.

The mechanism by which sphingolipids can act as activators of the EGF receptor tyrosine kinase activity is unclear. Activation of the receptor tyrosine kinase activity by EGF has been proposed to be mediated by an allosteric mechanism involving receptor dimerization (Schlessinger, 1988). Investigation of the effect of sphingosine, however, demonstrates that the receptor tyrosine kinase activity is stimulated in the absence of dimerization (Northwood and Davis, 1988). The mechanism of stimulation of the receptor tyrosine kinase by sphingosine is therefore distinct from that caused by EGF. Further studies are required to define the mechanism of regulation of the EGF receptor tyrosine kinase activity by these molecules.

An inhibition of the EGF receptor tyrosine kinase activity caused by sphingolipids has also been reported (Hakomori, 1990). Addition of ganglioside G_{M3} to the EGF receptor causes an inhibition of tyrosine kinase activity (Bremer *et al.*, 1986; Hanai *et al.*, 1988a,b; Igarashi *et al.*, 1990b; Song *et al.*, 1991). A smaller inhibition of tyrosine kinase activity was noted in experiments using ganglioside G_{M1} (Bremer *et al.*, 1986). Significantly, the inhibition of tyrosine kinase activity caused by G_{M3} could be detected in experiments using the EGF receptor in a lipid bilayer and did not require the addition of non-ionic detergent (Song *et al.*, 1991). Gangliosides therefore represent potential negative regulators of the EGF receptor tyrosine kinase activity (Hakomori, 1990). However, the mechanism of inhibition of the receptor tyrosine kinase by gangliosides is not understood. It is possible that the inhibition is accounted for by a specific regulatory mechanism, but a mechanism involving nonspecific processes is not excluded by the available evidence. Further studies of the mechanism of inhibition of the receptor tyrosine kinase activity by gangliosides are therefore warranted.

V. Physiological Relevance of Sphingolipid Regulation of the EGF Receptor

The alterations in ganglioside metabolism observed during oncogenic transformation, the cell cycle, and density-dependent growth inhibition indicate that sphingolipid metabolism may play a role in the regulation of cell growth (Hakomori, 1981, 1984). Consistent with this hypothesis are observations that sphingolipids are potent pharmacological regulators of cell growth and differentiation

(Hakomori, 1990). Direct addition of gangliosides to the tissue culture medium causes growth inhibition by extending the length of the G_1 phase of the cell cycle (Keenan *et al.*, 1975; Laine and Hakomori, 1973) and blocks cellular proliferation in the presence of fibroblast growth factor (Bremer and Hakomori, 1982), platelet-derived growth factor (Bremer *et al.*, 1984), and EGF (Bremer *et al.*, 1986). The molecular basis for the effects of gangliosides on cell growth is not understood. One possible mechanism is that the effects of gangliosides are mediated by an inhibition of protein kinase C caused by lysosphingolipids (Hannun and Bell, 1987, 1989). However, evidence that sphingolipids may interact directly with growth factor receptors has also been reported. Brenner *et al.* (1986) observed that gangliosides inhibit the tyrosine kinase activity of the EGF receptor. Gangliosides G_{M1} and G_{M3} were reported to inhibit the autophosphorylation of the EGF receptor by 26 and 64%, respectively, when present in an *in vitro* assay at a concentration of 0.35 m*M* (Bremer *et al.*, 1986).

Although sphingolipids are pharmacological regulators of the function of the EGF receptor, a significant question remains concerning whether the EGF receptor is physiologically regulated by sphingolipids. This is because the addition of exogenous gangliosides (or other sphingolipid) to the tissue culture media increases the incorporation of these molecules into cell membranes beyond the physiological range (Bremer and Hakomori, 1982; Bremer *et al.*, 1984, 1986; Laine and Hakomori, 1973). A direct investigation of the physiological relevance of the level of endogenous sphingolipids for the regulation of the EGF receptor is therefore required. One experimental strategy is to employ pharmacological inhibitors of sphingolipid metabolism. Recently, drugs that could be used for these experiments have been described: (1) fumonosins that inhibit sphingolipid biosynthesis (Wang *et al.*, 1991); (2) ceramide analogs that inhibit glycosphingolipid biosynthesis and increase the level of *N,N*-dimethylsphingosine (Felding-Habermann *et al.*, 1990; Okada *et al.*, 1988); and (3) inhibitors of serine palmitoyltransferase (such as β-chloroalanine) that block biosynthesis of long-chain sphingoid bases (Medlock and Merrill, 1988a). However, in view of the disadvantages of the use of inhibitors to study biological function, we have employed a genetic approach to investigate the role of sphingolipids in the control of EGF receptor (Weis and Davis, 1990).

The strategy that we used was to investigate a mutant CHO cell line (clone *ldl*D) that possesses a conditional defect in the biosynthesis of gangliosides (Kingsley *et al.*, 1986a,b). [These cells are also defective in "O"-linked glycosylation and in the biosynthesis of heparan-sulfate proteoglycans under the conditions employed for these experiments (Esko *et al.*, 1988; Kingsley *et al.*, 1986a,b)]. We prepared stable cell lines expressing the wild-type human EGF receptor by transfection with plasmid vectors (Weis and Davis, 1990). Under permissive conditions, these cells exhibit a normal phenotype and express levels of gangliosides (primarily G_{M3}) that are similar to control cells (Weis and Davis,

1990). However, under nonpermissive growth conditions, these cells do not express detectable levels of gangliosides (Weis and Davis, 1990). These cells can therefore be employed to test the hypothesis that the expression of gangliosides at physiological levels modulates signal transduction by the EGF receptor by comparing the properties of cells grown under permissive and nonpermissive conditions. It was observed that a reduction in the physiological level of gangliosides caused a potentiation of EGF receptor signal transduction (Weis and Davis, 1990). This result is consistent with the hypothesis that gangliosides represent physiological inhibitors of EGF receptor function (Hakomori, 1990). It is not, however, clear whether the effects observed can be directly attributed to ganglioside interaction with the EGF receptor. This is because of two considerations: (1) Evidence for a role of gangliosides in the process of cell adhesion, spreading, and other processes has been obtained (Kojima and Hakomori, 1991) and these changes may contribute to the observed perturbation of EGF receptor function. (2) The loss of ganglioside expression is likely to be coupled to alterations in the expression of other (potentially regulatory) sphingolipids. Further studies are therefore required to resolve these questions. However, as a result of this genetic analysis (Weis and Davis, 1990), it is established that sphingolipids are physiological regulators of EGF receptor function.

VI. Conclusions

The observation of a correlation between the level of ganglioside expression and the functional regulation of the EGF receptor in intact cells suggests that sphingolipid regulation of the EGF receptor is likely to be biologically significant (Weis and Davis, 1990). Goals for future research include (1) identification of specific sphingolipid regulatory molecules, (2) definition of the mechanism of action of these molecules to regulate the EGF receptor, and (3) performance of stringent tests of the hypothesis that sphingolipids are physiological regulators of EGF receptor function. It is likely that significant progress toward these goals will be achieved during the next few years.

References

Alvarez, E., Northwood, I. C., Gonzalez, F. A., Latour, D. A., Seth, A., Abate, C., Curran, T., and Davis, R. J. (1991). *J. Biol. Chem.* **266,** 15277–15285.

Borek, C., Ong, A., Stevens, V. L., Wang, E., and Merrill, A. H. (1991). *Proc. Natl. Acad. Sci. U.S.A.* **88,** 1953–1957.

Bowen, S., Stanley, K., Selva, E., and Davis, R. J. (1991). *J. Biol. Chem.* **266,** 1162–1169.

Bremer, E. G., and Hakomori, S. (1982). *Biochem. Biophys. Res. Commun.* **106,** 711–718.

Bremer, E. G., Hakomori, S., Bowen-Pope, D. F., Raines, E., and Ross, R. (1984). *J. Biol. Chem.* **259,** 6818–6825.
Bremer, E. G., Schlessinger, J., and Hakomori, S. (1986). *J. Biol. Chem.* **261,** 2434–2440.
Chinkers, M., and Garbers, D. L. (1986). *J. Biol. Chem.* **261,** 8295–8297.
Clark-Lewis, I., Sanghera, J. S., and Pelech, S. L. (1991). *J. Biol. Chem.* **266,** 15180–15184.
Cobb, M. H., Boulton, T. G., and Robbins, D. J. (1991). *Cell Regul.* **2,** 965–978.
Cochet, C., Gill, G. N., Meisenhelder, J., Cooper, J. A., and Hunter, T. (1984). *J. Biol. Chem.* **259,** 2553–2558.
Countaway, J. L., Northwood, I. C., and Davis, R. J. (1989a). *J. Biol. Chem.* **264,** 10828–10835.
Countaway, J. L., Gironès, N., and Davis, R. J. (1989b). *J. Biol. Chem.* **264,** 13642–13647.
Countaway, J. L., McQuilkin, P., Gironès, G., and Davis, R. J. (1990). *J. Biol. Chem.* **265,** 3407–3416.
Countaway, J. L., Nairn, A. C., and Davis, R. J. (1992). *J. Biol. Chem.* **267,** 1129–1140.
Cunningham, T. W., Kuppuswamy, D., and Pike, L. J. (1989). *J. Biol. Chem.* **264,** 15351–15356.
Davis, R. J. (1988). *J. Biol. Chem.* **263,** 9462–9469.
Davis, R. J., and Czech, M. P. (1984). *J. Biol. Chem.* **259,** 8545–8549.
Davis, R. J., and Czech, M. P. (1985a). *Proc. Natl. Acad. Sci. U.S.A.* **82,** 1974–1978.
Davis, R. J., and Czech, M. P. (1985b). *Proc. Natl. Acad. Sci. U.S.A.* **82,** 4080–4084.
Davis, R. J., and Czech, M. P. (1986). *EMBO J.* **5,** 653–658.
Davis, R. J., and Czech, M. P. (1987). *J. Biol. Chem.* **262,** 6832–6841.
Davis, R. J., Ganong, B. R., Bell, R. M., and Czech, M. P. (1985a). *J. Biol. Chem.* **260,** 1562–1566.
Davis, R. J., Ganong, B. R., Bell, R. M., and Czech, M. P. (1985b). *J. Biol. Chem.* **260,** 5315–5322.
Davis, R. J., Gironès, N., and Faucher, M. (1988). *J. Biol. Chem.* **263,** 5373–5379.
Decker, S. J., Ellis, C., Pawson, T., and Velu, T. (1990). *J. Biol. Chem.* **265,** 7009–7015.
Dobrowsky, R. T., and Hannun, Y. A. (1992). *J. Biol. Chem.* **267,** 5084–5051.
Downward, J., Waterfield, M. D., and Parker, P. (1985). *J. Biol. Chem.* **260,** 14538–14546.
Dressler, K. A., and Kolesnick, R. N. (1990). *J. Biol. Chem.* **265,** 14917–14921.
Dressler, K. A., Mathias, S., and Kolesnick, R. N. (1991). *Science* **255,** 1715–1718.
Endo, K., Igarashi, Y., Nisar, M., Zhou, Q. H., and Hakomori, S. (1991). *Cancer Res.* **51,** 1613–1618.
Enkvetchakul, B., Merrill, A. H., and Birt, D. F. (1989). *Carcinogenesis (London)* **2,** 379–381.
Erikson, R. L. (1991). *J. Biol. Chem.* **266,** 6007–6010.
Esko, J. D., Rostand, K. S., and Weinke, J. L. (1988). *Science* **241,** 1092–1096.
Faucher, M., Gironès, N., Hannun, Y., Bell, R. M., and Davis, R. J. (1988). *J. Biol. Chem.* **263,** 5319–5327.
Felding-Habermann, B., Igarashi, Y., Fenderson, B. A., Park, L. S., Radin, N. S., Inokuchi, J., Strassmann, G., Handa, K., and Hakormori, S. (1990). *Biochemistry* **29,** 6314–6322.
Friedman, B., Frackelton, A. R., Ross, A. H., Connors, J. M., Fujiki, H., Sugimura, T., and Rosner, M. R. (1984). *Proc. Natl. Acad. Sci. U.S.A.* **81,** 3034–3038.
Goldkorn, T., Dresser, K. A., Muindi, J., Radin, N. S., Mendelsohn, J., Menaldino, D., Liotta, D., and Kolesnick, R. N. (1991). *J. Biol. Chem.* **266,** 16092–16097.
Gonzalez, F. A., Raden, D. L., and Davis, R. J. (1991). *J. Biol. Chem.* **266,** 22159–22163.
Hakomori, S. (1981). *Annu. Rev. Biochem.* **50,** 733–764.
Hakomori, S. (1984). *Trends Biochem. Sci.* **9,** 453–456.
Hakomori, S. (1990). *J. Biol. Chem.* **265,** 18713–18716.
Hanai, N., Dohi, T., Nores, G. A., and Hakomori, S. (1988a). *J. Biol. Chem.* **263,** 6296–6301.
Hanai, N., Nores, G. A., Mcleod, C., Torres-Mendez, C., and Hakomori, S. (1988b). *J. Biol. Chem.* **263,** 10915–10921.
Hannun, Y. A., and Bell, R. M. (1987). *Science* **235,** 670–674.
Hannun, Y. A., and Bell, R. M. (1989). *Science* **243,** 500–507.

Hannun, Y. A., Loomis, C. R., Merrill, A. H., Jr., and Bell, R. M. (1986). *J. Biol. Chem.* **261,** 12604–12609.

Hausdorff, W. P., Caron, M. G., and Lefkowitz, R. J. (1990). *FASEB J.* **4,** 2881–2889.

Heisermann, G. J., and Gill, G. N. (1988). *J. Biol. Chem.* **263,** 13152–13158.

Heisermann, G. J., Wiley, H. S., Walsh, B. J., Ingraham, H. A., Fiol, C. J., and Gill, G. N. (1990). *J. Biol. Chem.* **265,** 12820–12827.

Hunter, T., Ling, N., and Cooper, J. A. (1984). *Nature (London)* **311,** 480–483.

Igarashi, Y., and Hakomori, S. (1989). *Biochem. Biophys. Res. Commun.* **164,** 1411–1416.

Igarashi, Y., Hakomori, S., Toyokuni, T., Dean, B., Fujita, S., Sugimoto, M., Ogawa, T., el-Ghendy, K., and Racker, E. (1989). *Biochemistry* **28,** 6796–6800.

Igarashi, Y., Kitamura, K., Toyokuni, T., Dean, B., Fenderson, B., Ogawass, T., and Hakomori, S. (1990a). *J. Biol. Chem.* **265,** 5385–5389.

Igarashi, Y., Kitamura, K., Zhou, Q. H., and Hakomori, S. (1990b). *Biochem. Biophys. Res. Commun.* **172,** 77–84.

Kahn, W. A., Dobrowsky, R., el Touny, S., and Hannun, Y. A. (1990). *Biochem. Biophys. Res. Commun.* **172,** 683–691.

Kahn, W. A., Mascarella, S. W., Lewin, A. H., Wyrick, C. D., Carroll, F. I., and Hannun, Y. A. (1991). *Biochem. J.* **278,** 387–392.

Keenan, T. W., Schmid, E., Franke, W. W., and Wiegandt, H. (1975). *Exp. Cell Res.* **92,** 259–270.

Kim, M. Y., Linardič, C., Obeid, L., and Hannun, Y. A. (1991). *J. Biol. Chem.* **266,** 484–489.

Kingsley, D. M., Kozarsky, K. F., Segal, M., and Krieger, M. (1986a). *J. Cell Biol.* **102,** 1576–1585.

Kingsley, D. M., Kozarsky, K. F., Hobbie, L., and Krieger, M. (1986b). *Cell (Cambridge, Mass.)* **44,** 749–759.

Koch, A. C., Anderson, D., Moran, M. F., Ellis, C., and Pawson, T. (1991). *Science* **252,** 668–674.

Kojima, N., and Hakomori, S. (1991). *J. Biol. Chem.* **266,** 17552–17558.

Kolesnick, R. N. (1987). *J. Biol. Chem.* **262,** 16759–16762.

Kolesnick, R. N. (1989a). *J. Biol. Chem.* **264,** 7617–7623.

Kolesnick, R. N. (1989b). *J. Biol. Chem.* **264,** 11688–11692.

Kolesnick, R. N., and Clegg, S. (1988). *J. Biol. Chem.* **263,** 6534–6537.

Lai, W. H., Cameron, P. H., Doherty, J. J., II, Posner, B. I., and Bergeron, J. J. M. (1989). *J. Cell Biol.* **109,** 2751–2760.

Laine, R. A., and Hakomori, S. (1973). *Biochem. Biophys. Res. Commun.* **54,** 1039–1045.

Lambeth, J. D., Burnham, D. N., and Tyagi, S. R. (1988). *J. Biol. Chem.* **263,** 3818–3822.

Livneh, E., Dull, T. J., Berent, E., Prywes, R., Ullrich, A., and Schlessinger, J. (1988). *Mol. Cell. Biol.* **8,** 2302–2308.

Lund, K. A., Lazar, C. S., Chen, W. S., Walsh, B. J., Welsh, J. B., Herbst, J. J., Walton, G. M., Rosenfeld, M. G., Gill, G. N., and Wiley, H. S. (1990). *J. Biol. Chem.* **265,** 20517–20523.

Mathias, S., Dressler, K. A., and Kolesnick, R. N. (1991). *Proc. Natl. Acad. Sci. U.S.A.* **88,** 10009–10013.

McCune, B. K., Propkop, C. A., and Earp, H. S. (1990). *J. Biol. Chem.* **265,** 9715–9721.

Medlock, K. A., and Merrill, A. H. (1988a). *Biochemistry* **18,** 7079–7084.

Medlock, K. A., and Merrill, A. H. (1988b). *Biochem. Biophys. Res. Commun.* **157,** 232–237.

Merrill, A. H. (1991). *J. Bioenerg. Biomembr.* **23,** 83–104.

Merrill, A. H., Wang, E., Mullins, R. E., Jamison, W. C., Nimkar, S., and Liotta, D. C. (1988). *Anal. Biochem.* **171,** 373–381.

Merrill, A. H., Nimkar, S., Menaldino, D., Hannun, Y. A., Loomis, C., Bell, R. M., Tyagi, S. R., Lambeth, J. D., Stevens, V. L., Hunter, R., and Liotta, D. C. (1989). *Biochemistry* **28,** 3138–3145.

Northwood, I. C., and Davis, R. J. (1988). *J. Biol. Chem.* **263,** 7450–7453.

Northwood, I. C., and Davis, R. J. (1989). *J. Biol. Chem.* **264,** 5746–5750.

Northwood, I. C., and Davis, R. J. (1990). *Proc. Natl. Acad. Sci. U.S.A.* **87,** 6107–6111.
Northwood, I. C., Gonzalez, F. A., Wartmann, M., Raden, D. L., and Davis, R. J. (1991). *J. Biol. Chem.* **266,** 15266–15276.
Okada, Y., Radin, N. S., and Hakomori, S. (1988). *FEBS Lett.* **235,** 25–29.
Okazaki, T., Bell, R. M., and Hannun, Y. A. (1989). *J. Biol. Chem.* **264,** 19076–19080.
Okazaki, T., Bielawska, A., Bell, R. M., and Hannun, Y. A. (1990). *J. Biol. Chem.* **265,** 15823–15831.
Schlessinger, J. (1988). *Trends Biochem. Sci.* **13,** 443–447.
Slife, C. W., Wang, E., Hunter, R., Wang, S., Burgess, C., Liotta, D. C., and Merrill, A. H. (1989). *J. Biol. Chem.* **264,** 10371–10377.
Song, W. X., Vacca, M. F., Welti, R., and Rintoul, D. A. (1991). *J. Biol. Chem.* **266,** 10174–10181.
Stevens, V. L., Winton, E. F., Smith, E. E., Owens, N. E., Kinkade, J. M., and Merrill, A. H. (1989). *Cancer Res.* **49,** 3229–3234.
Stevens, V. L., Nimkar, S., Jamison, W. C., Liotta, D. C., and Merrill, A. H. (1990). *Biochim. Biophys. Acta* **1051,** 37–45.
Takishima, K., Griswold-Prenner, I., Ingebritsen, T., and Rosner, M. R. (1991). *Proc. Natl. Acad. Sci. U.S.A.* **88,** 2520–2524.
Théroux, S. J., Taglienti-Sian, C., Nair, N., Countaway, J. L., Robinson, H. L., and Davis, R. J. (1992). *J. Biol. Chem.* **267,** 7967–7970.
Ullrich, A., and Schlessinger, J. (1990). *Cell (Cambridge, Mass.)* **61,** 203–212.
Van Veldhoven, P. P., Bishop, W. R., and Bell, R. M. (1989). *Anal. Biochem.* **183,** 177–189.
Wang, E., Norred, W. P., Bacon, C. W., Riley, R. T., and Merrill, A. H. (1991). *J. Biol. Chem.* **266,** 14486–14490.
Wedegaertner, P. B., and Gill, G. N. (1989). *J. Biol. Chem.* **264,** 11346–11353.
Weis, F. M. B., and Davis, R. J. (1990). *J. Biol. Chem.* **265,** 12059–12066.
Wilson, E., Wang, E., Mullins, R. E., Uhlinger, D. J., Liotta, D. C., Lambeth, J. D., and Merrill, A. H. (1988). *J. Biol. Chem.* **263,** 9304–9309.

Gangliosides and Glycosphingolipids as Modulators of Cell Growth, Adhesion, and Transmembrane Signaling

SEN-ITIROH HAKOMORI AND YASUYUKI IGARASHI

The Biomembrane Institute
Seattle, Washington 98119
and
Department of Pathobiology
University of Washington
Seattle, Washington 98195

I. Introduction: Evidence That Glycosphingolipids Regulate Cell Growth

During the mid 1960s to early 1970s, when biochemical studies were focused on structure and function of cell surface membranes, numerous studies showed that isolated plasma membranes are greatly enriched in glycosphingolipids (GSLs), particularly gangliosides, in comparison to intracellular membrane components (Yamakawa, 1966; Weinstein *et al.*, 1970; Renkonen *et al.*, 1970; Klenk and Choppin, 1970). Remarkable changes in ganglioside composition were observed in cultured cells upon viral transformation (Hakomori and Murakami, 1968; Mora *et al.*, 1969), suggesting that gangliosides in the cell surface membrane may be involved in cell growth regulation. This possibility was strengthened when we and others found that contact inhibition of cell growth is closely associated with GSL synthesis, and that this contact response of GSL synthesis is lost upon oncogenic transformation of cells (Hakomori, 1970; Sakiyama *et al.*, 1972). Other studies revealed cell cycle-dependent alteration of GSL organization and synthesis, and enhanced synthesis and exposure of GSLs at the G_1/G_0 phase (Chatterjee *et al.*, 1975; Gahmberg and Hakomori, 1975a,b; Lingwood and Hakomori, 1977). Monovalent Fab antibodies directed against G_{M3} [NeuAc(α2→3)Gal(β1→4)Glcβ1→Cer], but not those against globoside, inhibited growth of 3T3 and NIL cells, and an increase of G_{M3} mimicked contact response to G_{M3} synthesis (Lingwood and Hakomori, 1977). Exogenous addition of various GSLs incorporated into cell membrane inhibited cell growth through extension of the G_1 phase (Laine and Hakomori, 1973; Keenan *et al.*, 1975). All these studies, performed around 15–20 years ago, clearly suggested the essential functional role of GSLs in cell growth regulation.

The first indication that G_{M3} may control transmembrane signaling through fibroblast growth factor (FGF) receptor came from the following observations (Bremer and Hakomori, 1982): (1) BHK cells can be grown in medium containing insulin, transferrin, hydrocortisone, and 100 ng/ml FGF, and do not require other growth factors. This growth is inhibited when 50–100 μg/ml of G_{M3} (but not G_{M1} or other GSLs) is added in culture medium. However, G_{M3} itself does not bind to FGF; i.e., G_{M3} is not the FGF receptor and does not influence FGF binding to cells. (2) G_{M3} added to medium is quickly incorporated into cell membranes, and G_{M3}-fed cells become refractory to FGF stimulation. (3) Cells which are growth-inhibited by culture in the presence of G_{M3} (but not other GSLs) accumulate large quantities of ^{125}I-labeled FGF at the cell surface; internalization of this FGF is blocked.

At this time, the FGF receptor was not chemically defined, so no further studies were performed. However, these findings prompted us to consider the general role of sphingolipids and gangliosides in control of cell growth regulation through modulation of transmembrane signaling. A series of subsequent studies demonstrated the importance of this role.

II. Role of G_{M3} Ganglioside and Its Immediate Catabolite as Modulators of Transmembrane Signaling

Prompted by many startling discoveries concerning the structure and function of certain growth factor receptors, particularly the fact that the receptors are structurally as well as functionally associated with protein kinases (Cohen *et al.*, 1980; Ushiro and Cohen, 1980; Kasuga *et al.*, 1982c; Heldin *et al.*, 1983; Jacobs *et al.*, 1983; for a monograph, see Bradshaw and Prentis, 1987), we began a search for possible effects of gangliosides and sphingolipids on growth factor receptors. A few examples of our findings are summarized in this and subsequent sections.

G_{M3} is an ubiquitous component in many cell lines (e.g., BHK, 3T3, KB, and A431), and exogenous addition of anti-G_{M3} monoclonal antibodies (MAbs) arrested cell cycle at a defined stage (G_1 phase) (Lingwood and Hakomori, 1977).

Furthermore, FGF-dependent BHK cell growth was inhibited by exogenous G_{M3} addition (Bremer and Hakomori, 1982). Thus, G_{M3} may be a general modulator of cell proliferation and cell cycle.

These studies were subsequently extended to epidermal growth factor (EGF)-dependent cell growth and effects of gangliosides on EGF receptor. Growth of human ovarian epidermoid carcinoma KB and A431 cells was highly dependent on EGF, and was inhibited by exogenous addition of G_{M3}. The following results (Bremer *et al.*, 1986) indicated that the effect of G_{M3} on cell growth was due to modulation of transmembrane signaling through EGF receptor tyrosine kinase: (1) Neither G_{M3} nor G_{M1} had any effect on binding of ^{125}I-labeled EGF to its cell surface receptor. G_{M3} produced specific inhibition of EGF-stimulated tyrosine phosphorylation of the EGF receptor in membrane preparations from both KB and A431 cells. (2) The effect of G_{M3} on EGF-dependent tyrosine kinase was also observed in isolated EGF receptors after adsorption on antireceptor MAb/Sepharose complex. (3) Inhibition of tyrosine phosphorylation of EGF receptor by G_{M3} was further confirmed by phospho-amino acid analysis. Tyr (but not Ser) phosphorylation was affected by G_{M3}. (4) The inhibitory effect of G_{M3} on EGF-dependent receptor phosphorylation was reproduced in membranes isolated from G_{M3}-fed A431 cells. Further elaborate studies on EGF-dependent growth of an epimerase-less mutant of Chinese hamster ovary cells were performed by Weis and Davis (1990). These authors found that the mutant, which also lacks EGF receptor, was incapable of synthesizing LacCer or G_{M3} unless Gal was added to culture medium. The cells were transfected to express EGF receptor, and then showed EGF-dependent growth in the absence of Gal. In fact, the EGF-dependent growth of these transfected cells was *inhibited* by addition of Gal, which induces G_{M3} synthesis.

How does G_{M3} in cell membranes interact with growth factor receptors to modulate their function? Does G_{M3} directly and randomly interact with growth factor receptors, or does it act in some specific, organized fashion? Do G_{M3} derivatives exist in nature? If so, do they have distinct effects on transmembrane signaling, in comparison to native G_{M3}?

A recent study indicates that the inhibitory effect of G_{M3} on EGF receptor tyrosine kinase of A431 cells is greatly enhanced in the presence of lysophosphatidylcholine (lyso-PC), but not lysophosphatidylethanolamine (lyso-PE), lysophosphatidylserine (lyso-PS), or lysophosphatidylinositol (lyso-PI). In an *in vitro* assay system, lyso-PC (but not the other compounds listed above) greatly stimulated EGF-dependent tyrosine phosphorylation of EGF receptor. It is therefore assumed that lyso-PC promotes the inhibitory effect of G_{M3} in modulation of EGF receptor kinase (RK) (Igarashi *et al.*, 1990b); the possible cooperative effect between lyso-PC and G_{M3} is illustrated in Fig. 1.

Two derivatives of G_{M3} have been chemically detected, although only in minor quantities: (1) lyso-G_{M3}, in which the *N*-fatty acyl residue is eliminated to

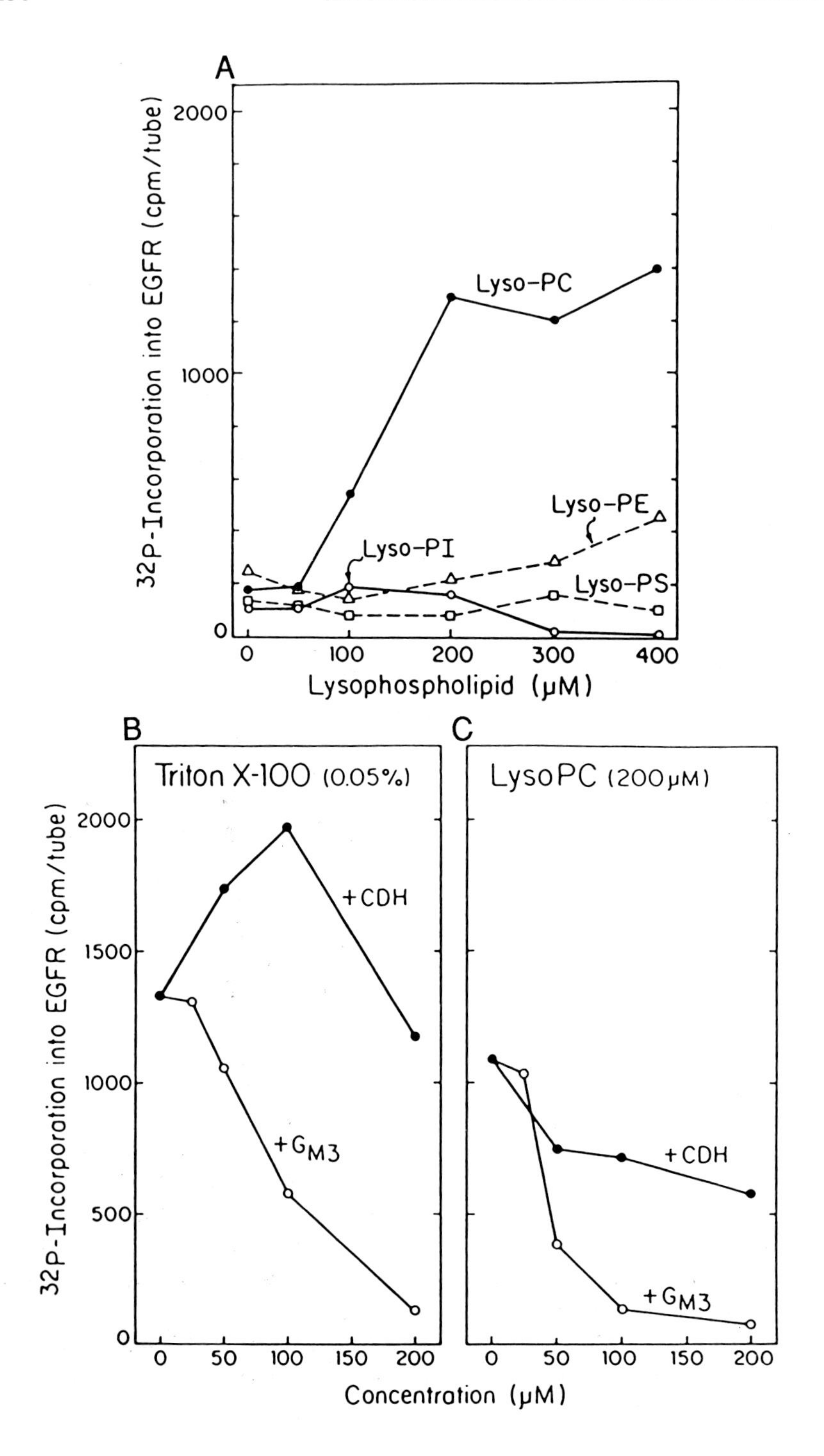
A
32P-Incorporation into EGFR (cpm/tube)
2000
1000
0
Lyso-PC
Lyso-PE
Lyso-PI
Lyso-PS
0
100
200
300
400
Lysophospholipid (μM)
B
Triton X-100 (0.05%)
+CDH
+GM3
C
LysoPC (200μM)
+CDH
+GM3
32P-Incorporation into EGFR (cpm/tube)
2000
1500
1000
500
0
0
50
100
150
200
Concentration (μM)

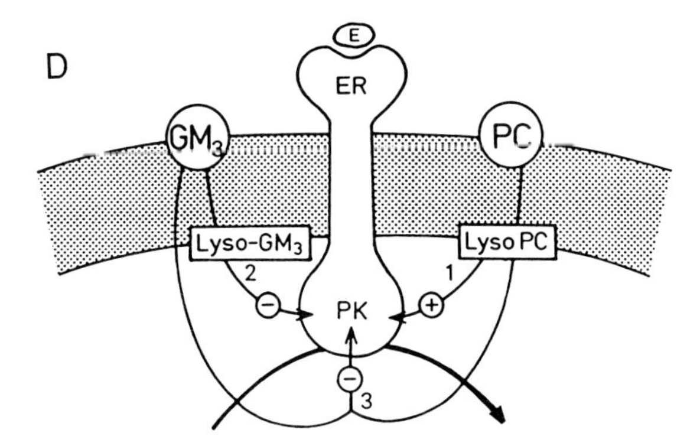

FIG. 1. Cooperative effect of lysophosphatidylcholine (lyso-PC) and G_{M3} on EGF-dependent receptor kinase activity. EGF-dependent receptor autophosphorylation was specifically stimulated by lyso-PC, but not by lyso-PE, lyso-PI, or lyso-PS (A). EGF-RK was inhibited in a dose-dependent manner by exogenous G_{M3} in the presence of 0.05% Triton X-100 (B), as well as in the presence of lyso-PC and absence of detergent (C). Other phospholipids and lysophospholipids had no effect on G_{M3}-dependent inhibition of EGF-RK. (D) EGF (E), when bound to its receptor (ER), activates receptor-associated protein kinase (PK) by an as-yet-unidentified mechanism (most plausibly through di- or oligomerization of receptor) (Schlessinger, 1988). EGF-dependent activation of PK is promoted by lyso-PC which is derived from PC (route 1). G_{M3} can inhibit PK through its degradation to the lyso form (route 2), or its synergetic effect with lyso-PC (route 3). However, free G_{M3}, which is not accessible to lyso-PC, is incapable of modulating PK activity. All four panels reprinted with permission from Igarashi *et al.* (1990b).

expose the free amino group of sphingosine (Sph), and (2) de-*N*-acetyl-G_{M3}, in which the *N*-acetyl group of sialic acid in G_{M3} is eliminated and the free amino group of sialic acid is exposed. Lyso-G_{M3} was detected in A431 cells (Hanai *et al.,* 1988b). De-*N*-acetyl-G_{M3} was detected by specific monoclonal antibody (MAb) DH5 in melanoma and A431 cells (Hanai *et al.,* 1988a), and in various human colonic cancers (E. Nudelman and S. Hakomori, unpublished) Lyso-G_{M3}, in comparison to G_{M3} (and other GSLs and sphingolipids), showed the strongest inhibitory effect on protein kinase C (PKC) (Igarashi *et al.,* 1989b). However, the effect of lyso-G_{M3} on EGF-RK is relatively weak (Y. Igarashi, unpublished).

De-*N*-acetyl-G_{M3} was found to have a stimulatory effect on EGF-RK, particularly in the presence of detergent Triton X-100 (Hanai *et al.,* 1988a). This effect varied considerably depending on detergent quality, but was always stimulatory, in contrast to the inhibitory effect of parent G_{M3}. A later study (Song *et al.,* 1991) showed that, while de-*N*-acetyl-G_{M3} had a stimulatory effect on EGF-RK in the presence of Triton X-100 (in agreement with the findings of Hanai *et al.*, 1988a), it had neither positive nor negative effects on EGF-RK in the absence of detergent (i.e., using membrane vesicles subjected to repeated freezing/thawing which increased ATP permeability). These authors observed no stimulation of cell growth by de-*N*-acetyl-G_{M3}.

Studies on the functional role of modified G_{M3}, and the concept of a cooperative effect on transmembrane signaling by endogenous G_{M3} and transiently formed catabolites of normal phospholipids (e.g., lyso-PC), are of great importance. Further extensive studies along this line should help clarify the functional roles of G_{M3} and its derivatives.

III. Role of Other Glycosphingolipids as Modulators of Receptor Function

A. G_{M1} as Modulator of PDGF Receptor and PDGF-Dependent Cell Growth

Growth of 3T3 cells in chemically defined medium requires platelet-derived growth factor (PDGF), and this growth (both in terms of cell number increase and thymidine incorporation) was found to be inhibited by exogenous G_{M1}, less so by G_{M3}, and not at all by other gangliosides. Kinetic studies showed clearly that (1) PDGF does not directly interact with gangliosides; (2) preincubation of cells with G_{M1} or G_{M3} alters the K_d of ^{125}I-labeled PDGF binding to cell surface receptors without alteration of receptor number; (3) ganglioside levels in membrane affect PDGF receptor tyrosine phosphorylation, i.e., tyrosine phosphorylation of PDGF receptor (M_r 170,000) associated with addition of PDGF was inhibited in a dose-dependent manner by G_{M1} and G_{M3}, but not by other GSLs (Bremer *et al.*, 1984).

Effects of exogenous gangliosides added in culture media on PDGF-stimulated growth of intact 3T3 cells have been reported recently. PDGF-dependent increase of intracellular Ca^{2+} concentration was inhibited by gangliosides in the order $G_{M1} \geq G_{T1b} > G_{M2} > G_{M3}$, whereas PDGF-stimulated tyrosine phosphorylation of PDGF receptor was inhibited in the order $G_{D1a} = G_{T1b} > G_{M1} > G_{M2} > G_{M3}$. None of these gangliosides bound to PDGF directly (Yates *et al.*, 1993). Thus, ganglioside-dependent modulation of PDGF receptor function has been clearly demonstrated *in situ*.

B. Sialosylparagloboside as a Specific Modulator of Insulin Receptor

Insulin-dependent cell growth, through the insulin receptor, is well characterized. The receptor consists of two 135-kDa insulin-binding subunits and two 95-kDa subunits showing tyrosine kinase activity. Tyrosine phosphorylation at the

95-kDa subunits occurs when insulin binds to its receptor (Kasuga *et al.*, 1982a,b; Czech, 1985). Insulin-dependent phosphorylation of the 95-kDa subunits was specifically inhibited by 2→3SPG (sialosylparagloboside (IV^{3}NeuAc-nLc$_4$)], but not by other gangliosides (Fig. 2). However, exogenous 2→3SPG did not alter binding of insulin to cellular receptors. When IM9 cells (which exhibit insulin-dependent growth) were preincubated with 2→3SPG and subsequently treated with insulin, *in situ* receptor phosphorylation as detected by immunoprecipitation with anti-receptor MAb was greatly reduced. Insulin-dependent growth of IM9, HL60, and K562 cells was inhibited by 2→3SPG, and 2→3SPG-treated HL60 cells differentiated into myelomonocytes, as evidenced by morphological and surface marker changes. We concluded that addition of 2→3SPG inhibits insulin-dependent cell growth and triggers differentiation in HL60 cells (Nojiri *et al.*, 1991).

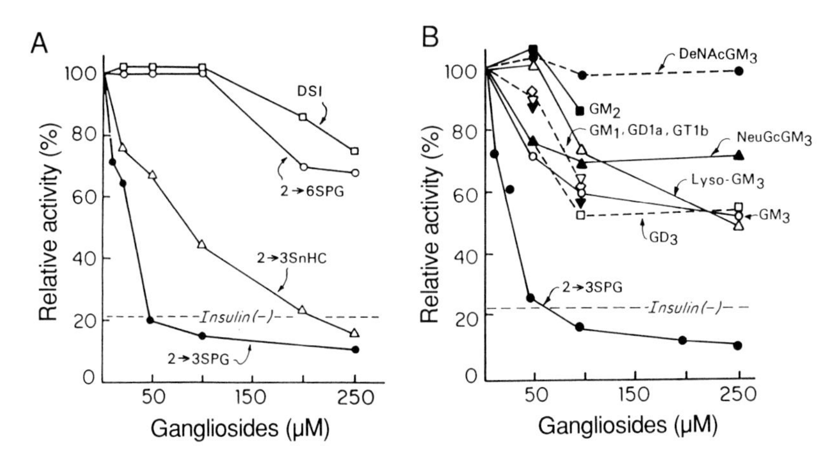

FIG. 2. Specificity of inhibitory effect of various gangliosides on RK activity: An example involving insulin receptor-associated tyrosine kinase. This example shows the effect of 13 gangliosides at various concentrations (abscissa) on activity of insulin RK, expressed as percentage relative activity (ordinate). Control value (no ganglioside added) is defined as 100%. Inhibitory effects of various lacto-series (A) and ganglio-series (B) gangliosides are compared with that of 2→3SPG. The important finding is that 2→3SPG was a far more potent inhibitor of insulin RK at low concentration than were the other gangliosides. Even 2→3 sialosylnorhexaosylceramide (2→3SnHC), which has identical terminal structure, had far less inhibitory effect at low concentrations. This example shows that the inhibitory effect of gangliosides on RK activity is highly specific to their carbohydrate structure. DSI, Disialosyl-I [Vi^{3}NeuAcIV^{6}NeuAc(α2→3)Gal(β1→4)GlcNAcnLc$_6$]; 2→6SPG, IV^{6}NeuAcnLc$_4$. Designations for ganglio-series gangliosides used in B are according to Svennerholm (1964). Both panels reprinted with permission from Nojiri *et al.* (1991).

C. Gangliotriaosylceramide as a Possible Modulator of Transferrin Internalization

Mouse T cell lymphoma L5178Y-AA12 is characterized by high expression of gangliotriaosylceramide (Gg3Cer; asialo-G_{M2}), whereas mutant clone L5178Y-AV27 does not express Gg3Cer. AA12 and AV27 both grow in serum-free medium supplemented by only a single factor: transferrin. Transferrin-dependent growth of AA12 cells was suppressed completely by addition of biotinylated anti-Gg3 MAb 2D4 and avidin, which induces extensive capping of Gg3Cer separated from transferrin receptor and other glycoprotein receptors (e.g., Con A receptor). Growth of AV27 cells was unaffected by the same treatment. Kinetic analysis of transferrin receptor internalization at various temperatures showed that such internalization in AA12 cells is functionally maintained by Gg3Cer (Okada *et al.*, 1985).

D. Modulation of Cell Adhesion by Gangliosides: Possible Modulation of Integrin and Other Receptor Functions

In an early study of membrane receptors able to recognize pericellular fibronectin (FN), it was observed that FN-dependent cell adhesion was inhibited by various gangliosides, particularly polysialogangliosides (Kleinman *et al.*, 1979). This finding suggested that a ganglioside could be the receptor for FN. Many related studies followed. We observed that G_{T1b} ganglioside produced equal inhibition of cell adhesion to not only FN but also gelatin and other substrates, indicating that this ganglioside may not be a specific receptor for FN (Rauvala *et al.*, 1981). The general idea of "integrin receptors" was not well established at the time of these studies.

Integrin receptors for various adhesive proteins (initially FN receptor, later vitronectin and many others) were identified between 1979 and 1985. The idea that gangliosides may modulate cell adhesion has been almost forgotten in the current literature. However, two findings should be noted: (1) G_{M3} has been found as a component of the "detergent-insoluble cell adhesion matrix," which represents a specific cell adhesion site morphologically known as "adhesion plaque." Thus, G_{M3} may play an important role in control of a functional protein present at the adhesion plaque, as evidenced not only by its presence in the plaque, but also by the ability of G_{M3} to block cell-to-substratum attachment and cell spreading in the presence of Ca^{2+} and Mg^{2+}. The association of G_{M3} with adhesion plaques in normal cells is much higher than in transformed cells, indicating that G_{M3} may have a specific function in control of cell adhesion (Okada *et al.*, 1984). (2) Integrin receptors isolated from human melanoma cells by affinity chromatography on Arg-Gly-Asp-Ser (RGDS)-containing peptide/Sepharose column contained G_{D3} ganglioside, and a close association was demonstrated between G_{D3} and the FN-dependent adhesion site in melanoma cells, suggesting that G_{D3} is a specific cofactor for integrin receptor (Cheresh *et al.*, 1987). However, G_{D3} is not widely present, i.e., most cells showing FN-dependent adhesion do not contain G_{D3}.

In more recent studies, mouse mammary carcinoma mutant cell line FUA169, characterized by high G_{M3} ganglioside content, was established from parent cell line FM3A/F28-7, which was high LacCer content but no G_{M3}. In contrast to F28-7 cells, FUA169 cells showed clear adhesion to FN. Evidence indicates that FUA169 adhesion to FN requires the presence of G_{M3}, which supports the function of integrin receptor: (1) Both FUA169 and F28-7 cells express the same quantity of FN integrin receptor, which consists of α5β1 (sensitive to RGDS peptide) and α4β1 (sensitive to CS1 peptide), although adhesion to FN-coated plates, regardless of type of FN, was much higher for FUA169 than for F28-7 cells. (2) F28-7 cells, which normally lack G_{M3} and adhere only weakly to FN, acquired G_{M3} during incubation in G_{M3}-containing medium, and subsequently adhered strongly to FN. (3) Cholesterol-lecithin liposomes (^{14}C-labeled cholesterol) incorporating α5β1 receptor isolated from human placenta showed clear adhesion to FN-coated plates, and this adhesion was completely inhibited by RGDS peptide and by anti-β_1 MAb ZH1. When liposomes included a moderate quantity of GM_3 (0.2–0.4 nmol per 55 μg phosphatidylcholine, 33 μg cholesterol, 5 μg α5β1 in liposome), adhesion was enhanced significantly. In contrast, adhesion was greatly reduced below control level for α5β1 liposomes containing a higher quantity (>2 nmol) of G_{M3}. Adhesion to FN was also inhibited, but never enhanced, for α5β1 liposomes with similar composition but containing 0.4 nmol (or other quantities) of LacCer or GlcCer, instead of G_{M3} (Zheng *et al.*, 1992, 1993). Thus, G_{M3} and other gangliosides may modulate not only membrane-associated RKs and PKC, but also functions of integrin receptors, some of which are involved in transmembrane signaling.

IV. Psychosine Derivatives and *N*-Methylsphingosines as Physiological and Pharmacological Modulators of Transmembrane Signaling

A. Naturally Occurring Psychosine Derivatives

Sphingosine (Sph), psychosine (galactosyl-Sph), and their derivatives are known to modulate transmembrane signaling (Hannun and Bell, 1987, 1989). We confirmed that psychosine is a strong inhibitor of PKC (Igarashi *et al.*, 1989b), in agreement with previous findings (Hannun and Bell, 1989). We also found that psychosine, but not glucosyl-Sph or lactosyl-Sph, had a marked stimulatory effect on *src*-kinase, but an inhibitory effect on *src*-kinase in the co-presence of psychosine and G_{M3}. Psychosine is therefore assumed to act as a "chaperon" on the substrate (Abdel-Ghany *et al.*, 1992).

In view of the interesting biological activities of this class of lipids, we performed an initial study on chemical quantity and identity of cationic GSLs in brain (Nudelman *et al.*, 1992). Sph and *N,N*-dimethyl-D-*erythro*-sphingosine (DMS) were isolated and characterized on mass spectrometry. Psychosine was undetectable in normal brain. In brain white matter (but not gray matter), we detected three major cationic GSL bands besides Sph and DMS. These bands were chemically identified as a new compound, termed "plasmalopsychosine," having different plasmal substitutions at different hydroxyl groups of the galactosyl moiety of psychosine (Fig. 3, A and B). Since plasmalopsychosine is the major GSL component of white matter, it presumably has functional significance in modulation of transmembrane signaling in axons. Plasmalopsychosine, in contrast to psychosine, was noncytotoxic and showed a weaker but still significant

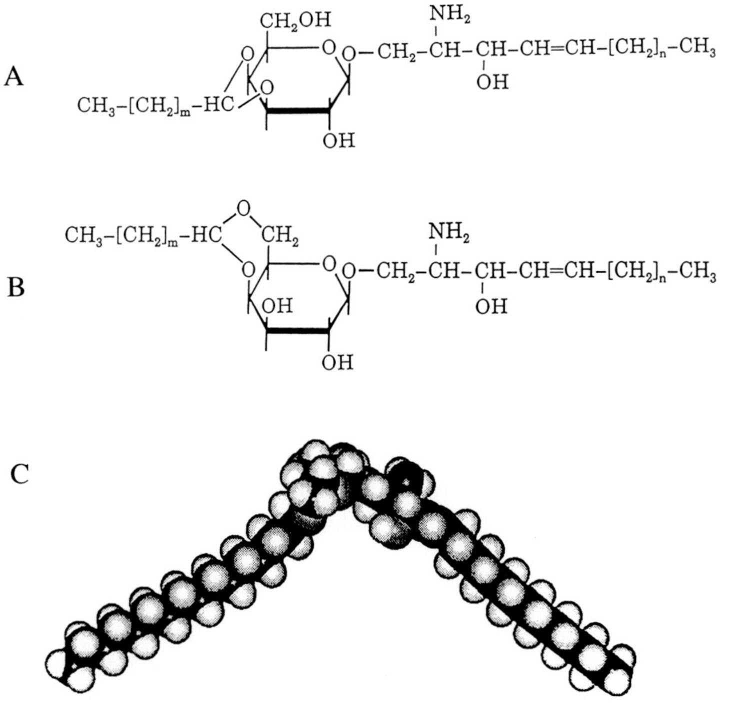

FIG. 3. Structure of plasmalopsychosine. (A and B) Structural formulas of two forms of plasmalopsychosine. (C) Molecular model of compound A showing novel structure (see text for details).

effect on PKC. It may be incorporated into cells and converted to psychosine, thereby regulating PKC and other protein kinases essential for specific neuronal functions.

A molecular model of one form of plasmalopsychosine (Fig. 3C) shows its novel structure. It shows two long aliphatic chains oriented in nearly opposite directions. In striking contrast, essentially all known GSLs and sphingolipids have two long aliphatic chains oriented parallel to each other. The novel structure of plasmalopsychosine could allow the two chains to be inserted into the lipid bilayers of two different cells. Alternatively, it could help anchor essential membrane proteins through hydrophobic interactions. However, these ideas are pure speculation at this point. Plasmalopsychosine may be present only in axons (myelin sheaths) and not in neuronal cell bodies, since it was found exclusively in white matter and not in gray matter of brain (Nudelman *et al.,* 1992).

B. Dimethylsphingosine and Trimethylsphingosine

A series of studies has shown that Sph acts a negative modulator (in contrast to diacylglycerol as a positive modulator) of PKC activity *in vitro,* and that exogenous addition of Sph produces a number of physiological responses. Along this line, we found that DMS significantly inhibits PKC activity of A431 cells, whereas *N,N*-dimethyl-D-*threo*- or -L-*erythro*-sphingosine, or unsubstituted Sph, show much weaker effects (Igarashi *et al.*, 1989a). Similarly, DMS (but not Sph) enhances EGF-dependent RK activity (Igarashi *et al.,* 1990a). *N*-Methylation of Sph was observed in brain homogenate (Igarashi and Hakomori, 1989), and labeling corresponding to DMS was observed in A431 and CTLL (mouse T cell line) cells metabolically labeled with [^{3}H]Ser (Igarashi *et al.,* 1990a; Felding-Habermann *et al.,* 1990). The PKC activity of various tumor cell lines was suppressed strongly by DMS, and even more so by its analog *N,N,N*-trimethylsphingosine (TMS). In *in vitro* assay, DMS and TMS inhibited PKC at concentrations of 20 and 5 μM, respectively, whereas Sph required 50–100 μM for similar effect. Similarly, DMS and TMS (relative to Sph) produced stronger suppression of growth of a variety of human tumor cell lines (e.g., MKN74, MKN45, LU65) in nude mice (Endo *et al.,* 1991).

An important factor in *in vivo* tumor progression is the ability of tumor cells to activate platelets, and to accelerate adhesion of tumor cells to microvascular endothelia. DMS and TMS showed clear inhibition of tumor cell-dependent as well as thrombin- or ADP-dependent platelet aggregation and activation. This phenomenon is presumably based on inhibition of platelet PKC, since phosphorylation of the 45-kDa platelet protein was inhibited strongly by TMS, moderately by DMS, and only weakly by Sph. Interestingly, well-known *in vitro* PKC inhibitors such as calphostin C and staurosporine, did not significantly affect platelet activation or aggregation, particularly that induced by tumor cells. These findings reflect a striking inhibitory effect of TMS on tumor cell metastasis, as further confirmed by suppression of spontaneous lung metastasis from subcutaneously injected BL6 tumor cells when TMS was injected subsequently (Okoshi *et al.*, 1991).

In the field of human tumor metastasis, there has been much interest in the role of P-selectin (GMP-140) expressed on activated platelets. P-Selectin has been shown to bind to the tumor-associated antigens sialosyl-Le^a and sialosyl-Le^x (Polley *et al.*, 1991; Handa *et al.*, 1991b), thereby causing tumor cell aggregation leading to microembolism, which in turn promotes metastatic deposition. DMS and TMS, but not Sph, inhibited P-selectin expression resulting from activation of platelets by thrombin or phorbol ester (Handa *et al.*, 1991a). Thus, the metastasis-inhibitory effect of DMS and TMS appears to be ascribable to suppression of platelet activation and tumor cell aggregation caused by tumor cell–platelet interaction (Okoshi *et al.*, 1991; Handa *et al.*, 1991b).

The comparative effects of TMS, DMS, and Sph on neutrophil function were distinct from those described above for tumor progression and platelet activation. The three compounds had roughly equal inhibitory effects on oxidative burst and transendothelial migration of neutrophils. Phagokinetic activity of neutrophils was inhibited more strongly by Sph than by DMS or TMS, possibly because this activity is controlled via a PKC-independent pathway (Kimura *et al.*, 1992). When radiolabeled Sph was added to culture media, it was rapidly taken up by cells and catabolized into various sphingolipids [e.g., ceramide, sphingosine-1-phosphate (Sph-1-P), GSLs, and sphingomyelin], and partially converted to phosphoethanolamine and phosphatidylethanolamine. In contrast, radiolabeled TMS was not catabolized, but stayed in its original form. The interesting pharmacological effects of TMS as described above may be due in part to the metabolic stability of this compound, and in part to its strong inhibitory effect on PKC. Our recent studies indicate that Sph-1-P inhibits tumor cell motility, transendothelial migration, and invasiveness of various tumors through a PKC-independent pathway (Sadahira *et al.*, 1992). Obviously, an as-yet-unknown transmembrane signaling mechanism could also be involved.

V. Conclusions and Perspectives

Cell growth is regulated by two major mechanisms: (1) signal transduction through plasma membrane receptors or transducers which are associated with protein kinases; (2) various cytoplasmic regulatory units, known as c-oncogenes, consisting of protein kinases and other functional units. Our studies have been focused on mechanism (1), particularly in terms of the modulatory functions of GSLs and sphingolipids. Preliminary studies indicate that c-*src*-kinase activity is enhanced in the presence of Sph and psychosine, and inhibited by G_{M3}. These effects seem to involve action on substrates or "chaperones" of the enzyme, rather than the enzyme itself. Further extensive studies will be necessary to clarify the role of GSLs in mechanism (2) above.

GSLs and sphingolipids have three types of modulatory function:

1. Effects of endogenous, intact GSL species: G_{M3}, G_{M1}, and SPG directly or indirectly affect RKs, although the mechanism of this action is unclear. Some evidence exists that G_{M3} interferes with EGF-dependent receptor–receptor interaction (dimerization or oligomerization). The effects of G_{M3} and SPG on EGF and insulin receptor (respectively) are highly specific, i.e., substitution of G_{M3} or SPG by closely related ganglioside species negates the effect.

2. Cooperative effects of GSLs with lysophospholipids: Interaction of GSLs with protein kinases is catalyzed by phospholipid catabolites. One example is the lyso-PC-mediated interaction of G_{M3} with EGF receptor. Lyso-PC itself specif-

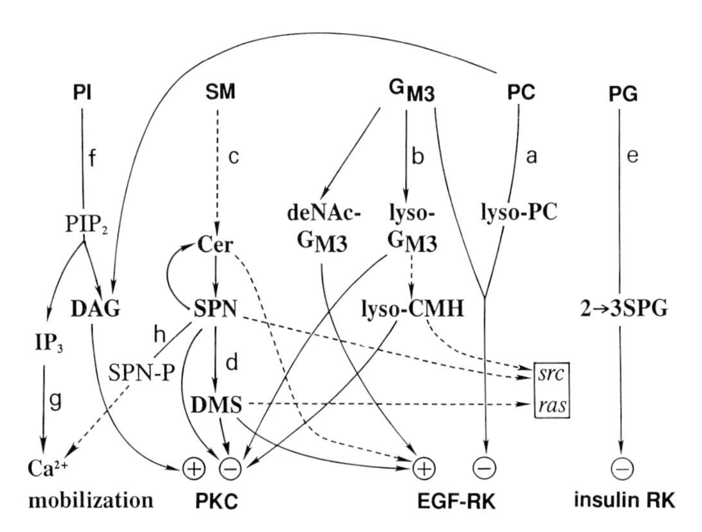

FIG. 4. Hypothetical scheme showing the cooperative modulatory effects of GSLs, phospholipids, and their metabolites on key regulators of cell proliferation. Membrane GSLs [G_{M3}, paragloboside (PG)], phospholipids [PC, sphingomyelin (SM), PI], and their derivatives [diacylglycerol (DAG), Sph, DMS, lyso-G_{M3}, de-*N*-acetyl-G_{M3}, 2→3SPG, ceramide (Cer)] function as modulators of PKC, EGF receptor kinase (RK), and insulin RK, some of them possibly affecting *src* and *ras* oncoproteins. Metabolic pathways (a), (b), (c), (d), and (e) are important in inhibition of EGF-RK, PKC, and insulin RK; therefore, factors (enzymes) involved in these pathways could be anti-oncogenic. In addition, the PI derivatives PIP_2 (phosphatidylinositol 4,5-bisphosphate) and IP_3 (inositol triphosphate), created by pathway (f), and Sph-P (Sph-1-phosphate), created by pathway (h), act to induce Ca^{2+} mobilization and, consequently, to promote cell growth. Dashed lines indicate processes that consist of multiple steps. Lyso-CMH (psychosine) in the presence of G_{M3} inhibits *src*- and *ras*-associated kinase activity, while Sph and DMS enhance this activity (Abdel-Ghany *et al.*, 1992). Reprinted with permission from Hakomori (1991).

ically enhances EGF-dependent RK activity, whereas lyso-PE, lyso-PS, and lyso-PI have no effect. This is merely one well-documented example; obviously, activity of various other RKs (or other kinases) may well be modulated by cooperative effects of GSLs and phospholipid catabolites.

3. Effects of catabolites or modified GSLs or sphingolipids: GSL degradation products are known to modulate RKs and PKC. Examples include ceramide, Sph, DMS, TMS, psychosine, lyso-G_{M3}, and de-*N*-acetyl-G_{M3}. Interestingly, modulation of RK or PKC activity is highly dependent on structure of lipid species. For example, PKC activity is strongly inhibited by psychosine (galactosyl-Sph), but not by glucosyl- or lactosyl-Sph. Similarly, it is strongly inhibited by lyso-G_{M3}, but unaffected by de-*N*-acetyl-G_{M3}. The enhancing effect on EGF RK was specifically produced by DMS, but not by its stereoisomers.

Modulatory functions of GSLs on various RKs, protein kinases, and other kinases are summarized in Fig. 4. Obviously, our current knowledge is fragmentary. Correlations and interrelationships between mechanisms shown in this figure are speculative or unknown. Nevertheless, it is clear that some of these lipid modulators are capable of inhibiting tumor cell growth or metastasis (e.g., through blocking of platelet activation or P-selectin expression, or suppression of oxidative burst and chemotaxis of neutrophils), and are also expected to display antiinflammatory effects. Thus, these lipids or modified GSLs have obvious potential pharmaceutical applicability. In-depth evaluation of this possibility remains to be accomplished.

Acknowledgment

The authors thank Stephen Anderson, Ph.D., for scientific editing and preparation of the manuscript.

References

Abdel-Ghany, M., Osusky, M., Igarashi, Y., Hakomori, S., Shalloway, D., and Racker, E. (1992). *Biochim. Biophys. Acta* **1137,** 349–355.

Bradshaw, R. A., and Prentis, S. (1987). "Oncogenes and Growth Factors." Elsevier, Amsterdam.

Bremer, E. G., and Hakomori, S. (1982). *Biochem. Biophys. Res. Commun.* **106,** 711–718.

Bremer, E. G., Hakomori, S., Bowen-Pope, D. F., Raines, E., and Ross, R. (1984). *J. Biol. Chem.* **259,** 6818–6825.

Bremer, E. G., Schlessinger, J., and Hakomori, S. (1986). *J. Biol. Chem.* **261,** 2434–2440.

Chatterjee, S. K., Sweeley, C. C., and Velicer, L. F. (1975). *J. Biol. Chem.* **250,** 61–66.

Cheresh, D. A., Pytela, R., Pierschbacher, M. D., Klier, F. G., Ruoslahti, E., and Reisfeld, R. A. (1987). *J. Cell Biol.* **105,** 1163–1173.

Cohen, S., Carpenter, G., and King, L. (1980). *J. Biol. Chem.* **255,** 4834–4842.

Czech, M. P. (1985). *Annu. Rev. Physiol.* **47,** 357–381.

Endo, K., Igarashi, Y., Nisar, M., Zhou, Q., and Hakomori, S. (1991). *Cancer Res.* **51,** 1613–1618.

Felding-Habermann, B., Igarashi, Y., Fenderson, B. A., Park, L. S., Radin, N. S., Inokuchi, J., Strassmann, G., Handa, K., and Hakomori, S. (1990). *Biochemistry* **29,** 6314–6322.
Gahmberg, C. G., and Hakomori, S. (1975a). *J. Biol. Chem.* **250,** 2438–2446.
Gahmberg, C. G., and Hakomori, S. (1975b). *J. Biol. Chem.* **250,** 2447–2451.
Hakomori, S. (1970). *Proc. Natl. Acad. Sci. U.S.A.* **67,** 1741–1747.
Hakomori, S. (1991). *Cancer Cells* **3,** 461–470.
Hakomori, S., and Murakami, W. T. (1968). *Proc. Natl. Acad. Sci. U.S.A.* **59,** 254–261.
Hanai, N., Dohi, T., Nores, G. A., and Hakomori, S. (1988a). *J. Biol. Chem.* **263,** 6296–6301.
Hanai, N., Nores, G. A., MacLeod, C., Torres-Mendez, C.-R., and Hakomori, S. (1988b). *J. Biol. Chem.* **263,** 10915–10921.
Handa, K., Igarashi, Y., Nisar, M., and Hakomori, S. (1991a). *Biochemistry* **30,** 11682–11686.
Handa, K., Nudelman, E. D., Stroud, M. R., Shiozawa, T., and Hakomori, S. (1991b). *Biochem. Biophys. Res. Commun.* **181,** 1223–1230.
Hannun, Y. A., and Bell, R. M. (1987). *Science* **235,** 670–674.
Hannun, Y. A., and Bell, R. M. (1989). *Science* **243,** 500–507.
Heldin, C.-H., Ek, B., and Ronnstrand, L. (1983). *J. Biol. Chem.* **258,** 10054–10061.
Igarashi, Y., and Hakomori, S. (1989). *Biochem. Biophys. Res. Commun.* **164,** 1411–1416.
Igarashi, Y., Hakomori, S., Yoyokuni, T., Dean, B., Fujita, S., Sugimoto, M., Ogawa, T., El-Ghendy, K., and Racker, E. (1989a). *Biochemistry* **28,** 6796–6800.
Igarashi, Y., Nojiri, H., Hanai, N., and Hakomori, S. (1989b). *In* "Methods in Enzymology" (V. Ginsburg, ed.), Vol. 179, pp. 521–541. Academic Press, Orlando, FL.
Igarashi, Y., Kitamura, K., Toyokuni, T., Dean, B., Fenderson, B. A., Ogawa, T., and Hakomori, S. (1990a). *J. Biol. Chem.* **265,** 5385–5389.
Igarashi, Y., Kitamura, K., Zhou, Q., and Hakomori, S. (1990b). *Biochem. Biophys. Res. Commun.* **172,** 77–84.
Jacobs, S., Kull, F. C., Jr., Earp, H. S., Svoboda, M. E., Van Wyk, J. J., and Cuatrecasas, P. (1983). *J. Biol. Chem.* **258,** 9581–9584.
Kasuga, M., Hedo, J. A., Yamada, K. M., and Kahn, C. R. (1982a). *J. Biol. Chem.* **257,** 10392–10399.
Kasuga, M., Karlsson, F. A., and Kahn, C. R. (1982b). *Science* **215,** 185–187.
Kasuga, M., Zick, Y., Blithe, D. L., Crettaz, M., and Kahn, C. R. (1982c). *Nature (London)* **298,** 667–669.
Keenan, T. W., Schmid, E., Franke, W. W., and Wiegandt, H. (1975). *Exp. Cell Res.* **92,** 259–270.
Kimura, S., Kawa, S., Ruan, F., Nisar, M., Sadahira, Y., Hakomori, S., and Igarashi, Y. (1992). *Biochem. Pharmacol.* **44,** 1585–1595.
Kleinman, H. K., Martin, G. R., and Fishman, P. H. (1979). *Proc. Natl. Acad. Sci. U.S.A.* **76,** 3367–3371.
Klenk, H.-D., and Choppin, P. W. (1970). *Proc. Natl. Acad. Sci. U.S.A.* **66,** 57–64.
Laine, R. A., and Hakomori, S. (1973). *Biochem. Biophys. Res. Commun.* **54,** 1039–1045.
Lingwood, C., and Hakomori, S. (1977). *Exp. Cell Res.* **108,** 385–391.
Mora, P. T., Brady, R. O., Bradley, R. M., and McFarland, V. W. (1969). *Proc. Natl. Acad. Sci. U.S.A.* **63,** 1290–1296.
Nojiri, H., Stroud, M. R., and Hakomori, S. (1991). *J. Biol. Chem.* **266,** 4531–4537.
Nudelman, E. D., Levery, S. B., Igarashi, Y., and Hakomori, S. (1992). *J. Biol. Chem.* **267,** 11007–11016.
Okada, Y., Mugnai, G., Bremer, E. G., and Hakomori, S. (1984). *Exp. Cell Res.* **155,** 448–456.
Okada, Y., Matsuura, H., and Hakomori, S. (1985). *Cancer Res.* **45,** 2793–2801.
Okoshi, H., Hakomori, S., Nisar, M., Zhou, Q., Kimura, S., Tashiro, K., and Igarashi, Y. (1991). *Cancer Res.* **51,** 6019–6024.
Polley, M. J., Phillips, M. L., Wayner, E. A., Nudelman, E. D., Singhal, A. K., Hakomori, S., and Paulson, J. C. (1991). *Proc. Natl. Acad. Sci. U.S.A.* **88,** 6224–6228.
Rauvala, H., Carter, W. G., and Hakomori, S. (1981). *J. Cell Biol.* **88,** 127–137.

Renkonen, O., Gahmberg, C. G., Simons, K., and Kaariainen, L. (1970). *Acta Chem. Scand.* **24,** 733–735.

Sadahira, Y., Ruan, F., Hakomori, S., and Igarashi, Y. (1992). *Proc. Natl. Acad. Sci. U.S.A.* **89,** 9686–9690.

Sakiyama, H., Gross, S. K., and Robbins, P. W. (1972). *Proc. Natl. Acad. Sci. U.S.A.* **69,** 872–876.

Schlessinger, J. (1988). *Trends Biochem. Sci.* **13,** 443–447.

Song, W., Vacca, M. F., Welti, R., and Rintoul, D. A. (1991). *J. Biol. Chem.* **266,** 10174–10181.

Svennerholm, L. (1964). *J. Lipid Res.* **5,** 145–155.

Ushiro, H., and Cohen, S. (1980). *J. Biol. Chem.* **255,** 8363–8365.

Weinstein, D. B., Marsh, J. B., Glick, M. C., and Warren, L. (1970). *J. Biol. Chem.* **245,** 3928–3937.

Weis, F. M. B., and Davis, R. J. (1990). *J. Biol. Chem.* **265,** 12059–12066.

Yamakawa, T. (1966). *Colloq. Ges. Physiol. Chem.* **16,** 87–111.

Yates, A. J., VanBrocklyn, J., Saqr, H. E., Guan, Z., Stokes, B. T., and O'Dorisio, M. S. (1993). *Exp. Cell Res.* **204,** 38–45.

Zheng, M., Tsuruoka, T., Tsuji, T., and Hakomori, S. (1992). *Biochem. Biophys. Res. Commun.* **186,** 1397–1402.

Zheng, M., Fang, H., Tsuruoka, T., Tsuji, T., Sasaki, T., and Hakomori, S. (1993). *J. Biol. Chem.* **268,** 2217–2222.

Part II

BIOLOGICAL AND PHARMACOLOGICAL FUNCTIONS

Gangliosides as Receptors for Bacterial Enterotoxins

PETER H. FISHMAN, TADEUSZ PACUSZKA,*
AND PALMER A. ORLANDI

Membrane Biochemistry Section
Laboratory of Molecular and Cellular Neurobiology
National Institute of Neurological Disorders and Stroke
National Institutes of Health
Bethesda, Maryland 20892

I. Introduction

Gangliosides, the sialic acid-containing glycosphingolipids, are a diverse class of biological compounds which vary in both their carbohydrate and lipid moieties (Wiegandt, 1982). Because of their structural diversity, their ubiquitous presence in almost all vertebrate cells and tissues, and their predominant localization on the plasma membrane, gangliosides have been implicated in various cell recognition and signaling phenomena (Fishman, 1988; Hakomori, 1990). As many of these aspects are addressed in other articles of this volume, we limit ourselves to the most well-characterized function of a ganglioside, namely that of G_{M1} as the receptor for cholera toxin (Holmgren, 1981; van Heyningen, 1983; Fishman, 1990). We emphasize the importance of the lipid moiety of G_{M1} and describe experiments in which G_{M1}-neoganglioproteins behave as nonfunctional toxin receptors. Finally, we consider the *Escherichia coli* heat-labile toxin,

*Present address: Department of Biochemistry, Medical Center of Postgraduate Education, Warsaw, Poland.

which is structurally and functionally similar to cholera toxin, but can bind to both G_{M1} and galactoproteins found in intestinal mucosal cells of certain species, including man.

II. Structure and Mechanism of Action of Cholera and *Escherichia coli* Heat-Labile Toxins

A. Structure

Colonization of the human small intestine by *Vibrio cholerae* or some enterotoxigenic strains of *E. coli* can result respectively in the severe diarrheal disease cholera (Finkelstein, 1973) or the milder travelers' diarrhea (Sack, 1975). The causative agents have been identified as heat-labile toxins secreted by the bacteria, which have been named cholera toxin (CT) and heat-labile enterotoxin (LT), and purified to homogeneity (Finkelstein, 1973; Kunkel and Robertson, 1979).

CT consists of an A subunit (CT-A) of 27 kDa and a homopentameric B subunit (CT-B) with each monomer of 11.6 kDa. The A subunit, which is synthesized as a single polypeptide, is proteolytically cleaved during bacterial processing to form A_1 (22 kDa) and A_2 (5 kDa) peptides linked by a single disulfide bridge. *Vibrio cholerae* also secretes a toxoid known as choleragenoid which is identical to CT-B. The pentameric nature of CT-B was indicated by cross-linking experiments (Gill, 1976), supported by physical chemical data (Dwyer and Bloomfield, 1982; Ludwig *et al.*, 1986; Reed *et al.*, 1987), and recently confirmed by the detailed crystal structure of LT (Sixma *et al.*, 1991).[1] From these latter studies, the five B monomers form a doughnut-shaped ring. On reduction, the A_1 peptide can be separated from the A_2–B complex, which indicates that CT-A is bound to CT-B through its A_2 peptide (Gill, 1976; Mekalanos *et al.*, 1979).

LT[2] is structurally similar to (Gill *et al.*, 1981) and highly homologous to CT, with almost 80% identity at both the nucleotide and amino acid levels (Mekalanos *et al.*, 1983; Yamamoto *et al.*, 1984). Thus, each B polypeptide contains 103 amino acids of which only 18 are different. Of the 240 amino acids in the A subunits, 46 are different. The A_1 peptides are even more highly conserved, with 87% homology, whereas the A_2 peptides exhibit the greatest divergence, with only 59% homology. Interestingly, the two A_2 peptides differ around the proteolytic cleavage site (Yamamoto *et al.*, 1984). This appears to have some practical consequences, as most of the CT used in research is proteo-

[1]A preliminary report indicates that CT has a crystal structure similar to LT [*Science* **253,** 382–383 (1991)].

[2]Sixma *et al.* (1991) reported on the crystal structure of the heat-labile enterotoxin produced by a strain of *E. coli* isolated from pig (LTp-I). See Section VI,A for further details on the types of toxins.

lytically nicked (Mekalanos *et al.*, 1979), whereas most of the LT is not nicked (Moss *et al.*, 1981) and thus its A subunit has to be cleaved by target cells during the intoxification process.

The crystal structure of LT at a resolution of 2.3 Å has been published by Sixma *et al.* (1991). The B pentamer has a diameter of ~64 Å and a height of ~40 Å. The monomers interact only between nearest neighbors and are highly buried within each other, which is consistent with the great stability of the B subunit. The central pore formed by the monomers is 30 Å long and varies in diameter from ~11 Å at the surface which interfaces with the A subunit to ~15 Å at the opposite surface. The former surface of the B pentamer is extremely flat and highly charged whereas the latter has exposed loops. Trp^{88}, which has been implicated as part of the receptor binding site of both LT and CT (Fishman *et al.*, 1978; Moss *et al.*, 1981; De Wolf *et al.*, 1981; Ludwig *et al.*, 1985), is at the bottom of a small cavity about 23 Å from the flat surface of the pentamer and very close to Gly^{33} of the adjacent monomer. This residue also has been implicated in receptor binding (Tsuji *et al.*, 1985) and may indicate that the binding site is formed by two adjacent monomers, there being five binding sites per pentamer. The association of the A subunit with the B pentamer is mediated almost entirely by the A_2 peptide, with its C-terminus forming a hairpin loop which is buried in the central pore of the pentamer. The N-terminus of the A_2 peptide forms an extended α-helix which interacts with the A_1 peptide. The latter has minimal interactions with the B pentamer and appears triangular in shape with a base of 57 Å and sides of ~35 Å. A structural representation is shown in Fig. 1.

B. Mechanism of Action

The mechanism of action of both CT and LT are essentially identical and most of the details are well established (Fishman, 1990). The normal target cell for the toxins is the mucosal epithelium cell of the small intestine. The intoxicated enterocyte secretes electrolytes and water, a process mediated by cyclic AMP

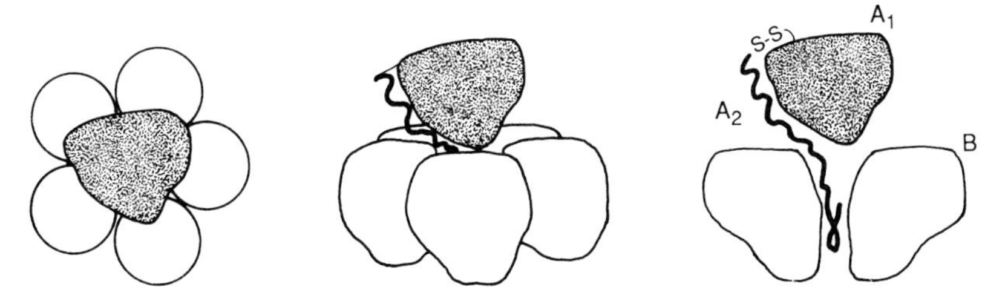

FIG. 1. Conceptual model of the structure of cholera toxin and *E. coli* heat-labile toxin. See text for more details.

(Field, 1971). The toxins bind to the intestinal cells and activate adenylyl cyclase (Kimberg *et al.*, 1971; Sharp and Hynie, 1971; Evans *et al.*, 1972). As is the case for other toxins with A–B structures (Eidels *et al.*, 1983), the two subunits perform different functions. Whereas neither of the resolved A or B subunits is toxic to intact cells, the B subunits will block the secretory effects of their respective holotoxins (Holmgren *et al.*, 1982). Thus, as is described below, the B subunits bind to specific receptors on the cell surface while the A subunits activate adenylyl cyclase (Gill, 1977).

The activation process is rather complex and requires several cofactors (Moss and Vaughan, 1979; Bobak *et al.*, 1990). The A_1 peptide is an ADP-ribosyltransferase (Moss and Vaughan, 1979) and transfers ADP-ribose from NAD to the stimulatory guanine nucleotide-binding protein G_s of adenylyl cyclase (Cassel and Pfeuffer, 1978; Gill and Meren, 1978; Gill and Richardson, 1980). G_s is a heterotrimer consisting of α, β, and γ subunits (Gilman, 1987). During the normal cycle of hormone stimulation, $G_{s\alpha}$ binds GTP, is activated, and in turn stimulates the adenylyl cyclase catalyst. The cycle is turned off by hydrolysis of the bound GTP by the intrinsic GTPase activity of $G_{s\alpha}$. The target of the toxins is $G_{s\alpha}$; an ADP-ribose is transferred to Arg^{187} (Freissmuth and Gilman, 1989), which inhibits the GTPase activity and results in a persistent activation of the adenylyl cyclase.

In addition to NAD and GTP, the A_1 peptide requires small GTP-binding protein(s) in order to ADP-ribosylate $G_{s\alpha}$ (Bobak *et al.*, 1990). Several of these ~20-kDa proteins, known as ADP-ribosylation factors (ARF), have been purified and cloned. In the current model of activation, ARF with bound GTP forms a complex with the A_1 peptide and activates the latter, which then catalyzes the ADP-ribosylation of $G_{s\alpha}$. ARF recently has been identified as one of the coat proteins of Golgi-derived non-clathrin-coated transport vesicles and its normal function may involve vesicular transport (Serafini *et al.*, 1991).

It is still unclear as to how the A_1 peptide gains access to $G_{s\alpha}$ which is located at the cytoplasmic side of the plasma membrane. One model envisions the A subunit penetrating directly across the plasma membrane bilayer and being reduced to A_1 on the cytoplasmic side (Fishman, 1990). Others have proposed that CT undergoes endocytosis through non-coated invaginations which form non-coated vesicles or endosomes (Tran *et al.*, 1987). The low endosomal pH then promotes the generation of A_1 and its translocation across the endosomal membrane into the cytosol (Janicot *et al.*, 1991). The latter authors detected A_1 peptide in endosomes purified from rat liver following *in vivo* injection of CT and found that chloroquine inhibited the endocytosis of CT and its activation of hepatic adenylyl cyclase. This effect of chloroquine may be unique to hepatocytes as the drug has no or minimal effects on the response of many cultured cells to CT (Table I).

Table I
EFFECT OF CHLOROQUINE ON STIMULATION OF VARIOUS CELLS BY CHOLERA TOXIN[a]

Cell line	Origin	Treatment conditions		Cyclic AMP response (% of control)[b]
		CT (n*M*)	Time (hr)	
HeLa	Human carcinoma	10	1.5	91
SK-N-MC	Human neurotumor	1	2	113
CaCo-2	Human intestinal	1	2	97
Fibroblasts	Human skin	5	1	101
NB41A	Murine neuroblastoma	10	1	99
Friend	Murine erythroleukemic	25	1	94
C6	Rat glioma	10	1.5	102[c]

[a]T. Pacuszka, P. A. Orlandi, and P. H. Fishman, unpublished observations. Cells were incubated in the absence and presence of 0.1 m*M* chloroquine for 30 minutes, then exposed to CT as indicated, and assayed for cyclic AMP and protein content.

[b]Values are expressed as the response in chloroquine-treated cells as a percentage of control cells.

[c]The effects of chloroquine were somewhat variable on C6 cells, where it sometimes caused a slight inhibition and other times up to a 50% potentiation of CT stimulation (Pacuszka *et al.*, 1991; Pacuszka and Fishman, 1992).

III. G_{M1} as the Natural Receptor for Cholera Toxin

CT binds with high affinity (0.1–1 n*M*) to intestinal cells from various species, including man (Walker *et al.*, 1974; Holmgren *et al.*, 1975; Critchley *et al.*, 1981). As CT binds with similar affinity to most vertebrate cells and activates their adenylyl cyclase (Fishman and Atikkan, 1980), the toxin receptor would appear to be widespread and not specific to intestinal cells. The nature of the receptor was first suggested by van Heyningen *et al.* (1971), who reported that mixtures of gangliosides are potent inhibitors of CT activity. It was subsequently shown that ganglioside G_{M1} (Fig. 2) is the most effective inhibitor and is able to form a complex with the toxin (Cuatrecasas, 1973; Holmgren *et al.*, 1973; King and van Heyningen, 1973). The proposal that G_{M1} is the receptor for CT initially was met with some skepticism (Donta, 1976; Kanfer *et al.*, 1976; King and van Heyningen, 1975). Although G_{M1} is readily found in nervous tissue, it is only a minor component in other cells and tissues (Wiegandt, 1982). Using the crude methods available 20 years ago, G_{M1} was not detected in certain cells such as rat adipocytes, which were employed as model systems to study the mechanism of action of CT (Kanfer *et al.*, 1976).

The evidence that G_{M1} is the receptor for CT is now extensive and convincing (reviewed in Fishman, 1982, 1986, 1990). First, there is a close correlation between the amount of G_{M1} and the number of toxin-binding sites in various

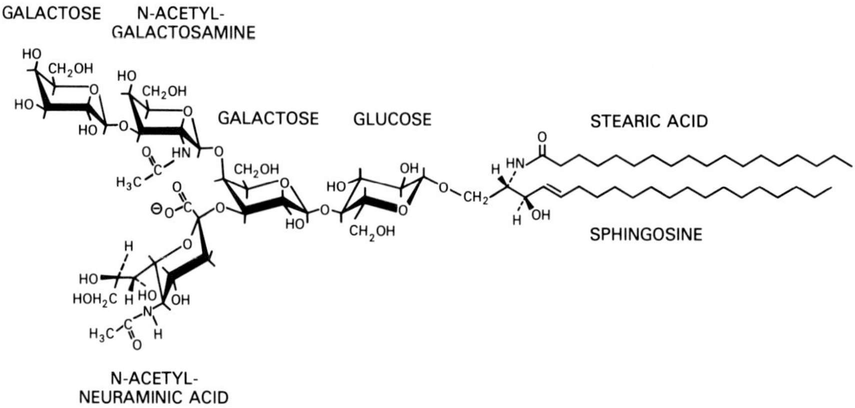

FIG. 2. Structure of ganglioside G_{M1}.

cells and tissues (Holmgren *et al.*, 1975; Hansson *et al.*, 1977; Miller-Podraza *et al.*, 1982). Treatment with bacterial sialidase results in a corresponding increase in both parameters, as does treatment with sodium butyrate, an inducer of cell differentiation (Fishman and Atikkan, 1979). When highly quantitative methods are used, the ratio of G_{M1} to CT bound is close to 6 : 1 over a 1000-fold range in a variety of control and sialidase-treated cells (Table II). This is consistent with the pentavalent nature of CT-B. Of particular interest is the presence of small but sufficient amounts of G_{M1} to account for all of the toxin-binding sites in human small intestinal mucosal cells, the normal target for CT. Even cells such as rat adipocytes, which have very few toxin-binding sites, have adequate amounts of G_{M1} when very sensitive detection techniques are used (Pacuszka *et al.*, 1978).

Second, all CT binding activity is removed from cells and tissues, including intestine, by delipidation, whereas binding activity is resistant to protease treatment (Critchley *et al.*, 1981; Holmgren *et al.*, 1982, 1985; Griffiths *et al.*, 1986). Third, CT specifically protects cell surface G_{M1} but not other glycolipids from being labeled by the galactose oxidase/NaB^3H_4 method (Moss *et al.*, 1977; Critchley *et al.*, 1981; Miller-Podraza *et al.*, 1982; Fishman *et al.*, 1984). Fourth, CT–receptor complexes have been isolated from cells or membranes labeled with the above method or from cells metabolically labeled with [^{3}H]galactose by an immunoadsorption procedure and analyzed (Critchley *et al.*, 1981; Miller-Podraza *et al.*, 1982). Briefly, the labeled cells or membranes are incubated with CT, washed to remove any unbound CT, and extracted with a nonionic detergent. The soluble complexes are then adsorbed with anti-CT antibodies and protein A-agarose. After denaturation, the labeled receptors are analyzed either by

Table II
COMPARISON OF G_{M1} CONTENT AND CHOLERA TOXIN RECEPTORS IN VARIOUS CELLS

Cell	Molecules/cell $\times 10^{-3}$		Ratio
	G_{M1} content	Toxin receptors	
Human small intestinal mucosal[a]	90	15	6.0
Murine thymocytes[a]	870	140	6.2
Murine Leydig tumor[b]	5,200	862	6.0
Murine NB41A neuroblastoma[c]	8,160	1,370	6.0 (5.5)[d]
Murine Friend erythroleukemic[c]	17,200	2,990	5.75 (6.3)
Murine N18 neuroblastoma[c]	83,100	14,150	5.9 (5.9)

[a]Data from Hansson *et al.* (1977).
[b]Data from Fishman *et al.* (1984).
[c]Data from Miller-Podraza *et al.* (1982).
[d]Values in parentheses are for cells treated with sialidase, which causes a corresponding 4- to 5-fold increase in both G_{M1} content and CT binding.

sodium dodecyl sulfate–polyacrylamide gel elecrophoresis (SDS-PAGE) or thin-layer chromatography (TLC) on silica gel. Only G_{M1} is specifically recovered by this method. Fifth, using sensitive overlay or blotting techniques, G_{M1} is the only toxin-binding component detected in various cells and tissues (Critchley *et al.*, 1981; Spiegel *et al.*, 1985). Thus, membrane components are separated by SDS-PAGE and electrophoretically transferred to nitrocellulose sheets, which are overlaid with ^{125}I-labeled CT (analogous to Western blotting); or lipid extracts are separated by TLC and the chromatograms overlaid with ^{125}I-labeled CT.

Finally, G_{M1} has been shown to function as the receptor for CT using G_{M1}-deficient cells such as NCTC-2071A murine fibroblasts (Moss *et al.*, 1976). These cells do not bind or respond to CT, but, when [^{3}H]G_{M1} is added to the culture medium, the cells take up the ganglioside and can bind CT and accumulate cAMP in response to the toxin. Nanomolar concentrations of G_{M1} are sufficient to sensitize the cells to CT, whereas other gangliosides are ineffective. Thus, maximal stimulation is observed in cells exposed to 2.3 nM G_{M1} for 18 hours, at which time 120,000 molecules of ganglioside have been taken up per cell. Similar results are obtained with rat glioma C6 cells (Fishman, 1980; Fishman *et al.*, 1980). These cells bind trace amounts of CT, exhibit a slight response to CT, but are much more sensitive to the toxin after treatment with G_{M1}. Exposure to G_{M1} causes a corresponding increase in toxin binding. Using [^{3}H]G_{M1}, the uptake of the ganglioside by C6 cells parallels the increase in CT binding with an initial ratio of 7 : 1. C6 cells treated with other gangliosides exhibit no increase in either CT binding or responsiveness.

IV. Neogangliolipids as Receptors for Cholera Toxin

The first indication that the lipid moiety of G_{M1} is important for CT action involved studies with an analog in which the fatty acid (>90% stearic) had been replaced with an acetate group (Fishman *et al.,* 1980). The analog, DA-G_{M1}, is taken up by the cells more rapidly and extensively than G_{M1}. Presumably, this is due to its higher critical micellar concentration and the increased ability of a glycolipid with a single hydrocarbon tail to insert into lipid bilayers. Cells exposed to DA-G_{M1} are more responsive to CT than cells exposed to G_{M1} even under conditions where equal amounts of each are taken up by the cells and the cells bound equal amounts of CT (Table III). The sphingoid long-chain base of G_{M1} may also play a role in modulating the response of cells to CT (Masserini *et al.,* 1990). Native G_{M1} is a mixture of 56% $C_{18:1}$, 37% $C_{20:1}$, 4% $C_{18:0}$, and 3% $C_{20:0}$. G_{M1} containing $C_{18:1}$ sphingosine was found to be taken up by HeLa cells more rapidly than the other molecular species of G_{M1}. HeLa cells treated with the $C_{18:1}$ molecular species also were observed to respond more rapidly to CT than cells treated with the other G_{M1} species. Unfortunately, no toxin-binding studies were done in parallel in order to determine whether the variously treated cells bound the same amount of CT.

In order to explore further the influence of the ceramide of G_{M1} on CT action, a series of G_{M1}-neogangliolipids were synthesized and tested as toxin receptors (Pacuszka *et al.,* 1991). Their structures are shown in Fig. 3. The simplest analogs consisted of G_{M1}-oligosaccharide coupled by reductive amination to aliphatic amines of increasing chain length. Whereas the C_{10} and C_{12} derivatives are taken up very poorly or not at all by rat glioma C6 cells, the C_{14}, C_{16}, and C_{18} derivatives are taken as well as or even better than G_{M1}. Cells treated with each of these derivatives are stimulated more by CT than are G_{M1}-treated cells for a given amount of bound toxin (Fig. 4). The cholesterol analog also is a more effective receptor than G_{M1}, whereas the phospholipid analogs are less effective.

Table III

COMPARISON OF G_{M1} AND DA-G_{M1} AS RECEPTORS FOR CHOLERA TOXIN[a]

Cell treatment	^{125}I-CT bound	Cyclic AMP response (-fold increase)	Adenylyl cyclase activity
None	1	1	1
G_{M1}	50	5.7	3.7
DA-G_{M1}	50	26.6	6.1

[a]Data extrapolated from Fishman *et al.* (1980). Rat glioma C6 cells were incubated for 1 hour in serum-free medium containing G_{M1} or DA-G_{M1}, washed extensively, and assayed either for ^{125}I-labeled CT binding, CT-stimulated cyclic AMP accumulation, or CT-activated adenylyl cyclase activity

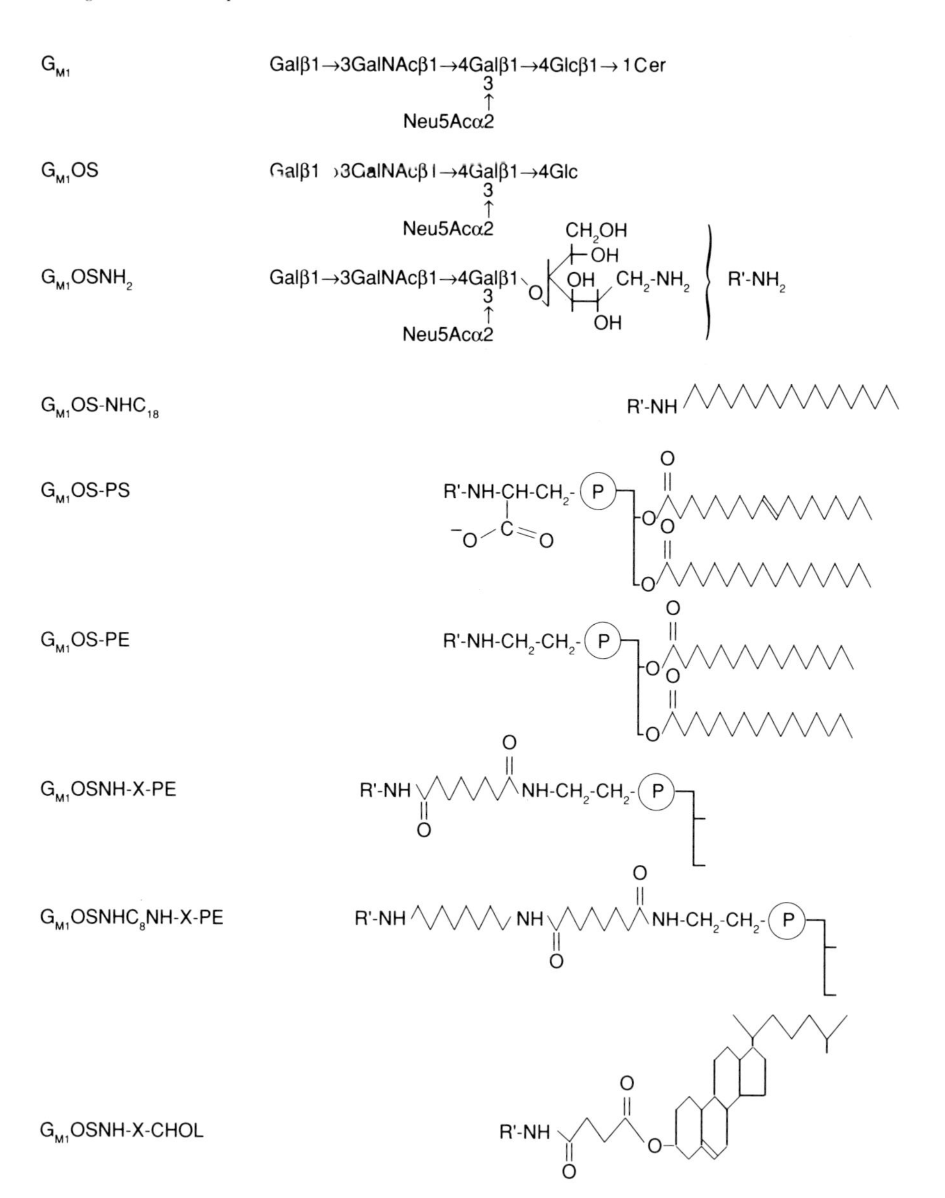

FIG. 3. Structures of neogangliolipid analogs of G_{M1}. Reproduced from Pacuszka *et al.* (1991).

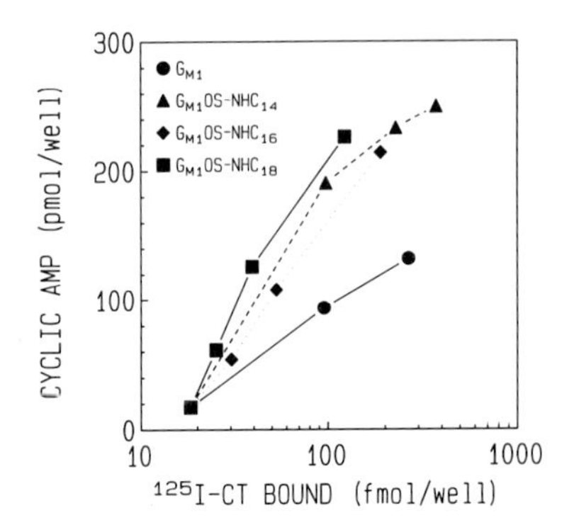

FIG. 4. Cholera toxin binding and cyclic AMP accumulation in rat glioma C6 cells treated with G_{M1} and aliphatic amine derivatives of G_{M1}-oligosaccharide. Data expressed from Pacuszka *et al.* (1991).

Furthermore, the addition of a spacer between the oligosaccharide and the phospholipid decreases the effectiveness of these analogs (Table IV).

The neogangliolipids appear to fall into two classes: DA-G_{M1}, and the analogs based on long-chain aliphatic amines or cholesterol that are more efficient toxin receptors than G_{M1}, but with the same lag period. Analogs based on phospholipids are less efficient than G_{M1} and produce a longer lag period. These differences are not due to any differences in the affinity of CT for the G_{M1} and the various neoglycolipids; the concentration required to yield half-maximal binding is between 0.37 and 0.76 n*M* for C6 cells treated with G_{M1}, DA-G_{M1}, and the other neoglycolipids (Fishman *et al.*, 1980; Pacuszka *et al.*, 1991). As is the case with G_{M1}-treated C6 cells, cells treated with the various neogangliolipids

Table IV

CHOLERA TOXIN BINDING AND CYCLIC AMP RESPONSE IN RAT GLIOMA C6 CELLS TREATED WITH G_{M1} AND PHOSPHOETHANOLAMINE DERIVATIVES OF G_{M1} OLIGOSACCHARIDE[a]

Treatment	^{125}I-CT bound (fmol/well)	Cyclic AMP response (pmol/well)
None	6.9	26.8
G_{M1}	385	277
G_{M1}OS-PE	375	209
G_{M1}OS-X-PE	374	96.5
G_{M1}OS-NHC$_8$NH-X-PE	393	58.8

[a]Rat glioma C6 cells were incubated with the indicated glycolipid for 1 hour at 37°C, washed, and assayed for either specific ^{125}I-labeled CT binding or CT stimulation of cyclic AMP accumulation. For further details, see Pacuszka *et al.* (1991).

exhibit similar multivalent binding of the toxin based on the molar ratio of ^{3}H-labeled glycolipid taken up to ^{125}I-labeled CT bound (Pacuszka and Fishman, 1991). Interestingly, C6 cells treated with the phospholipid derivatives of G_{M1}-oligosaccharide are more responsive to CT in the presence of chloroquine (Pacuszka *et al.*, 1991).

V. Neoganglioproteins as Nonfunctional Receptors for Cholera Toxin

The above studies with G_{M1}-neogangliolipids suggest that the nature of the moiety which anchors the G_{M1}-oligosaccharide to the cell surface may influence the action of CT. As glycosphingolipids and glycoproteins can share similar oligosaccharide structures, the possibility that cell surface glycoproteins may function as receptors for CT had not been eliminated, even though such entities had not been detected (as indicated in Section III). To explore this further, a method has been developed to attach G_{M1}-oligosaccharide specifically to surface proteins of viable rat glioma C6 cells (Pacuszka and Fishman, 1990). Briefly, G_{M1}-oligosaccharide is reductively aminated and coupled to a heterobifunctional cross-linker. The resulting derivative reacts with free sulfhydral groups and covalently attaches to cell surface proteins under mild conditions which maintain cell viability.

The covalent attachment is monitored by measuring the increase in ^{125}I-labeled CT binding. Generation of toxin-binding sites is dependent on time of exposure and concentration of cross-linker, optimum conditions being 50 μM for 30 minutes. Prior treatment of the cells with dithiothrèitol increases binding an additional 4-fold. The nature of these toxin-binding sites has been confirmed by Western blotting of cell membranes components with ^{125}I-labeled CT. Whereas no toxin-binding components are detected in control membranes, a wide spectrum of proteins (30 to 200 kDa) are found to bind CT in membranes from the derivatized cells, some of which are trypsin resistant. When the cells are exposed to dithiothreitol prior to the cross-linker, even more extensive toxin binding is detected by this method. The affinity of CT for cells expressing these surface "neoganglioproteins" is the same as that for cells treated with G_{M1}. By contrast, G_{M1}-treated cells exhibit an increased response to CT whereas the cells treated with cross-linker are no more responsive than control C6 cells (Fig. 5). This lack of effect is not due an impairment of the adenylyl cyclase as the control and treated cells responded equally to the β-adrenergic agonist isoproterenol.

Although cell surface G_{M1}-neoganglioproteins bind CT with high affinity, they fail to act as functional receptors for the toxin. We can only speculate on the reason(s) for this failure. It may be that the G_{M1}-oligosaccharide chains are too far from the cell surface and the bound toxin is unable to come in contact

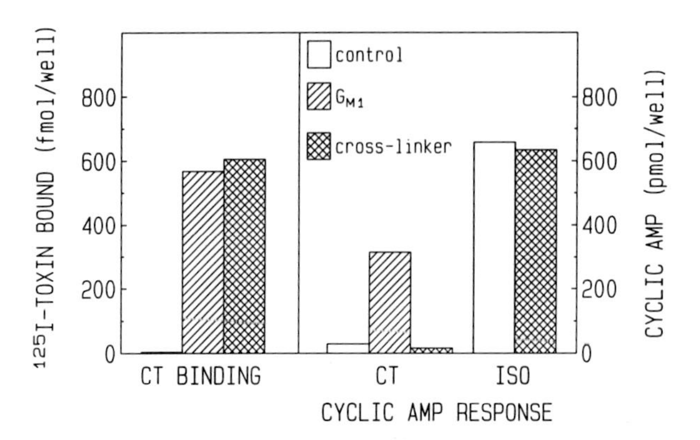

FIG. 5. Cholera toxin (CT) binding and cyclic AMP accumulation in rat glioma C6 cells treated with G_{M1} and cross-linker derivative of G_{M1}-oligosaccharide. Data extrapolated from Pacuszka and Fishman (1990, 1992). Cells were treated with no addition, G_{M1}, or cross-linker, washed extensively, and assayed for specific ^{125}I-labeled CT binding or stimulation of cyclic AMP by CT or isoproterenol (ISO).

with the membrane lipid bilayer. It is believed that such contact or proximity with the membrane is important in promoting the entry of CT or its A subunit into the cell (Fishman, 1990). More recently, it was found that C6 cells treated with the cross-linking derivative of G_{M1}-oligosaccharide could be induced to respond to CT in the presence of chloroquine (Pacuszka and Fishman, 1992). Based on these results, it was suggested that CT bound to neoganglioproteins on the cell surface may enter the cells through endocytosis and be spared from lysosomal degradation by chloroquine, which is known to elevate lysosomal pH.

In order to explore this possibility more directly, we covalently coupled reductively aminated G_{M1}-oligosaccharide to human transferrin (Pacuszka and Fishman, 1992). Transferrin enters cells through a well-defined pathway of receptor-mediated endocytosis. The G_{M1}-oligosaccharide-modified transferrin binds with high affinity to transferrin receptors on HeLa cells and supports the binding of CT (Table V). For these studies, the HeLa cells are first treated with B subunit to block endogenous G_{M1} and minimize the stimulation by CT through these endogenous receptors. Although CT is able to bind to HeLa cells through the modified transferrin and transferrin receptors, it is unable to activate adenylyl cyclase. In the presence of chloroquine, however, the cells respond to the toxin. Chloroquine has little if any effect on stimulation of control or G_{M1}-treated HeLa cells by CT. Interestingly, the lag period is more than twice as long as is observed when CT is bound to G_{M1} on HeLa cells, suggesting that the toxin is entering the cell through an alternative pathway. The ability of excess unmodified transferrin to inhibit toxin binding and stimulation of cyclic AMP

Table V
EFFECT OF CHLOROQUINE ON STIMULATION OF CYCLIC AMP IN HELA CELLS BY CHOLERA TOXIN BOUND TO G_{M1} OS-MODIFIED TRANSFERRIN[a]

Treatment	^{125}I-CT bound (-fold increase)	Cyclic AMP response (pmol/mg protein)[b] – Chloroquine	Cyclic AMP response (pmol/mg protein)[b] + Chloroquine	Lag period (min)
None	1	125	129	30
G_{M1}	14.6	388	398	15
G_{M1}OS-transferrin	8	44	273	75

[a]Data extrapolated from Pacuszka and Fishman (1992). Control HeLa cells and cells treated with G_{M1} or G_{M1}OS-transferrin were assayed for specific ^{125}I-labeled CT binding and CT-stimulated cyclic AMP accumulation. The latter cells were incubated first with B subunit to block endogenous CT receptors. Chloroquine was present at 0.1 m*M* where indicated.

[b]Cyclic AMP response was determined at 45 minutes for control and G_{M1}-treated cells, and at 4 hours for cells treated with G_{M1}OS-transferrin. Basal levels were 10–20 pmol/mg protein.

(albeit in the presence of chloroquine) confirmed that CT is binding and entering the cell through the transferrin receptor.

VI. Galactoproteins as Additional Receptors for Heat-Labile Toxin

A. VARIANTS OF HEAT-LABILE ENTEROTOXIN WITH DIFFERENT BINDING SPECIFICITIES

Heat-labile enterotoxins of *E. coli* and *V. cholerae* are classified into two antigenically distinct groups. The type I toxins include CT and type I heat-labile toxin of *E. coli* (LT-I), which are immunologically, structurally, and functionally similar (see Section II). There are two antigenic variants of heat-labile toxin produced by strains of *E. coli* isolated from humans (LTh-I) and pigs (LTp-I), respectively (Honda *et al.*, 1981). All type I toxins are neutralized by antisera raised against CT and each will cross-react with others within the serogroup while not necessarily being structurally identical. Serogroup II contains the type II *E. coli* heat-labile toxin LT-II (Holmes *et al.*, 1986), of which two antigenic variants, LT-IIa and LT-IIb, have been identified (Guth *et al.*, 1986). Although LT-II is composed of A and B subunits with molecular weights similar to the corresponding subunits of CT and LT-I, antiserum against LT-II does not neutralize CT or LT-I or vice versa. In addition, DNA probes for the structural genes for LTp-I and LT-II do not hybridize with each other under low-stringency conditions. Like the type I toxins, LT-II also activates adenylyl cyclase by ADP-ribosylation of $G_{s\alpha}$ (Chang *et al.*, 1987).

While it is well established that G_{M1} is the functional receptor for CT (see Section III) and can serve as a functional receptor for LT-I (see Section VI,B), LT-II appears to have a different receptor specificity. Thus, G_{M1} is a much less effective inhibitor of LT-II than LT-I, whereas mixed brain gangliosides are more potent inhibitors (Holmes *et al.*, 1986). Fukata *et al.* (1988) compared the binding specificity of the type I and II toxins using either a solid-phase radioimmunoassay or a TLC immunostaining method. The latter involves separating gangliosides by TLC and sequentially overlaying the chromatogram with the toxin, a monoclonal antibody against the toxin, and ^{125}I-labeled anti-mouse IgG. Type I toxins (CT, LTh-1, and LTp-1) all bind predominantly to G_{M1} and weakly to G_{D1b}. LT-1 also binds very weakly to asialo-G_{M1} and G_{M2}. This may represent a limitation of these detection methods, as ^{125}I-labeled CT fails to bind to G_{D1b} using the direct TLC overlay procedure (Fishman, 1986) or when taken up by rat glioma C6 cells (Fishman, 1980). Different binding specificities, however, are seen with type II toxins. LT-IIa binds more strongly to G_{D1b} than to G_{M1} but also recognizes G_{D1a}, G_{T1b}, G_{Q1b}, and G_{D2}. LT-IIb does not recognize G_{M1} but binds preferentially to G_{D1a}, G_{T1b}, and weakly to G_{M3}. These results suggest that differences in oligosaccharide structures are required for efficient binding of the different toxins. Thus, CT and LT-I are highly selective for the sequence Galβ1–3GalNAcβ1–4[NeuAcα2–3]Gal . . . , with additional sialyl residues inhibiting binding. LT-IIa prefers Ga1β–3GalNAc . . . , with a disialyl group (as in G_{D1b}) enhancing binding and a terminal sialyl residue (as in G_{D1a}) not affecting binding, whereas LT-IIb prefers NeuAcα2–3Galβ1–3GalNAc . . . and is the only toxin which does not appear to bind G_{M1}.

B. Evidence That G_{M1} Is a Functional Receptor for LT-I

Both Holmgren (1973) and Pierce (1973) observed that, whereas CT-B completely blocks the biological effects of CT on rabbit intestine, it only partially inhibits LT-I. In addition, brain gangliosides and G_{M1} are less effective inhibitors of LT-I compared to CT. Of the various gangliosides tested, however, G_{M1} is the most effective at interacting with LT-I (Holmgren, 1973). These results suggested that LT-I may bind to other receptors besides G_{M1}. By contrast, secretion in dog intestinal loops by either CT or LT-I was found to be inhibited by the B subunits of either of the toxins (Nalin and McLaughlin, 1978).

To address this issue, the ability of LT-I to bind to and stimulate G_{M1}-deficient NCTC-2071A and rat glioma C6 cells was explored. As had been found for CT, LT-I does not stimulate control NCTC-2071A cells to accumulate cyclic AMP, whereas G_{M1}-treated cells do respond (Moss *et al.*, 1979). Control C6 cells bind only in trace amounts of ^{125}I-labeled LT-I, whereas cells treated with G_{M1} bind 30-fold more LT-I and cells treated with other gangliosides bind only

2-fold more toxin (Moss *et al.*, 1981). We have directly compared the effect of different gangliosides on the ability of LT-I to bind to and stimulate C6 cells (Table VI). There was a striking parallel between the two parameters which clearly indicates that G_{M1} can serve as a functional receptor for LT-I.

C. Identification of Other Receptors for LT-I

Even though G_{M1} can serve as a functional receptor for LT-I, there is a growing body of evidence that LT-I binds to additional receptors in intestinal epithelial cells of various species. Holmgren *et al.* (1982) reported that rabbit intestinal epithelial cells and brush-border membranes bind up to 13-fold more LT-I than CT. Whereas all of the CT-binding sites are removed by delipidation of the membranes, a substantial fraction of the LT-I-binding sites remain in the delipidated residue. The lipid-extractable binding activity for both toxins was found to be similar, predominantly associated with the monosialoganglioside fraction, and accounted for by the amount of mucosal G_{M1} present. Griffiths *et al.* (1986) obtained similar results. In addition, they used the sensitive, direct overlay techniques to identify the toxin receptors further. With the TLC overlay, both toxins were found to bind to G_{M1} extracted from rabbit intestinal brush-border membranes. With the Western blot-like method, LT-I was found to bind to proteins with mobilities identical to the major brush-border galactoproteins.

Similar, additional binding sites for LT-I are also present in rat intestinal brush-border membranes (Zemelman *et al.*, 1989; Griffiths and Critchley, 1991). Most of the binding activity is associated with proteins of 130–140 kDa with lesser binding to smaller proteins. That these proteins are galactoproteins is supported by the ability of ricin, a galactose-specific lectin, to inhibit LT-I binding

Table VI

LT Binding and Cyclic AMP Stimulation in Rat Glioma C6 Cells Treated with Different Gangliosides[a]

Ganglioside treatment	-Fold increase	
	^{125}I-LT bound	Cyclic AMP response
None	1	1
G_{M2}	2.42	1.23
G_{M1}	27.4	27.8
G_{D1a}	2.32	1.64
G_{D1b}	2.57	1.43

[a]P. A. Orlandi and P. H. Fishman, unpublished observations. Rat glioma C6 cells were incubated with 0.5 μ*M* of the indicated gangliosides for 1 hour at 37°C, washed extensively, and assayed for specific ^{125}I-labeled LT-I binding (2.5 n*M* for 50 minutes at 25°C with 1 μ*M* LT-I-B for nonspecific binding) and LT-I stimulation of cyclic AMP accumulation (10 n*M* for 75 minutes at 37°C).

to them. Furthermore, LT-I is able to bind to the sucrase–isomaltase complex purified from the rat brush-border membranes (Griffiths and Critchley, 1991). The LT-I-binding proteins appear to be developmentally regulated, as their levels increase with age in both rats (Zemelman *et al.,* 1989) and rabbits (Griffiths and Critchley, 1991). There also appear to be variations between strains and species. Thus, intestinal brush-border membranes from Sprague-Dawley rats bind 8.3-fold more LT-I than CT, whereas the corresponding ratio is 2.2-fold for membranes from Wistar rats (Griffiths and Critchley, 1991). Whereas the ratio is highest in rabbit followed by dog and rat, it appears to be the lowest in human intestine, ranging between 1 and 1.44 (Holmgren *et al.,* 1982; Griffiths and Critchley, 1991).

D. LT-I Receptors in a Human Intestinal Cell Line

Our laboratory has recently begun to characterize the receptors for LT-I in the human intestinal epithelial cell line CaCo-2 (Orlandi *et al.,* 1991, and unpublished observations of P. Orlandi, D. R. Critchley, and P. H. Fishman). These cells, isolated from a primary colonic tumor, differentiate in culture into high polarized epithelial cells with characteristic brush-border and basolateral membranes. CaCo-2 cells bound about 20 times as much LT-I as CT, as did membranes prepared from the cells (Table VII). Whereas 30 n*M* CT-B completely inhibited

Table VII

Binding of ^{125}I-Labeled LT-I and ^{125}I-Labeled CT to CaCo-2 Epithelial Cell Membranes[a]

	Binding (% of control)	
Condition	^{125}I-LT-I	^{125}I-CT
Untreated membranes	100 (45.2)[b]	100 (2.0)[b]
+30 n*M* CT-B	67.6	2.5
Delipidated membranes	23.6	7.7
+30 n*M* CT-B	19.0	—
Trypsin-treated membranes	65.2	120
+30 n*M* CT-B	52.3	—
Lipid extract[c]	100	100
+30 n*M* CT-B	12.0	0.4

[a]P. A. Orlandi and P. H. Fishman, unpublished observations. Membranes were prepared from CaCo-2 cells and portions were treated with trypsin or delipidated. The control and treated membranes were then assayed for specific binding of either 30 n*M* ^{125}I-labeled LT-I or 1 n*M* ^{125}I-labeled CT in the absence and presence of 30 n*M* CT-B.

[b]The amount of bound toxin in pmol/mg of membrane protein is shown in parentheses for LT-I and CT at concentrations of 30 and 1.0 n*M*, respectively, which reflect saturable binding.

[c]Portions of the lipid extract were used to coat the wells of microtiter plates; after the wells were blocked with bovine serum albumin, they were incubated with the iodinated toxins as indicated. For details of the method, see Griffiths *et al.* (1986).

^{125}I-labeled CT binding (97.5%), it only partially inhibited ^{125}I-labeled LT-I binding (33%). CT binding was not decreased by trypsin treatment, whereas LT-I binding was decreased by 35%. By contrast, pronase and papain digestion of rabbit and rat intestinal membranes does not appear to reduce LT-I binding (Holmgren *et al.*, 1982; Griffiths *et al.*, 1986; Griffiths and Critchley, 1991). Such treatment, however, does solubilize LT-I binding activity, including the sucrase–isomaltase complex. As expected, most of the CT-binding sites were removed from CaCo-2 membranes by lipid extraction; delipidation also removed a significant fraction of the LT-I-binding sites, a phenomenon not observed with rabbit membranes (Holmgren *et al.*, 1982; Griffiths *et al.*, 1986). All of the CT-binding activity in the lipid extracts from CaCo-2 membranes was inhibited by CT-B, whereas some of the LT-I-binding activity was not so inhibited (see below).

CaCo-2 cells were metabolically labeled with [^{3}H]galactose and detergent extracts of LT-I/receptor complexes were prepared and immunoprecipitated. Analysis by SDS-PAGE revealed the presence of several putative LT-1-binding galactoproteins with apparent masses of 25, 35, 100, 150, and 190 kDa. These proteins appeared to be specifically recognized by LT-I, as they were not immunoprecipitated from extracts of labeled membranes containing bound CT-B. Similar proteins were detected when the Western blot-like method was used with ^{125}I-labeled LT-I. Preliminary analysis of the lipid extracts indicated the presence of a 25-kDa protein, which may explain the residual LT-I binding which was insensitive to CT-B (Table VII).

Griffiths and Critchley (1991) had shown that the binding of LT-I to rat intestinal brush-border membranes is effectively inhibited by both galactose and fucose, two principle sugars in the oligosaccharides of intestinal epithelial glycoproteins. Partial characterization of the oligosaccharide determinants required for efficient LT-I binding to CaCo-2 membrane galactoproteins was accomplished by treatment with specific glycosidases. The binding of ^{125}I-labeled LT-I to electroblotted membrane proteins was found to be sensitive to several specific glycosidases. In particular, treatment with *V. cholerae* neuraminidase drastically enhanced LT-I binding to a family of glycoproteins of 90–200 kDa. This suggested the presence of cryptic glycoprotein receptors for LT-I which are masked by terminal sialic acid residues. Treatment with β-galactosidase from *Diplococcus pneumoniae,* which preferentially hydrolyzes β1–4 linkages, destroyed LT-I-binding activity, whereas limited digestion with a bovine testes β-galactosidase, a general galactosidase with preference for β1–3 linkages, did not appreciably affect toxin binding. LT-I-binding assays with intact membrane preparations following glycosidase treatments gave similar results (Table VIII). It is especially interesting to note the effects of *D. pneumoniae* β-galactosidase and endo-β-galactosidase from *Bacteroides fragilis* on LT-I binding. Neither glycosidase had any significant effect on CT binding while destroying LT-I binding. These results indicated that LT-I binding to its glycoprotein receptors has an absolute requirement for terminal β-galactosyl residues, and it also suggested that

Table VIII
EFFECT OF GLYCOSIDASES ON BINDING OF ^{125}I-LABELED LT-I AND ^{125}I-LABELED CT TO CACO-2 MEMBRANES[a]

Glycosidase treatment	Binding (% of control)	
	^{125}I-LT-I	^{125}I-CT
Control	100[b]	100[b]
+ 30 n*M* CT-B	56.3	4
V. cholerae neuraminidase	634	388
+ 30 n*M* CT-B	491	33
D. pneumoniae β-galactosidase	16.7	87.2
+ 30 n*M* CT-B	6.1	—
Endo-β-galactosidase	37.9	119
+ 30 n*M* CT-B	0.5	—

[a]P. A. Orlandi and P. H. Fishman, unpublished observations. Details are similar to those described in Table VII except the membranes were incubated with the indicated glycosidases before being assayed for binding.
[b]At saturation, control membranes bound 16.3-fold more LT-I than CT.

these receptors may belong to the poly-*N*-lactosamine series (of the lacto-*N*-neotetraosyl type) with the common structure:

Gal(β1–4)GlcNAc(β1–3)Gal(β1–4)GlcNAc–R

LT-I was found to bind to asialofetuin and asialotransferrin, which provided unique model glycoproteins to differentiate LT-I binding specificities. Asialotransferrin contains a biantennary complex structure with terminal Gal(β1–4) GlcNAc(β1–3)–R. Asialofetuin contains both O-linked oligosaccharide determinants and triantennary N-linked complex chains with the terminal disaccharide Gal(β1–4)GlcNAc(β1–3)–R. As expected, binding to asialofetuin or asialotransferrin is dependent on terminal galactose residues and was abrogated either by *D. pneumoniae* β-galactosidase or peptide-*N*-glycosidase F treatments. Removal of the O-linked oligosaccharides of asialofetuin or treatment with β1–3 galactosidase had no effect on LT-I binding.

E. ARE INTESTINAL GALACTOPROTEINS FUNCTIONAL RECEPTORS FOR LT-I?

The galactoprotein receptors for LT-I appear to be found only in intestinal cells. There is some evidence that these receptors may be functional. As described above, excess CT-B is able to block completely the secretory response of rabbit intestinal loops to CT but not to LT-I (Holmgren, 1973; Pierce, 1973;

Holmgren *et al.*, 1982). Similar results were obtained with rat intestinal loops (Zemelman *et al.*, 1989). The latter investigators also found that CT-B only partially inhibits the activation of the intestinal adenylyl cyclase by LT-I. We have pursued this problem with the CaCo-2 cells as a model system. Pretreatment of CaCo-2 cells with a concentration of CT-B which inhibited the CT-stimulated cyclic AMP response by more than 95% only caused a 50–60% inhibition of the response to LT-I. By contrast, other human nonintestinal cell lines such SK-N-MC neurotumor and HeLa cervical carcinoma exhibited no such differential response to the two toxins. Thus, CT-B as well as LT-I-B was effective in inhibiting the cyclic AMP response induced by either CT or LT-I.

F. Neoganglioproteins Are Nonfunctional Receptors for LT-I

Because the above studies indicate that both G_{M1} and certain intestinal galactoproteins are functional receptors for LT-I, we decided to determine whether G_{M1}-neoganglioproteins would act as functional receptors for this toxin. As shown in Fig. 6A, treatment of rat glioma C6 cells with the cross-linker derivative of G_{M1}-oligosaccharide did not sensitize the cells to LT-I. If the cells were stimulated in the presence of chloroquine, however, they became responsive, albeit at a slower rate. Similar results were observed with HeLa cells treated with G_{M1}-oligosaccharide-modified transferrin (Fig. 6B). The presence of chloroquine dramatically enhanced the stimulation of cyclic AMP by LT-I bound to the cells through the transferrin receptor, whereas chloroquine had no effect on the response to LT-I bound to endogenous or exogenously incorporated G_{M1}. We confirmed that LT-I was able to bind to the neoganglioproteins (data not shown). We also found that chloroquine had no effect on the response of CaCo-2 cells to LT-I. In addition, the lag time before cyclic AMP levels began to rise was not increased when LT-I was bound to CaCo-2 cell surface galactoproteins.

Thus, LT-I behaved the same as CT in terms of its ability to bind to these G_{M1}-neoganglioproteins but remain biological inert unless chloroquine is present. These results also indicate that LT-I bound to G_{M1} is processed by the cell and activates adenylyl cyclase in a a manner analogous to that described for CT (see Section II,B); they also appear to raise some paradoxes. When bound to cell surface G_{M1}, both toxins are able to enter the cell and activate adenylyl cyclase through an apparently identical pathway. When bound to neoganglioproteins, both toxins remain inactive unless chloroquine is present, which permits some of each toxin to escape from a nonproductive pathway and slowly activate adenylyl cyclase. Yet, when bound to endogenous galactoproteins present on the intestinal enterocyte, LT-I is able to enter the cells and activate adenylyl cyclase through a productive, chloroquine-insensitive pathway. Whether this pathway is the same as that used by either LT-I or CT bound to G_{M1} is still unknown.

VII. Conclusions

To date, the ganglioside G_{M1} is the only natural, functional receptor for CT yet identified. Although the binding determinants for CT reside in the oligosaccharide moiety, the ceramide appears to be important for the biological activity of the toxin. Thus, changes in the lipid moiety can lead to either an enhanced or a decreased ability of the neogangliolipid to sensitize G_{M1}-deficient cells to CT. The extreme case occurs when the G_{M1}-oligosaccharide is attached to proteins. These G_{M1}-neoganglioproteins behave as nonfunctional receptors for CT. It appears that toxin bound to such receptors is taken into the cell through a nonproductive pathway distinct from the normal pathway of toxin entry which leads to activation of adenylyl cyclase. In the presence of chloroquine, some of the CT is salvaged from this nonproductive pathway and is able to stimulate the cells. LT-I, the heat-labile toxin produced by certain enterotoxigenic strains of *E. coli,* can also use G_{M1} as a functional receptor, and the latter appears to be the natural receptor found on most if not all nonintestinal cells. Intestinal epithelial cells of

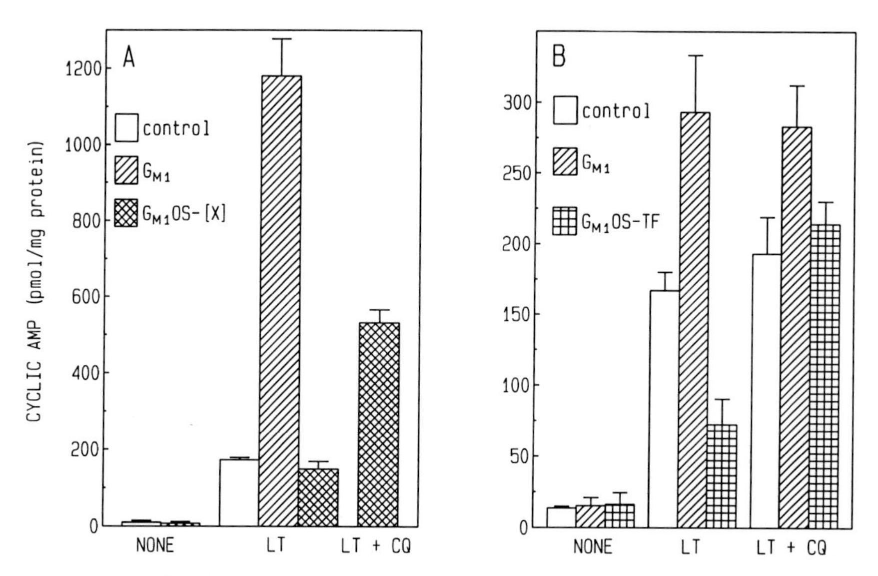

FIG. 6. Ability of LT-I to utilize G_{M1}-neoganglioproteins as functional receptors (T. Pacuszka and P. H. Fishman, unpublished observations). (A) Stimulation of rat glioma C6 cells by LT-I. Details are similar to those described in the legend to Fig. 5 except the cells were stimulated with 20 n*M* LT-I and 0.1 m*M* chloroquine (CQ) was present as indicated. G_{M1}OS-[X], Cross-linker derivative of G_{M1}-oligosaccharide. (B) Stimulation of HeLa cells by LT-I. Details are the same as those described in Table V except the cells were stimulated for 90 minutes with 20 n*M* LT-I. G_{M1}OS-TF, Transferrin modified with G_{M1}-oligosaccharide.

most species, including human, however, have additional receptors for LT-I. These have been identified as a group of galactoproteins found on the brush-border membranes of the enterocyte and may include the sucrase–isomaltase enzyme complex. Some of these galactoproteins are functional receptors for LT-I, as CT-B which binds to cell surface G_{M1} can completely block the response to CT but not to LT-I. Preliminary analyses indicate the oligosaccharide-binding determinants on these galactoproteins are different from G_{M1}-oligosaccharide. Thus, while the two enterotoxins exhibit remarkable similarities in structure and mechanism of action, subtle structural differences in their B subunits must exist to account for the variation in binding specificities. Finally, it is unclear as to how LT-I bound to these intestinal brush-border galactoproteins enters the cells and whether it utilizes the same pathway as LT-I or CT bound to G_{M1}. In this regard, the latter pathway has not been fully elucidated. Further studies will be required to resolve these questions.

References

Bobak, D. A., Tsai, S.-C., Moss, J., and Vaughan, M. (1990). *In* "ADP-Ribosylating Toxins and G Proteins: Insights into Signal Transduction" (J. Moss and M. Vaughan, eds.), pp. 439–456. Am. Soc. Microbiol., Washington, DC.

Cassel, D., and Pfeuffer, T. (1978). *Proc. Natl. Acad. Sci. U.S.A.* **75,** 2669–2673.

Chang, P. P., Moss, J., Twiddy, E. M., and Holmes, R. K. (1987). *Infect. Immun.* **55,** 1854–1858.

Critchley, D. R., Magnani, J. L., and Fishman, P. H. (1981). *J. Biol. Chem.* **256,** 8724–8731.

Cuatrecasas, P. (1973). *Biochemistry* **12,** 3558–3566.

De Wolf, M. J. S., Fridkin, M., and Kohn, L. D. (1981). *J. Biol. Chem.* **256,** 5489–5496.

Donta, S. (1976). *J. Infect. Dis.* **133** Suppl., S115–S119.

Dwyer, J. D., and Bloomfield, V. A. (1982). *Biochemistry* **21,** 3227–3231.

Eidels, L., Proia, R. L., and Hart, D. A. (1983). *Microbiol. Rev.* **47,** 596–620.

Evans, D. J., Chen, L. C., Curlin, G. T., and Evans, D. G. (1972). *Nature (London)* **236,** 137–138.

Field, M. (1971). *N. Engl. J. Med.* **284,** 1137–1144.

Finkelstein, R. A. (1973). *CRC Crit. Rev. Microbiol.* **2,** 553–623.

Fishman, P. H. (1980). *J. Membr. Biol.* **54,** 61–72.

Fishman, P. H. (1982). *J. Membr. Biol.* **69,** 85–97.

Fishman, P. H. (1986). *Chem. Phys. Lipids* **42,** 137–151.

Fishman, P. H. (1988). *In* "New Trends in Ganglioside Research: Neurochemical and Neuroregenerative Aspects" (R. W. Ledeen, E. L. Hogan, G. Tettamanti, A. J. Yates, and R. K. Yu., eds.), Fidia Res. Ser., Vol. 14, pp. 183–201. Liviana Press/Springer-Verlag, Padova/Berlin.

Fishman, P. H. (1990). *In* "ADP-Ribosylating Toxins and G Proteins: Insights into Signal Transduction" (J. Moss and M. Vaughan, eds.), pp. 127–140. Am. Soc. Microbiol., Washington, DC.

Fishman, P. H., and Atikkan, E. E. (1979). *J. Biol. Chem.* **254,** 4342–4344.

Fishman, P. H., and Atikkan, E. E. (1980). *J. Membr. Biol.* **54,** 51–60.

Fishman, P. H., Moss, J., and Osborne, J. C., Jr. (1978). *Biochemistry* **17,** 711–716.

Fishman, P. H., Pacuszka, T., Hom, B., and Moss, J. (1980). *J. Biol. Chem.* **255,** 7657–7664.

Fishman, P. H., Bradley, R. M., Rebois, R. V., and Brady, R. O. (1984). *J. Biol. Chem.* **259,** 7963–7989.

Freissmuth, M., and Gilman, A. G. (1989). *J. Biol. Chem.* **264,** 21907–21914.

Fukuta, S., Magnani, J. L., Twiddy, E. M., Holmes, R. K., and Ginsburg, V. (1988). *Infect. Immun.* **56,** 1748–1753.

Gill, D. M. (1976). *Biochemistry* **15,** 1242–1248.

Gill, D. M. (1977). *Adv. Cyclic Nucleotide Res.* **8,** 85–118.

Gill, D. M., and Meren, R. (1978). *Proc. Natl. Acad. Sci. U.S.A.* **75,** 3050–3054.

Gill, D. M., and Richardson, S. H. (1980). *J. Infect. Dis.* **141,** 64–70.

Gill, D. M., Clements, J. D., Robertson, D. C., and Finkelstein, R. A. (1981). *Infect. Immun.* **33,** 677–682.

Gilman, A. G. (1987). *Annu. Rev. Biochem.* **56,** 615–649.

Griffiths, S. L., and Critchley, D. R. (1991). *Biochim. Biophys. Acta* **1075,** 154–161.

Griffiths, S. L., Finkelstein, R. A., and Critchley, D. R. (1986). *Biochem. J.* **238,** 313–322.

Guth, B. E., Twiddy, E. M., Trabulsi, L. R., and Holmes, R. A. (1986). *Infect. Immun.* **54,** 529–536.

Hakomori, S. (1990). *J. Biol. Chem.* **265,** 18713–18716.

Hansson, H.-A., Holmgren, J., and Svennerholm, S. (1977). *Proc. Natl. Acad. Sci. U.S.A.* **74,** 3782–3786.

Holmes, R. K., Twiddy, E. M., and Pickett, C. L. (1986). *Infect. Immun.* **53,** 464–473.

Holmgren, J. (1973). *Infect. Immun.* **8,** 851–859.

Holmgren, J. (1981). *Nature (London)* **292,** 413–417.

Holmgren, J., Lönnroth, I., and Svennerholm, S. (1973). *Infect. Immun.* **8,** 208–214.

Holmgren, J., Lönnroth, I., Månsson, J.-E., and Svennerholm, S. (1975). *Proc. Natl. Acad. Sci. U.S.A.* **72,** 2520–2524.

Holmgren, J., Fredman, P., Lindbad, M., Svennerholm, A.-M., and Svennerholm, L. (1982). *Infect. Immun.* **38,** 424–433.

Holmgren, J., Lindbad, M., Fredman, P., Svennerholm, L., and Myrvold, H. (1985). *Gastroenterology* **89,** 27–35.

Honda, T., Tsuji, T., Takeda, Y., and Miwatani, T. (1981). *Infect. Immun.* **34,** 337–340.

Janicot, M., Fouque, F., and Desbuquois, B. (1991). *J. Biol. Chem.* **266,** 12858–12865.

Kanfer, J. N., Carter, T. P., and Katzen, H. M. (1976). *J. Biol. Chem.* **251,** 7610–7619.

Kimberg, D. V., Field, M., Johnson, J., Henderson, A., and Gershon, E. (1971). *J. Clin. Invest.* **50,** 1218–1231.

King, C. A., and van Heyningen, W. E. (1973). *J. Infect. Dis.* **127,** 639–647.

King, C. A., and van Heyningen, W. E. (1975). *J. Infect. Dis.* **131,** 643–648.

Kunkel, S. L., and Robertson, D. C. (1979). *Infect. Immun.* **25,** 586–596.

Ludwig, D. S., Holmes, R. K., and Schoolnik, G. K. (1985). *J. Biol. Chem.* **260,** 12528–12534.

Ludwig, D. S., Ribi, H. O., Schoolnik, G. K., and Kornberg, R. D. (1986). *Proc. Natl. Acad. Sci. U.S.A.* **83,** 8585–8588.

Masserini, M., Palestini, P., Pitto, M., Chigorna, V., Tomasi, M., and Tettamanti, G. (1990). *Biochem. J.* **271,** 107–111.

Mekalanos, J. J., Collier, R. J., and Romig, W. R. (1979). *J. Biol. Chem.* **254,** 5855–5861.

Mekalanos, J. J., Swartz, D. J., Pearson, G. D. N., Harford, N., Groyne, F., and de Wilde, M. (1983). *Nature (London)* **306,** 551–557.

Miller-Podraza, H., Bradley, R. M., and Fishman, P. H. (1982). *Biochemistry* **21,** 3260–3265.

Moss, J., and Vaughan, M. (1979). *Annu. Rev. Biochem.* **48,** 581–600.

Moss, J., Fishman, P. H., Manganiello, V. C., Vaughan, M., and Brady, R. O. (1976). *Proc. Natl. Acad. Sci. U.S.A.* **73,** 1034–1037.

Moss, J., Manganiello, V. C., and Fishman, P. H. (1977). *Biochemistry* **16,** 1876–1881.

Moss, J., Garrison, S., Fishman, P. H., and Richardson, S. H. (1979). *J. Clin. Invest.* **64,** 381–384.

Moss, J., Osborne, J. C., Fishman, P. H., Nakaya, S., and Robertson, D. C. (1981). *J. Biol. Chem.* **256,** 12861–12865.
Nalin, D. R., and McLaughlin, J. C. (1978). *J. Med. Microbiol.* **11,** 177–186.
Orlandi, P. A., Critchley, D. R., and Fishman, P. H. (1991). *J. Cell Biol.* **115,** 203a.
Pacuszka, T., and Fishman, P. H. (1990). *J. Biol. Chem.* **265,** 7673–7678.
Pacuszka, T., and Fishman, P. H. (1991). *Biochim. Biophys. Acta* **1083,** 153–160.
Pacuszka, T., and Fishman, P. H. (1992). *Biochemistry* **31,** 4773–4778.
Pacuszka, T., Moss, J., and Fishman, P. H. (1978). *J. Biol. Chem.* **253,** 5103–5108.
Pacuszka, T., Bradley, R. M., and Fishman, P. H. (1991). *Biochemistry* **30,** 2563–2570.
Pierce, N. F. (1973). *J. Exp. Med.* **137,** 1009–1023.
Reed, R. A., Mattai, J., and Shipley, G. G. (1987). *Biochemistry* **26,** 824–832.
Sack, R. B. (1975). *Annu. Rev. Microbiol.* **29,** 333–353.
Serafini, T., Orci, L., Amherdt, M., Brunner, M., Kahn, R. A., and Rothman, J. E. (1991). *Cell (Cambridge, Mass.)* **67,** 239–253.
Sharp, G. W. G., and Hynie, S. (1971). *Nature (London)* **229,** 266–269.
Sixma, T. K., Pronk, S. E., Kalk, K. H., Wartna, E. S., van Zanten, B. A. M., Witholt, B., and Hol, W. G. J. (1991). *Nature (London)* **351,** 371–377.
Spiegel, S., Fishman, P. H., and Weber, R. J. (1985). *Science* **230,** 1285–1287.
Tran, D., Carpentier, J.-L., Sawano, F., Gordon, P., and Orci, L. (1987). *Proc. Natl. Acad. Sci. U.S.A.* **84,** 7957–7961.
Tsuji, T., Honda, T., Miwatani, T., Wakabayashi, S., and Matsubara, H. (1985). *J. Biol. Chem.* **260,** 8552–8558.
van Heyningen, S. (1983). *Curr. Top. Membr. Transp.* **18,** 445–471.
van Heyningen, W. E., Carpenter, C. C. J., Pierce, N. F., and Greenough, W. B., III (1971). *J. Infect. Dis.* **124,** 415–471.
Walker, W. A., Field, M., and Isselbacher, K. J. (1974). *Proc. Natl. Acad. Sci. U.S.A.* **71,** 320–324.
Wiegandt, H. (1982). *Adv. Neurochem.* **4,** 149–223.
Yamamoto, T., Nakazawa, T., Miyata, T., Kaji, A., and Yokota, T. (1984). *FEBS Lett.* **169,** 241–246.
Zemelman, B. V., Chu, S.-H. W., and Walker, W. A. (1989). *Infect. Immun.* **57,** 2947–2952.

Verotoxins and Their Glycolipid Receptors

CLIFFORD A. LINGWOOD

Department of Microbiology
The Hospital for Sick Children
Toronto, Ontario, Canada M5G 1X8
and
Departments of Clinical Biochemistry,
Biochemistry, and Microbiology
University of Toronto
Toronto, Ontario, Canada

I. Verotoxins

The *Escherichia coli* derived verotoxins (VTs) were first described in 1977 by Konowalchuk (Konowalchuk *et al.*, 1977) as being a novel cytotoxic activity toward African green monkey kidney cells (vero cells, hence the name verotoxin). The prototype toxin (VT1) was subsequently purified (O'Brien and LaVeck, 1983; Petric *et al.*, 1987) and found to be highly homologous with the well-studied toxin from *Shigella dysenteriae* (Shiga toxin) (O'Brien and LaVeck, 1983). Later cloning studies confirmed the identity of Shiga toxin with VT1 (Takao *et al.*, 1988), with the possible exception of one amino acid in the A subunit (Jackson *et al.*, 1987; Calderwood *et al.*, 1987; DeGrandis *et al.*, 1987; Strockbine *et al.*, 1988). Thus, the VTs are also referred to as Shiga-like toxins (SLT). Several additional variants in the VT "family" have since been described (Linggood and

Thompson, 1987; Scotland *et al.*, 1985), purified (O'Brien and LaVeck, 1983; Dickie *et al.*, 1989; Head *et al.*, 1988b; Downes *et al.*, 1988), and cloned (Newland *et al.*, 1985, 1987; Gyles *et al.*, 1988; Ito *et al.*, 1990). VT2 is approximately 60% homologous to VT1 but shows reduced cytotoxicity *in vitro* (Head *et al.*, 1988b, 1991). SLTII (Downes *et al.*, 1988) was originally thought to be synonymous with VT2 but has now been shown to be a distinct (97% homologous) toxin (Head *et al.*, 1988a). As a result of this confusion, VT2 has also been called VT2 variant (Ito *et al.*, 1990; Schmitt *et al.*, 1991). Indeed, many human *E. coli* strains produce both SLTII and VT2 (Schmitt *et al.*, 1991). The fourth member (VTE) is from porcine *E. coli* strains (Linggood and Thompson, 1987; Newland *et al.*, 1987; Gyles *et al.*, 1988) and is the cause of the pathology of the pig edema disease. VTE is 90% homologous to SLTII and has been referred to as SLTIIv (variant) (Marques *et al.*, 1987).

The nomenclature of these toxins has been clarified recently (M. A. Karmali, A. D. O'Brien, and S. M. Scotland, personal communication). The terms SLT and VT become synonymous and four toxins are recognized: VT1, SLTI (~ Shiga toxin); VT2, SLTII (previously SLTII); VT2c, SLTIIc (previously VT2, VT2v, SLTvh, SLTIIc); and VT2e, SLTIIe (previously VTE or SLTIIv, SLTIIvp).

Verotoxins are subunit toxins comprising a ~30-kDa A (toxic) subunit which inhibits intracellular protein synthesis via a specific RNA *N*-glycohydrolase activity (Obrig *et al.*, 1987; Endo *et al.*, 1988; Saxena *et al.*, 1989) and five noncovalently associated receptor-binding B subunits (Donohue-Rolfe *et al.*, 1984), which facilitate the entry of the toxic subunit into susceptible cells. The B subunits, comprising ~70 amino acids, contain a single disulfide bond necessary for function.

II. Tissue Targeting

A. Clinical Features

Great interest in the study of the pathogenesis of these toxins has been fostered by their implication in the etiology of two serious human pathologies, hemorrhagic colitis (HC) (Riley *et al.*, 1983) and the hemolytic uremic syndrome (HUS) (Karmali *et al.*, 1985b), the leading cause of pediatric renal failure. The role of systemic toxemia in these pathologies subsequent to infection with VT-producing *E. coli* (VTEC) in relation to HUS and HC has been reviewed elsewhere (Karmali, 1989). In both these pathologies, the primary site of damage is the blood vasculature (of the large bowel in HC and of the capillaries within the glomerulus in the HUS) (Richardson *et al.*, 1988). The endothelial cells lining these small blood vessels are damaged, resulting in the production of a fibrin clot and the local occlusion of the vasculature, which in turn results in infarct and rupture of the small blood vessels. VT binding sites were detected in infant glomerulii (Lingwood, 1993). In HUS, thrombocytopenia and hemolysis of red cells are also characteristic clinical features (Richardson *et al.*, 1988). However, damage to the renal endothelium is the primary site of pathology.

B. Cytotoxicity *in Vitro*

Verotoxin has been shown to be cytotoxic to human umbilical vein endothelial cells (HUVEC) in culture (Obrig *et al.*, 1988; Tesh *et al.*, 1991). However, sensitivity has been shown to vary markedly between different endothelial cell preparations. The level of Gb_3 in such cells was estimated to be at least 30-fold lower than that extracted from human kidney and VT sensitivity was proportionally reduced (Tesh *et al.*, 1991). Our recent studies have found that cultured human renal endothelial cells contain 50- to 100-fold higher levels of Gb_3 than do HUVECs (Obrig *et al.*, 1993). It is also likely that endothelial cell cultures prepared from different vascular beds exhibit markedly different properties. HUVEC sensitivity was found to be a function of cell growth (Obrig *et al.*, 1988), stationary phase cells being resistant to VT cytotoxicity. Similar results have also been found for vero cells (Pudymaitis and Lingwood, 1992). This may relate to the observed higher incidence of HUS cases following VTEC infection in the very young infant (Richardson *et al.*, 1988), when presumably there is a higher rate of cell growth. A wide variety of cultured cells are sensitive to VTs (Eiklid and Olsnes, 1980), including HeLa cells (Marques *et al.*, 1987), MDCK cells (Sandvig *et al.*, 1991), Daudi lymphoblastoma (Cohen *et al.*, 1987), and HEp2 cells (Tyrrell *et al.*, 1992).

C. Animal Models

Studies with the purified toxin injected into rabbits (Richardson *et al.*, 1992; Zoja *et al.*, 1993) have shown that the toxin localizes to those tissues in which damage is primarily observed, i.e., blood vessels of the gastrointestinal tract (primarily the cecum) and the central nervous system. In the rabbit, even long-term administration of toxin from subcutaneous implanted capsules failed to result in any renal damage (Barrett *et al.*, 1989) due to the lack of the VT receptor (the glycolipid Gb_3, see below) in the rabbit kidney (Boyd and Lingwood, 1989; Zoja *et al.*, 1992). Similarly, toxin administration to mice results in tubular necrosis with no evidence of glomerular damage (Wadolkowski *et al.*, 1990), presumably due to toxin receptor in the tubules but not in the renal endothelium. Recent studies by Karmali's group have shown that the exogenously administered VT binds selectively to endothelial cells in a subset of small venules within the rabbit gastrointestinal tract and capillaries in the central nervous system (Bitzan *et al.*, 1992; Richardson *et al.*, 1992). Thus, although there is no animal model of HUS, such studies strongly infer a direct role of toxin binding to target tissues in HUS and HC. Evidence for cytokine augmentation of VT2 cytopathology is available (Tesh *et al.*, 1991; Barrett *et al.*, 1990). Tumor necrosis factor α (TNF-α) has been found to potentiate the effect of Shiga toxin on human vascular endothelial cells *in vitro* (Tesh *et al.*, 1991; Louise and Obrig, 1991), via stimulation of Gb_3 synthesis (van de Kar *et al.*, 1993). This may relate to the finding that TNF-α was elevated for macrophages after VT2 treatment and macrophage-defective mice were found to be resistant to VT pathology (Barrett *et al.,* 1990). Such results would imply "pathological amplification," whereby VT enhances TNF-α production which increases endothelial Gb_3 to increase sensitivity to VT to exacerbate endothelial damage.

The studies of Sansonetti's group (Fontaine *et al.,* 1988) showed that experimental infection of monkeys with a mutant strain of *S. dysenteriae,* in which the Shiga toxin genes had been deleted, resulted in pathology in which the diarrheal sequelae of infection were still present but the renal pathology was not now evident. This provides a persuasive argument for a direct role of the toxin in the renal pathology observed with HUS.

Differences have been observed between the pathogenesis of VT1 and VT2c following parental administration to rabbits. Both cause microangiopathic lesions but VT1 has a lower LD_{50} due to paralysis, whereas VT2c causes less severe neurological symptoms but more severe hemorrhagic cecitis, bloody diarrhea, and mucous secretion (Head *et al.,* 1988c). Such differences, however, have not been observed for SLTII (Barrett *et al.,* 1989).

The tissue tropism of the toxin demonstrated *in vivo* using animal models and the selective pathological effect on different tissues following natural infection *in vivo* strongly indicate the presence of a specific VT receptor on target tissues.

III. Glycolipid Receptor

A. Specificity

The glycolipid-binding specificities of Shiga toxin, SLT1, and VT1 were first described by Jacewicz *et al.* (1986), Lindberg *et al.* (1987), and Lingwood *et al.* (1987), respectively. Specific binding to the ceramide trihexoside glycolipid globotriaosylceramide (Galα1–4Galβ1–4glucosylceramide, Gb_3) was observed. The binding specificity was determined by thin-layer chromatography (TLC) overlay procedures using iodinated toxin or specific antitoxin antibodies. The terminal Galα1-4Gal residue was found to be crucial for binding and it was shown that the P_1 blood group glycolipid, a pentaosyl ceramide (Galα1–4Galβ1–4GlcNAcβ1–4Galβ1–4GlcCer), was also bound, as was galabiosylceramide (Cohen *et al.,* 1987; Lindberg *et al.,* 1987). Digestion of such glycolipids with α-galactosidase resulted the loss of toxin binding (Lingwood *et al.,* 1987). Interestingly, it was determined early on that the ceramide moiety played a crucial role in the binding specificity of VT. Unlike galabiosylceramide, the plant glycolipid digalactosyldiglyceride, containing the same carbohydrate sequence but attached to a glycerol rather than a ceramide moiety, was not recognized by VT (Lingwood *et al.,* 1987). This implied that either that there were two binding sites, one recognizing the Galα1–4Gal sequence and another involved in lipid recognition, or, alternatively, that the lipid moiety played a role in the presentation of the carbohydrate sequence in an appropriate conformation for toxin binding.

Ceramide-based carbohydrate may orient in a manner semiparallel to the plane of the membrane (Strömberg *et al.*, 1991) due to a hydrogen bond between the sphingosine amide nitrogen and the glycosidic oxygen-linking glucose (Pascher and Sundell, 1977). Such an interaction cannot occur in glycoglycerolipids and might result in a different relative orientation of the sugar moiety to effect ligand binding.

The Gb_3 oligosaccharide alone does not bind to VT and cannot prevent the binding of VT to susceptible cells. This is in part due to the requirement for polyvalency, since immobilization of the galabiosyl disaccharide on BSA results in the generation of an effective though weak inhibitor of cytotoxicity (Lindberg *et al.*, 1987). This inhibition depends on the degree of substitution of the BSA, which suggests the requirement for a specific three-dimensional configuration of several galabiosyl moieties. Since "multivalent" Gb_3 is necessary for optimum VT binding, such configuration might well be best presented when the oligosaccharide is conjugated to ceramide.

The Galα1–4Gal must be terminal on the carbohydrate sequence to permit binding. This thus distinguishes VT binding from that of PapG adhesins, which recognize both internal and terminal Galα1–4Gal carbohydrate sequences (Lund *et al.*, 1987, 1988; Bock *et al.*, 1985; Lindstedt *et al.*, 1989; Strömberg *et al.*, 1990). Isoglobotriaosylceramide, containing Galα1–3 as opposed to Galα–4 linked to galactose was not recognized by VT (Lindberg *et al.*, 1987). The Galα1–4Gal linkage is not found on glycoproteins, with the exception of the P glycoprotein from hydatid cyst fluid (Cory *et al.*, 1974) and salivary glycoproteins of the Chinese swiftlet (Wieruszeski *et al.*, 1987). The former has been very effectively used by Keusch's group for the affinity purification of Shiga toxin (Donohue-Rolfe *et al.*, 1989), since, unlike the binding to Gb_3, this can be reversed in the presence of chaotropes. The pig edema disease toxin, VT2e, binds Gb_4 in addition to Gb_3 (DeGrandis *et al.*, 1989; Samuel *et al.*, 1990) (see below). Galactosylgloboside (SSEA 3; Shivensky *et al.*, 1982) is also bound by this toxin (Samuel *et al.*, 1990).

B. Affinity

1. Toxin Heterogeneity

VT1 and VT2c show approximately 60% homology in the B receptor-binding subunit (Newland *et al.*, 1987), although the cytotoxicity of VT2c is a thousandfold less than VT1 (Head *et al.*, 1988b). Both these toxins have been shown to bind specifically to Gb_3 (Lingwood *et al.*, 1987; DeGrandis *et al.*, 1989; Samuel *et al.*, 1990; Waddell *et al.*, 1988).

By separation of subunits and reannealing to form chimeric toxins, the reduced cytotoxicity of VT2c as compared with VT1 was shown to be a function of the B subunit, likely mediated by the reduced binding affinity of the VT2c B subunit for Gb_3 (Head *et al.*, 1991). The A subunits of VT1 and VT2c were found to be equipotent inhibitors of cell-free protein synthesis (Head *et al.*, 1991). Chimeras containing the B subunit of VT2c showed *in vitro* cytotoxicity similar to VT2c irrespective of the A subunit type (Head *et al.*, 1991). Using a VT1 displacement microtiter Gb_3-binding assay, a thousand-fold excess of VT2cB was required for displacement equivalent to that of VT1 B subunit. Thus, receptor binding affinity, in addition to specificity, is of prime importance in determining pathogenesis.

It is interesting to note that, although there is a high degree of homology in the B subunits and both toxins bind to the same receptor glycolipid, there is no cross-neutralization between antisera raised against these two species of toxin (Perera *et al.*, 1988), indicating that the glycolipid binding site is likely not an immunodominant epitope.

The VT1 B subunit was found to pentamerize in the absence of the A, and the affinity of the VT1 B subunit for Gb_3 was equal to that of the holotoxin (Head *et al.*, 1991; Ramotar *et al.*, 1990), indicating that the A subunit does not "organize" the B subunits nor in any way contribute to the binding affinity. The VT1 B subunit alone was, on a molar basis, equally effective to displace ^{125}I-labeled VT1 from Gb_3 as was the holotoxin (Head *et al.*, 1991).

Unlike VT2c, SLTII shows a cytotoxic potential *in vitro* equivalent to SLTI (Downes *et al.*, 1988). This may reflect a higher affinity for SLTII/Gb_3 binding as compared with VT2c (Samuel *et al.*, 1990). VT2c may show higher affinity binding for galabiosyl ceramide as opposed to VT1 and SLTII (Samuel *et al.*, 1990). Thus, VT2c and VT1/SLTII, VT2e may recognize different epitopes on the Galα1–4Gal sequence which become available as the structure is found in longer chain globo series glycolipids, in much the same way as variants of the PapG adhesins have been shown to discriminate between globo series glycolipids (Strömberg *et al.*, 1991).

2. *Gb_3 Microheterogeneity*

While the presence of Gb_3 is necessary to confer susceptibility to VT cytotoxicity, the cellular concentration of Gb_3 is not necessarily proportional to that sensitivity (Cohen *et al.*, 1990), although within variants of the same cell line, this correlation appears to hold (Jacewicz *et al.*, 1989). Thus, it is likely that other factors, in addition to receptor concentration, determine sensitivity to VT. One such possible factor is the microheterogeneity of Gb_3 itself. Although the carbohydrate sequence of Gb_3 is by definition constant, there is considerable variation in the lipid moiety, both in terms of sphingosine base and fatty acid heterogeneity. While no studies on the effect of variation of the sphingosine base on VT binding have yet been reported, data with regard to fatty acid differences are available. Fatty acids may vary in chain length, degree of saturation, and hydroxylation. These variations are typified by the doublet, or sometimes triplet, observed in the separation of shorter chain glycolipids by TLC. Shiga toxin has been reported to bind by TLC overlay equally well to hydroxylated and nonhydroxylated fatty acid-containing Gb_3 species (Lindberg *et al.*, 1987). However, human renal Gb_3 contains very little, if any, hydroxylated fatty acid-containing Gb_3 (Makita, 1964; Martensson, 1966; Pellizzari *et al.*, 1992).

The effect of heterogeneity of fatty acid chain length of human renal Gb_3 on VT binding has recently been investigated (Pellizzari *et al.*, 1992). $C_{16:0}$, $C_{22:0}$, $C_{24:0}$, and $C_{24:1}$-containing Gb_3s were found to be the major human renal species. Semisynthetic Gb_3s were prepared, homogeneous for these fatty acids. No gross difference in VT1 binding as monitored by TLC overlay was detected. In addition, Gb_3s without fatty acid (lyso-Gb_3) or containing a C_6 or C_{12} (nonphysiological fatty acids) were also efficiently bound by TLC overlay. However, on Scatchard analysis of binding in a microtiter plate, whereby the glycolipid is immobilized in the presence of cholesterol and lecithin, significant differences in binding parameters were observed. The first important difference was that the lyso-Gb_3 and C_6 and C_{12} receptors were barely recognized, indicating that there is a lower limit for fatty acid chain length for Gb_3 to function as a toxin receptor in a lipid environment. The second important observation was that mixtures of Gb_3 containing different fatty acids bound with higher affinity and capacity than any of the homogeneous preparations. The data indicated that such Gb_3 "isoforms" could interact cooperatively to generate a new binding site of higher affinity than any of the individual components alone. The binding capacity and affinity were dependent on the ratio of the Gb_3 isoforms within the mixture, suggesting that the affinity of VT for the more heterogeneous Gb_3 in a natural membrane may be the result of an extremely complex series of interactions. The individual Gb_3 isoforms in VT-sensitive cells may interact in a cell type-specific manner to generate a wide diversity of receptor affinities and thus markedly vary the ability of a given concentration of cell surface Gb_3 to mediate VT cytopathology.

To explain these effects, it was hypothesized that the individual Gb_3-binding sites on the B subunit pentamer might not lie in a single plane and that Gb_3 fatty acid heterogeneity might better accommodate such nonequivalent sites (Pellizzari *et al.*, 1992). The "split washer" solution to the crystal structure of the VT1 B subunit pentamer (Stein *et al.*, 1992) would be consistent with this explanation.

The third interesting point to emerge was that the lipid content of Gb_3 may actually influence the conformation of the carbohydrate moiety for ligand binding. Comparison of the ability of the different Gb_3 isoforms to act as VT receptors or substrates for galactose oxidase when immobilized in a microtiter plate showed that there was an overall increase in binding capacity as a function of chain length up until the length of C_{22}. For the C_{24} and $C_{24:1}$ fatty acid Gb_3, however, there was a dramatic decrease in the ability to function as a galactose oxidase substrate. In contrast, VT was able to recognize these species effectively. Since galactose oxidase modifies the C-6 position of the terminal galactose moiety, whereas VT requires an unsubstituted C-3 position of the terminal galactose, these data can be interpreted to suggest that, for the C_{24} fatty acid Gb_3, the conformation of the carbohydrate has changed in relation to the other Gb_3 species in that the C-3 position of the terminal galactose is exposed, whereas the C-6 position is not. Therefore, it is possible that the fatty acid can directly influence the orientation of the carbohydrate moiety of a glycolipid. This would have wide implications in the regulation of cellular glycolipid receptor function. It is possible that a particular combination of Gb_3 fatty acid isoforms might interact in the renal glomerulus of, for example, infants as opposed to adults, to form a particularly effective receptor for VT and thus predispose such individuals to the HUS following VTEC infections.

C. Receptor ELISA for VT

As a result of the high biological potency of VT (CD_{50} ~10pg), diagnosis of infections with VT-producing *E. coli* (VTEC) has been difficult. The gold standard has been the assay of cell cytotoxicity using vero cells and the neutralization of this pathology with appropriate antisera (Karmali *et al.*, 1985a; Karmali, 1987). This is a somewhat subjective, time-consuming, and expensive procedure and is performed only in major centers. We made use of glycolipid receptor binding to develop a rapid diagnostic assay for VT. Although the fatty acid of Gb_3 modifies receptor function in a lipid matrix, the fatty acid does not limit binding per se. Since the affinity of binding for lyso-Gb_3 immobilized in the absence of auxiliary lipid is so high, we were able to develop a receptor-based ELISA (RELISA) for VT1 which was able to detect <1 pg (Basta *et al.*, 1989; Boulanger *et al.*, 1990), a sensitivity which is in excess of a receptor ELISA in which Gb_3 was directly immobilized in microtiter wells (Ashkenazi and Cleary, 1989) and traditional ELISA formats (Kongmuang *et al.*, 1987; Donohue-Rolfe *et al.*, 1986; Strockbine *et al.*, 1985), which are not sufficiently sensitive for clinical use. Thus, the affinity of VT for lyso-Gb_3 is greater than that for Gb_3 under these conditions, illustrating the importance of the aglycone in the recognition of lipid-bound carbohydrate. In a blind study, this assay correlated well with the detection of VT1 by the cytotoxicity assay. Because of the extreme sensitivity, we were able to modify the assay to detect neutralizing antibodies in convalescent sera (Boulanger *et al.*, 1990). Only antibodies capable of neutralizing VT binding to lyso-Gb_3 were found to be able to neutralize cytotoxicity *in vitro*. These assays thus facilitate investigation of the epidemiology of these infections and detection of the toxin in clinical samples before renal damage is apparent.

D. Chemical Modification of VT Binding Specificity

There was some discrepancy in the literature as to whether Gb_4 (globotetraosylceramide), containing an additional terminal β1–3-linked GalNac residue, could be recognized by this toxin in addition to Gb_3. Our studies indicated that it could not be bound, whereas others showed significant but reduced binding for this glycolipid (Jacewicz *et al.*, 1986; Lindberg *et al.*, 1987).

We have recently suggested an explanation for this discrepancy (Yiu and Lingwood, 1992). The TLC overlay procedure commonly used to screen for glycolipid binding specificity involves the use of a compound, polyisobutylmethacrylate (PIBM), to pretreat the TLC plate prior to overlay with the ligand in question (Magnani *et al.*, 1980). The rationale behind this procedure is that PIBM prevents the loss of silica from the plate during the various washing steps necessary in the binding assay and that the treatment somehow orients the carbohydrate on the silica gel in a suitable fashion to expose the carbohydrate for ligand binding (Magnani *et al.*, 1980). We have not found PIBM treatment to be necessary (Cohen *et al.*, 1987; Boyd and Lingwood, 1989; Lingwood *et al.*, 1987) and have data to indicate that treatment of glycolipids with this amphipathic compound can result in artifactual glycolipid binding specificity (Yiu and Lingwood, 1992). VT binding to Gb_3 and the P1 glycolipid is decreased, while binding to Gb_4 is induced.

We have previously reported that certain amphipathic cyclic peptides (polymyxin B or melittin) were capable of directly interacting with Gb_4 to render it bound by VT (Head *et al.*, 1990). The VT binding to Gb_4 in the presence of amphipathic cyclic peptides required the presence of the lipid moiety, since immobilized globotetraose was not recognized by VT in the presence or absence of such peptides. This implies that the mechanism of induction of Gb_4 binding may involve hydrophobic interactions between the lipid moiety and the hydrophobic region of the cyclic peptide and a hydrophilic interaction between the carbohydrate and the hydrophilic part of the peptide, such that the carbohydrate sequence is "distorted" to expose the internal Galα1–4Gal residue for VT binding.

These studies are based only on solid-phase binding assays in the presence of nonphysiological modifying agents. However, they demonstrate the potential to alter carbohydrate conformation and thereby alter receptor-binding function. It may well be that this precedent set *in vitro* may have an *in vivo* counterpart such that the conformation of carbohydrates of glycolipids within the membrane of eukaryotic cells may be liable to "distortion" or alteration when the glycolipid is associated with other amphipathic components (e.g., membrane proteins) in the plasma membrane. This would add another degree of complexity to cell surface receptors and provide a novel mechanism by which carbohydrate receptor function might be regulated.

E. Site-Specific Mutagenic Modification of Binding Specificity

While VT1 (and VT2) can only bind to Gb_4 under such artifactual conditions as described above, one member of the VT family is indeed able to recognize the native configuration of Gb_4.

The pig edema toxin (VT2e) is only found in strains of *E. coli* which colonize pigs. The B subunit of this toxin is 85% homologous with VT2 (Newland *et al.*, 1987) and antibodies against these toxins do show some degree of cross-neutralization (Marques *et al.*, 1987; MacLeod and Gyles, 1990). A difference in cytotoxicity between these VTs was noted. VT2e showed reduced cytotoxicity for HeLa cells (Marques *et al.*, 1987; Gannon *et al.*, 1988). Although the homology between VT2 and VT2e is higher than that between VT2 and VT1, the glycolipid binding specificity of VT2e is different from that of VT2, in that VT2e preferentially recognizes Gb_4 (DeGrandis *et al.*, 1989; Samuel *et al.*, 1990). Significant, but reduced binding to Gb_3 is also observed. HeLa cells have only Gb_3, whereas vero cells have both Gb_3 and Gb_4. The high degree of homology between the B subunits of these two toxins and their different glycolipid binding specificities have allowed us to pinpoint some of the amino acids which play a role in the discrimination between Gb_3 and Gb_4 binding. Site-specific mutagenesis was used to alter those amino acids of the VT2e B subunit which were different from those in VT2 and VT1 (Tyrrell *et al.*, 1992). The amino acid was changed to that found in VT2. Similarly, amino acids conserved in VT1 and VT2 but different in VT2e were altered in VT1 by site-directed mutagenesis to the VT2e residue. These amino acids were mutated singly, in pairs, and, in some instances, in triplets and the effect on glycolipid binding specificity was determined (Tyrrell *et al.*, 1992). Previous site-specific mutagenesis studies on the Shiga toxin B subunit had established that changing the aspartic acids at positions 16 and 17 to histidine resulted in the complete loss of glycolipid binding specificity (Jackson *et al.*, 1990). Chemical mutagenesis of VT2 identified three residues in the B subunit important for Gb_3 binding (Perera *et al.*, 1991), the arginine at position 32, alanine at 42, and glycine at 59. Exchanging the single tryptophan residue at position 33 for glutamine resulted in a loss of receptor binding, but replacement with tyrosine had no effect (Jackson *et al.*, 1990).

In the mutagenesis studies of our group (Tyrrell *et al.*, 1992), alteration of the aspartic acid at position 18 of VT1 to asparagine found in VT2e resulted in the generation of a toxin able to bind both Gb_3 and Gb_4. [This mutation was also performed with the holotoxin by O'Brien's group with no reported effect, but an assay to detect Gb_4 binding was not employed (Jackson *et al.*, 1990).] Other mutations had no effect on glycolipid binding. This thus changed the glycolipid binding phenotype of VT1 to that of VT2e, providing the first example in which an alteration in the carbohydrate binding specificity of a ligand was achieved by a single amino acid substitution. Several mutations of VT2e were of significance. Changing the isoleucine at position 57 to the lysine found in VT1 resulted in the complete loss of glycolipid binding. However, this may have in addition resulted in a defect in the assembly of the toxin, since probable degradation products of the A subunit were detected. This suggests that the A subunit of the mutant toxin is hypersensitive to proteolytic digestion, perhaps due to a failure of subunits to assemble correctly. VT2e has two additional amino acids at the C-terminal end of the B subunit. When the five C-terminal amino acids were removed by mutagenesis, the resulting truncated toxin was also unable to bind glycolipid. Truncation of Shiga toxin by five amino acids at the C-terminus was previously shown to cause toxin inactivation (Jackson *et al.*, 1990).

The most interesting mutant was a double mutation in which amino acids Gln-64 and Lys-66 were changed to their corresponding VT1 residues (Glu and Gln, respectively). This was found to result in the selective loss of Gb_4 binding, while Gb_3 binding remained intact (enhanced in fact). This mutant toxin, termed GT3, was subsequently purified and a detailed analysis of its glycolipid binding specificity and *in vitro* cytotoxicity carried out. Cytotoxicity was found to correlate with the ability to bind Gb_3 alone. Cell lines containing Gb_4 but little Gb_3, e.g., HEp2 cells, were found to be sensitive to the wild-type toxin but resistant to GT3 (Tyrrell *et al.*, 1992). The amino acid changes which primarily affected glycolipid binding in VT2e were for the most part localized in the C-terminal region of the B subunit (GT3 and the truncation mutant).

These studies are consistent with the recent solution of the crystal structure for the VT1 B subunit pentamer (Stein *et al.*, 1992). In this structure, similar to the structure of the *E. coli* heat-labile enterotoxin (Sixma *et al.*, 1991), the interaction between the B subunits to maintain the pentameric form is via the C-terminal region and a β-sheet of the adjacent B subunit. This cleft between two B subunits was hypothesized to be the glycolipid binding site. Furthermore, the triple aspartic acid sequence at residues 31 to 34 is found near the top of this putative glycolipid-binding cleft. The availability of the crystal structure will aid in selecting future mutations and provide a unique model for determination of the molecular basis of carbohydrate recognition.

In addition, when injected into pigs, the mutant toxin was found to show an altered tissue distribution which in turn was found to result in alteration of the sites of tissue damage (Boyd *et al.*, 1993). Most notably, red blood cells were found to bind extensively to the wild-type VT2e, whereas no or little binding was detected for GT3 and VT1, suggesting that Gb_4 on pig red cells is an effective ligand, whereas Gb_3 is masked. Binding of the mutant to gastrointestinal tissue was reduced, as were the gastrointestinal lesions. Binding to brain tissue was increased, which correlated with edema at neurological sites not affected by the wild-type toxin. These results clearly demonstrate that the glycolipid binding specificity determines the site of tissue damage following infection with VT2e-producing *E. coli*.

IV. Role of Gb_3 in VT Cytotoxicity

A. Gb_3 VERSUS VT SENSITIVITY

It was not initially apparent that the binding specificity of VT for Gb_3 observed using the TLC overlay procedure was relevant to the mechanism of VT-induced cytotoxicity for cells *in vivo* and in culture. Early studies had provided evidence for a glycoprotein receptor for Shiga toxin (Keusch *et al.,* 1986), and kinetic evidence supported the existence of different classes of receptors with different affinities for toxin (Jacewicz *et al.,* 1989). Nevertheless, Gb_3 was found to be present in high concentrations in vero cells and HeLa cells used for the routine assay of VT cytotoxicity (Lingwood *et al.,* 1987). Gb_3 was shown to be a major component of the neutral glycolipid fraction of the human kidney (Boyd and Lingwood, 1989), and levels were found to be higher in the cortex than in the medulla (Boyd and Lingwood, 1989), correlating with the observed higher incidence of pathology at this site in HUS. However, analysis of the Gb_3 content as a function of age showed that levels of Gb_3 in the infant and elderly kidney were, if anything, lower than that found in the adult. Thus, the level of Gb_3 did not correlate with the higher prevalence of HUS following VTEC infection observed for infants and the elderly.

The role of Gb_3 in VT-induced cytotoxicity for cells in culture was most easily addressed. Initially, mutant Daudi (human Burkitt's lymphoma) cells (Cohen *et al.,* 1987) and, later, vero cells (Pudymaitis *et al.,* 1991), were selected for resistance for VT cytotoxicity *in vitro*. The Daudi cell mutant (VT20) selected for resistance to 20 ng of VT/ml was found to be 100,000-fold more resistant to VT than the wild-type Daudi cell *in vitro* (Cohen *et al.,* 1987). This correlated with a greater than 95% selective deletion in the level of Gb_3 in this mutant cell line. Similarly, a selected VT-resistant vero cell line was found to lack Gb_3 completely (Pudymaitis *et al.,* 1991). Such mutant cell lines offered the opportunity to investigate the ability of Gb_3 alone to mediate the cytopathology of VT. Daudi VT20 mutant cells were reconstituted with exogenous purified Gb_3 using a liposomal fusion technique (Waddell *et al.,* 1990). VT20 cells were unable to bind and internalize toxin as monitored by fluorescence-activated cell sorting. Cells reconstituted with Gb_3 were able to bind (though not to the same extent as wild-type Daudi cells) and internalize exogenous toxin when incubated at 37°C. Such cells became sensitive once again to the cytotoxic effects of added VT1. Cells reconstituted with the glycolipid DGDG, containing the galabiosyl carbohydrate sequence, but attached to a glycerolipid, which does not bind VT *in vitro*, remained resistant to VT cytopathology. Gb_3 was similarly incorporated into a human fetal B cell line which lacked detectable Gb_3 and was highly resistant to VT *in vitro*. Once again, the Gb_3 incorporated cells became sensitive to VT killing. Interestingly, cells in which Gb_4 was incorporated remained resistant to VT1, indicating that Gb_4 cannot function as a receptor to mediate the cytopathology of VT1. These studies clearly demonstrate that Gb_3 alone is able to mediate the cytopathology of VT *in vitro*. A similar conclusion was obtained in recent studies using a cell line in which Gb_3 synthesis was inducible in the presence of butyric acid (Sandvig *et al.*, 1991).

In the rabbit model, development of intestinal sensitivity to Shiga toxin during ontogeny was found to correlate precisely to the time when Gb_3 was first detectable in this tissue (Mobassaleh *et al.*, 1988). However, the situation is different in the human gastrointestinal tract in that Gb_3 is not present in the mucosal epithelial cells (Holgersson *et al.*, 1991; Falk *et al.*, 1979). Although glycolipid containing the Galα1–4Gal sequence was not detected chemically, human intestinal epithelium is reactive with a monoclonal anti-Gb_3 (Oosterwijk *et al.*, 1991). Earlier studies, however (Kasai *et al.*, 1985), had failed to detect intestinal epithelial anti-Gb_3 reactivity. It may be that systemic toxin is a result of some noncytotoxic and, as yet, unidentified mechanism of mucosal transit, resulting in subsequent damage to the gastrointestinal vasculature.

B. P Blood Group Status

It has also been postulated that P blood group status might determine susceptibility to HUS, in much the same way P blood group influences susceptibility to urinary tract infections with uropathogenic *E. coli* (Lomberg and Svanborg Eden, 1989) which also bind to Galα1–4Gal residues. The P blood group comprises the p^k antigen (Gb_3), P antigen (Gb_4), and P_1 antigen (Ln_5-Galα1–4Galβ–4GlcNAcβ1–3Galβ1–4GlcCer). Both p^k and P_1 are bound by VT (Lindberg *et al.*, 1987). Epidemiological evidence however, indicates a negative correlation for susceptibility to HUS, in that individuals containing the P_1 blood group antigen show lower incidence of HUS than p individuals lacking Galα1–4Gal terminal glycolipids or P_2 individuals who lack P_1 antigen (Taylor *et al.*, 1990). It was suggested that the P_1 blood group determinants on circulating erythrocytes might protect such individuals from renal damage by adsorbing any circulating VT following VTEC infection and preventing attachment to glomerular receptors. While this is an attractive hypothesis, we have been unable to demonstrate significant ^{125}I-labeled VT binding to any human red blood cells irrespective of blood group status; thus, the biochemical basis of the epidemiological observation remains obscure.

C. Toxin Internalization

The cytopathology of VT is mediated by A subunit inhibition of protein synthesis at the ribosome (Saxena *et al.*, 1989). The A subunit alone has no effect on cells and thus cytotoxicity is a function of the ability of the B subunit to translocate the toxin into receptor positive cells. This translocation is achieved by *receptor-mediated endocytosis* from clathrin-coated pits (Sandvig *et al.*, 1989). For lymphoid cells, we have found that VT internalization is preceded by surface receptor patching and subsequent capping. In the standard format for this process (Smythe and Warren, 1991), cell surface receptors are cross-linked by multivalent ligand binding to form discrete patches on the cell surface. This process is temperature dependent, not occurring at 4°C, but is not energy dependent. Such patches are then capped to a single cell pole at 37°C via an energy-dependent mechanism. During receptor-mediated endocytosis, the receptor-bound ligand accumulates in clathrin-coated pits which then internalize to form vesicles which target the GERL. For protein receptors, the clustering in coated pits has been shown to involve organization of submembrane adaptor elements, mediated via the cytoplasmic domain of the transmembrane receptor (Smythe and Warren, 1991).

Shiga toxin was the first lipid receptor binding ligand shown to be endocytosed via coated pits (Sandvig *et al.*, 1989, 1991). In contrast, cholera toxin, which binds G_{M1} ganglioside, was found to be internalized via noncoated pits (Montesano *et al.*, 1982). Capping and endocytosis of lipid receptors into coated pits cannot occur via direct transmembrane signaling. It is likely that either directional lipid flow (Bretscher, 1976) or glycolipid association with transmembrane proteins provides the mechanism under such circumstances. We have found that VT1 capping, internalization, and cell cytotoxicity *in vitro* are prevented by dansyl cadaverine (Khine and Lingwood, 1993), an inhibitor of clathrin-mediated endocytosis (Davies *et al.*, 1980) which prevents the action of cytosolic transglutaminase, suggesting that the latter mechanism is operational.

V. Physiology of Gb_3

Glycolipid function is a poorly understood cellular process. Changes occur during cell differentiation (Hakomori, 1986), growth, and neoplastic transformation (Hakomori and Kannagi, 1983), but no overall pattern can be discerned. This may be due in part to the lack of sufficiently selective probes to perturb such functions. Antiglycolipid antibodies are usually of the IgM class and therefore of low affinity.

G_{M1} ganglioside is perhaps the best studied glycolipid in this field and such studies are greatly facilitated by the availability of cholera toxin to target this glycolipid (Hansson *et al.*, 1977). An alternative viewpoint is that this glycolipid is targeted by the *Vibrio cholera* organism because it is involved in unique pathways of normal cell physiology. I would suggest that the same rationale applies for VT and that Gb_3 is targeted by VTEC because of the unique role played by this glycolipid in certain aspects of cellular physiology.

A. Cell Growth

VT can be used to target Gb_3 to study the role of this glycolipid in various normal processes of cellular physiology. As mentioned above for fibroblastic cells, growing cells are more sensitive to VT than stationary phase cells. Vero cell sensitivity to VT was found to be reduced by more than 10-fold for stationary phase as opposed to log phase cell cultures (Pudymaitis and Lingwood, 1992). Human endothelial cells in the stationary phase were found to be resistant to Shiga toxin (Obrig *et al.*, 1988). An additional variation in sensitivity to VT was found for vero cells during the cell cycle. Synchronized cell cultures of vero cells were found to be 10-fold more sensitive to the cytopathology of this toxin at the G_1/S boundary of the cell cycle (Pudymaitis and Lingwood, 1992). Metabolic labeling of Gb_3 at this point was also enhanced, as was the exposure of Gb_3 on the plasma membrane as monitored by galactose oxidase and fluorescent toxin labeling. In contrast, the total cellular level of Gb_3 remained essentially unchanged during the cell cycle. This indicates that modulation of Gb_3 organization within the plasma membrane occurs as a function of cell cycle progression. The G_1/S boundary is a major point of control within the cell cycle (Norbury and Nurse, 1989), when environmental stimuli can determine whether or not a cell will enter S and divide. Perhaps Gb_3 is involved in the transduction of such signals.

In this regard, it is of interest to note that Gb_3 is specifically elevated in several cancers when compared to the normal counterpart (Li *et al.*, 1986; Mannori *et al.*, 1990) and has been proposed as a marker for testicular tumors (Ohyama *et al.*, 1990). Surface exposure of Gb_3 can vary vastly (Wiels *et al.*, 1984) and has been correlated with metastatic potential (Mannori *et al.*, 1990; Junqua *et al.*, 1989).

While Gb_3 cannot be essential for growth, it may modify growth parameters. A VT-resistant Gb_3-deficient clone of vero cells was isolated and found to exhibit a remarkably different growth pattern as compared to the parental cell line (Pudymaitis and Lingwood, 1992). Growth rate was reduced and, on reaching confluency, the Gb_3-deficient mutant showed a cobblestone morphology unlike the flat integrated monolayer seen for wild-type vero cells.

It was interesting to note that this resistant cell clone contained no α-galactosyltransferase activity, which explained the lack of Gb_3 in these cells (Pudymaitis *et al.*, 1991). Such a phenotype is similar to that of fibroblasts from p blood group individuals which lack Gb_3 and detectable α-galactosyltranferase activity (Kijimoto-Ochiai *et al.*, 1977). In contrast, the VT-resistant mutant selected from Daudi lymphoma cells, although similarly defective in Gb_3, showed wild-type levels of α-galactosyltransferase activity *in vitro* (Pudymaitis *et al.*, 1991). This is similar to the phenotype of lymphocytes from p individuals which also contain no Gb_3 yet express normal levels of α-galactosyltransferase activity (Iizuka *et al.*, 1986). It is therefore possible that alternative mechanisms for thc regulation of Gb_3 synthesis exist in lymphocytes as compared to fibroblastoid cells. Such a hypothesis is consistent with a role for Gb_3 in human B cell differentiation.

B. Human B Cell Differentiation

Human B cell precursors undergo a complex differentiation pathway to acquire immunocompetence to produce antibody. Much of this process has been defined at the genetic level, and the availability of monoclonal antibodies to cell differentiation (CD) antigens, which are sequentially expressed on the surface of B cells during this developmental process, has greatly added to the understanding of this pathway (Knapp *et al.*, 1989). It was shown early on that globo series glycolipids were primarily restricted to human lymphocytes of the B cell series (Schwarting, 1980; Lee *et al.*, 1981), and several of these monoclonal antibodies against differentiation antigens turned out to recognize globo series glycolipids, primarily Gb_3 (Fyfe *et al.*, 1987), which has been named CD77 (Knapp *et al.*, 1989; Mangeney *et al.*, 1991).

Gb_3 was also determined to be the antigen reactive with a monoclonal antibody, raised against the Daudi human lymphoma cell line (Wiels *et al.*, 1981), which defined the BLA (Burkitt lymphoma antigen) proposed as a marker of early B cell differentiation (Balana *et al.*, 1985). This antigen also defined a unique subset of normal germinal center human B lymphocytes (Mangeney *et al.*, 1991; Murray *et al.*, 1985; Gregory *et al.*, 1987). A monoclonal antibody against a differentiation antigen appearing later in B cell ontogeny was found to be specific for Gb_4 (Madassery *et al.*, 1991). Thus, the development of human B cells may require a sequential expression of longer chain globo series glycolipids (Wiels *et al.*, 1991).

Independent of these studies, we had been determining the effect of VT on human B cells. We had originally shown that Daudi cells were highly sensitive to this toxin (Cohen *et al.*, 1987) and were interested to determine the spectrum of sensitivity in other lymphoid cells and cell lines. Our studies showed that, *in vitro,* VT was able to prevent B cell, but not T cell, responses (Cohen *et al.*, 1990). Interestingly, we found that a subset of human B cells, those committed to the production of IgG as opposed to IgM, were primarily sensitive to VT. These results suggested that patients suffering from these infections may be, at least temporarily, partially immunocompromised. Indeed, toxin-neutralizing antibodies found in convalescent human sera are restricted to the IgM class (Keusch *et al.*, 1976), consistent with a selective loss of IgG-producing cell clones. The mechanism of VT-induced cytotoxicity to Gb_3-containing human B lymphocytes is of particular interest. VT cytotoxicity is generally mediated by the A subunit which inhibits protein synthesis by an endoRNA-*N*-glycohydrolase activity (Saxena *et al.*, 1989). However, one of the earliest events which could be detected following treatment of human B lymphocytes with VT was the activation of an endogenous endonuclease which cleaved DNA into discrete fragments, giving a ladder of DNA fragments when separated by agarose gel electrophoresis (Mangeney *et al.*, 1993). Such an activity is characteristic of apoptosis, or programmed cell death, in which a series of events encoded within the cell result in the self-destruction of the cell. This process requires *de novo* protein synthesis, as cells can be protected from apoptosis by inhibitors of protein synthesis. It had been previously suggested that Gb_3 was a marker for B cells capable of undergoing apoptosis (Mangeney *et al.*, 1991) and that this provided a mechanism for the selection of high-affinity-antibody producing clones.

Interestingly, this effect on the induction of apoptosis by VT1 was mimicked by the VT1 B subunit alone. For vero cells, the holotoxin is approximately nine orders of magnitude more cytotoxic than is the VT1 B subunit alone. However, for activation of apoptosis in human B cells, the B subunit is only 100-fold less effective than the holotoxin (Mangeney *et al.*, 1993). These studies indicate that binding to cell surface Gb_3 itself in these cells is sufficient to activate the apoptosis pathway and is consistent with a specific role for Gb_3 in this process.

C. Role of Gb_3 in α2-Interferon Signal Transduction

α2-Interferon is produced by lymphocytes following viral infection and is involved in the regulation of differentiation of these cells. Several genes contain upstream interferon-responsive elements (Williams, 1991), most notably that encoding 2′5′-ADP synthase (Lengyel, 1982), the induction of which is often used as a marker of α2-interferon sensitivity. In addition, such cells are growth inhibited in the presence of higher interferon concentrations. Susceptibility of cells in culture to growth inhibition by α2-interferon was found to correlate with susceptibility to VT cytotoxicity. Mutant cells selected for resistance to VT and deficient in Gb_3 were found to be cross-resistant to α2-interferon (Cohen *et al.*, 1987). Such cells had loss the high-affinity component in the Scatchard analysis of interferon binding. This component was thought to mediate growth inhibition by α2-interferon since most cells contain low-affinity interferon-binding sites (Grossberg *et al.*, 1989). It was therefore postulated that an interaction between the interferon receptor and Gb_3 in the plane of the plasma membrane resulted in a conformational change in the interferon receptor which generated a higher affinity for exogenous α2-interferon and in turn mediated growth inhibition (Cohen *et al.*, 1987). With the cloning of the human α2-interferon receptor (Uzé *et al.*, 1990), it became possible to determine whether there was a structural basis for such a hypothetical interaction.

Comparison of the amino acid sequence of the B subunit with the sequence of the α2-interferon receptor identified three regions near the N-terminus of striking homology (Lingwood and Yiu, 1992) which may indeed provide the structural basis for a glycolipid binding, "allosteric" site on the α2-interferon receptor. The homologous regions include an RWN sequence, the tryptophan residue of which had been previously hypothesized from infrared binding studies (Surewicz *et al.*, 1989) to reside within the VT1 Gb_3-binding site, and the DDN sequence of VT, which site-specific mutagenesis studies have also implicated in the binding specificity (Tyrrell *et al.*, 1992; Jackson *et al.*, 1990). In the α2-interferon receptor sequence these amino acid sequences are contiguous, whereas in the VT1 B subunit they are separated by some 10 amino acids. However, analysis of the recent crystal structure of the VT B subunit shows that these two regions of amino acid sequences are in fact closely apposed in VT1 B subunit and at the head of the putative glycolipid binding cleft between the B subunit monomers (Stein *et al.*, 1992); they could thus provide the structural basis of the postulated interaction between the α2-interferon receptor and Gb_3 (Cohen *et al.*, 1987; Lingwood and Yiu, 1992).

Thus, the α2-interferon receptor may be modulated via a lateral interaction with Gb_3 to effect signal transduction. Further evidence to support this pathway is provided by our finding that Gb_3-deficient Daudi cell mutants, resistant to growth inhibition by α2-interferon, are resensitized to α2-interferon when such cells are reconstituted with exogenous Gb_3 (Maloney and Lingwood, unpublished results). In a manner similar to VT resensitization (Waddell *et al.*, 1990), sensitivity is not restored following reconstitution with Gb_4.

In addition, recent studies in Fish's laboratory have established that Gb_3 deficient cells are defective in nuclear factor binding to interferon response gene elements (Ghislain *et al.*, 1992).

The role of membrane Gb_3 in α2-interferon signaling may relate to its function as a differentiation antigen in human B cell ontogeny (Fyfe *et al.*, 1987; Gregory *et al.*, 1987, 1988; Schwartz-Albeiz *et al.*, 1991). B cells are responsive to β-interferon (Sehgal *et al.*, 1987), which shares the α2-interferon receptor (Pestka *et al.*, 1987), and α2-interferon induction of B cell lymphoma differentiation has been demonstrated *in vitro* (Exley *et al.*, 1987a,b). Other lymphokines may, however, be more relevant. TNF potentiates VT cytotoxicity on endothelial cells (Tesh *et al.*, 1991; Barrett *et al.*, 1990; van de Kar, *et al.*, 1992) by increasing Gb_3 synthesis, and plays a role in regulation of lymphocyte cytokines (Camussi *et al.*, 1991). Perhaps the α2-interferon/Gb_3 interaction provides a precedent for the transduction of other signals involved in the regulation of the immune response.

In conclusion, the study of VT/Gb_3 binding includes a wide spectrum of disciplines since it provides an opportunity to study the molecular basis of carbohydrate recognition, the influence of the lipid moiety on glycolipid receptor function, the role of glycolipid binding in tissue targeting and toxin pathogenesis, and the function of Gb_3 in B cell development, endothelial physiology, α2-interferon signal transduction, and cell growth. Receptor binding has also allowed the rapid diagnosis of VTEC infection and may yet provide the basis of novel modalities for the treatment of such infections.

ACKNOWLEDGMENTS

I would like to thank Dr. Maloney for his critical review of this manuscript. Studies from our laboratory discussed in this review have been supported by Canadian MRC Program Grant # 11123.

References

Ashkenazi, S., and Cleary, T. G. (1989). *J. Clin. Microbiol.* **27**(6), 1145–1150.

Balana, A., Wiels, J., Tetaud, C., Mishal, Z., and Tursz, T. (1985). *Int. J. Cancer* **36,** 453–460.

Barrett, T. J., Potter, M. E., and Wachsmuth, I. K. (1989). *J. Infect. Dis.* **159,** 774–777.

Barrett, T. J., Potter, M. E., and Strockbine, N. A. (1990). *Microb. Pathog.* **9,** 95–103.

Basta, M., Karmali, M., and Lingwood, C. A. (1989). *J. Clin. Microbiol.* **127**(7), 1617–1622.

Bitzan, M., Wiebe, M., Martino, T., Huang, C., Petrič, M., Karmali, M., and Richardson, S. (1992). *Am. Soc. Microbiol. Annu. Meet., New Orleans,* p. 76.

Bock, E., Breimer, M. E., Brignole, A., Hansson, G. C., Karlsson, K.-A., Larson, G., Leffler, H., Samuelsson, B. E., Strömberg, N., and Svanborg Edén, C. (1985). *J. Biol. Chem.* **260**(14), 8545–8551.

Boulanger, J., Petrič, M., Lingwood, C. A., Law, H., Roscoe, M., and Karmali, M. (1990). *J. Clin. Microbiol.* **28**(12), 2830–2833.

Boyd, B., and Lingwood, C. A. (1989). *Nephron* **51,** 207–210.

Boyd, B., Tyrrell, G., Maloney, M., Gyles, C., Brunton, J., and Lingwood, C. A. (1993). *J. Exp. Med.* (in press).

Bretscher, M. S. (1976). *Nature (London)* **260,** 21–23.

Calderwood, S. B., Auclair, F., Donohue-Rolfe, A., Keusch, G. T., and Mekalanos, J. J. (1987). *Proc. Natl. Acad. Sci. U.S.A.* **84,** 4364–4368.

Camussi, G., Albano, E., Tetta, C., and Bussolini, F. (1991). *Eur. J. Biochem.* **202,** 3–14.

Cohen, A., Hannigan, G. E., Williams, B. R. G., and Lingwood, C. A. (1987). *J. Biol. Chem.* **262**(35), 17088–17099.

Cohen, A., Madrid-Marina, V., Estrov, Z., Freedman, M., Lingwood, C. A., and Dosch, H.-M. (1990). *Int. Immunol.* **2,** 1–8.

Cory, H. T., Yates, A., Donald, A., Watkins, W., and M., W. T. (1974). *Biochem. Biophys. Res. Commun.* **61,** 1289–1296.

Davies, P. J., Davies, D. R., Levitski, A., Maxfield, F. R., Milhaud, P., Willingham, M., and Pastan, I. (1980). *Nature (London)* **283,** 162–167.

DeGrandis, S. A., Ginsberg, J., Toone, M., Climie, S., Friesen, J., and Brunton, J. (1987). *J. Bacteriol.* **169,** 4313–4319.

DeGrandis, S. A., Law, H., Brunton, J., Gyles, C., and Lingwood, C. A. (1989). *J. Biol. Chem.* **264,** 12520–12525.

Dickie, N., Speirs, J. I., Akhtar, M., Johnson, W. M., and Szabo, R. A. (1989). *J. Clin. Microbiol.* **27**(9), 1973–1978.

Donohue-Rolfe, A., Keusch, G. T., Edson, C., Thorley-Lawson, D., and Jacewicz, M. (1984). *J. Exp. Med.* **160,** 1767–1781.

Donohue-Rolfe, A., Kelley, N. A., Bennish, M., and Keusch, G. T. (1986). *J. Clin. Microbiol.* **24**(1), 65–68.

Donohue-Rolfe, A., Acheson, D. W. K., Kane, A. V., and Keusch, G. T. (1989). *Infect. Immun.* **57**(12), 3888–3893.

Downes, F. P., Barrett, T. J., Green, J. H., Aloisio, C. H., Spika, J. S., Strockbine, N. A., and Wachsmuth, I. K. (1988). *Infect. Immun.* **56**(8), 1926–1933.

Eiklid, K., and Olsnes, S. (1980). *J. Recept. Res.* **1**(2), 199–213.

Endo, Y., Tsurugi, K., Yutsudo, T., Takeda, Y., Ogasawara, T., and Igarashi, K. (1988). *Eur. J. Biochem.* **171,** 45–50.

Exley, R., Gordon, J., and Clemens, M. J. (1987a). *Proc. Natl. Acad. Sci. U.S.A.* **84,** 6467–6470.

Exley, R., Gordon, J., Nathan, P., Walker, L., and Clemens, M. J. (1987b). *Int. J. Cancer* **40,** 53–57.

Falk, K.-E., Karlsson, K.-A., Leffler, H. J., and Samuelsson, B. E. (1979). *FEBS Lett.* **101**(2), 273–276.

Fontaine, A., Arondel, J., and Sansonetti, P. J. (1988). *Infect. Immun.* **56**(12), 3099–3109.

Fyfe, G., Cebra-Thomas, J. A., Mustain, E., Davie, J. M., Alley, C. D., and Nahm, M. H. (1987). *J. Immunol.* **139,** 2187–2194.

Gannon, V. P. J., Gyles, C. L., and Friendship, R. W. (1988). *Can. J. Vet. Res.* **52,** 331–337.

Ghislain, J., Lingwood, C., Maloney, M., Penn, L., and Fish, E. (1992). *J. Interferon Res.* **12,** 5114.

Gregory, C. D., Turz, T., Edwards, C. F., Tetaud, C., Talbot, M., Caillou, B., Rickenson, A. B., and Lipinski, M. (1987). *J. Immunol.* **139**(1), 313–318.

Gregory, G. D., Edwards, C. F., Milner, A., Wiels, J., Lipinski, M., Rowe, M., Tursz, T., and Rickenson, A. B. (1988). *Int. J. Cancer* **42,** 213–220.

Grossberg, S. E., Taylor, J. L., and Kushnaryov, V. M. (1989). *Experientia* **45,** 508–513.

Gyles, C. L., DeGrandis, S. A. D., MacKenzie, C., and Brunton, J. L. (1988). *Microb. Pathog.* **5,** 419–426.

Hakomori, S. (1986). *Sci. Am.* **245**(5), 40–53.

Hakomori, S., and Kannagi, R. (1983). *JNCI, J. Natl. Cancer Inst.* **71,** 231–251.

Hansson, H. A., Holmgren, J., and Svennerholm, L. (1977). *Proc. Natl. Acad. Sci. U.S.A.* **74,** 3782–3786.

Head, S. C., Karmali, M. A., Roscoe, M. E., Petrič, M., Strockbine, N. A., and Wachsmuth, I. K. (1988a). *Lancet* **1,** 751.

Head, S. C., Petrič, M., Richardson, S. E., Roscoe, M. E., and Karmali, M. (1988b). *FEMS Microbiol Lett.* **51,** 211–216.

Head, S. C., Richardson, S., Petrič, M., and Karmali, M. (1988c). *Am. Soc. Microbiol. Annu. Meet.,* p. 39, Miami, FL, Abst B61.

Head, S. C., Ramotar, K., and Lingwood, C. A. (1990). *Infect. Immun.* **58,** 1532–1537.

Head, S. C., Karmali, M., and Lingwood, C. A. (1991). *J. Biol. Chem.* **266**(6), 3617–3621.

Holgersson, J., Jovall, P.-A., and Breimer, M. E. (1991). *Biochem. J.* **110,** 120–131.

Iizuka, C., Chen, S.-H., and Yoshida, A. (1986). *Biochem. Biophys. Res. Commun.* **137**(3), 1187–1195.

Ito, H., Terai, A., Kurazono, H., Takeda, Y., and Nishibuchi, M. (1990). *Microb. Pathog.* **8,** 47–60.

Jacewicz, M., Clausen, H., Nudelman, E., Donohue-Rolfe, A., and Keusch, G. T. (1986). *J. Exp. Med.* **163,** 1391–1404.

Jacewicz, M., Feldman, H. A., Donohue-Rolfe, A., Balasubramanian, K. A., and Keusch, G. T. (1989). *J. Infect. Dis.* **159**(5), 881–889.
Jackson, M. P., Neill, R. J., O'Brien, A. D., Holmes, R. K., and Newland, J. W. (1987). *FEMS Microbiol. Lett.* **44,** 109–114.
Jackson, M. P., Wadolkowski, E. A., Weinstein, D. L., Holmes, R. K., and O'Brien, A. D. (1990). *J. Bacteriol.* **172**(2), 653–658.
Junqua, S., Larsen, A. K., Wils, P., Mishal, Z., Wiels, J., and Le Pecq, J.-B. (1989). *Cancer Res.* **49,** 6480–6486.
Karmali, M. A. (1987). *Clin. Microbiol. Newsl.* **9,** 65–70.
Karmali, M. A. (1989). *Clin. Microbiol. Rev.* **2,** 15–38.
Karmali, M. A., Petrič, M., Lim, C., Cheung, R., and Arbus, G. S. (1985a). *J. Clin. Microbiol.* **22,** 614–619.
Karmali, M. A., Petrič, M., Lim, C., Fleming, P. C., Arbus, G. S., and Lior, H. (1985b). *J. Infect. Dis.* **151,** 775–782.
Kasai, K., Galton, J., Terasaki, P., Wakisaka, A., Kawahara, M., Root, T., and Hakomori, S.-I. (1985). *J. Immunogenet.* **12,** 213–220.
Keusch, G., Jacewicz, M., Levine, M., Hornick, R., and Kochwa, S. (1976). *J. Clin. Invest.* **57,** 194.
Keusch, G. T., Jacewicz, M., and Donohue-Rolfe, A. (1986). *J. Infect. Dis.* **153**(2), 238–248.
Khine, A. A., and Lingwood, C. A. (1993). Submitted for publication.
Kijimoto-Ochiai, S., Naiki, M., and Makita, A. (1977). *Proc. Natl. Acad. Sci. U.S.A.* **74**(12), 5407–5410.
Knapp, W., Dörken, P., Rieber, P., Schmidt, R. E., Stein, H., and von dem Borne, A. E. (1989). *Int. J. Cancer* **44**(1), 190–191.
Kongmuang, U., Honda, T., and Miwatani, T. (1987). *J. Clin. Microbiol.* **25**(1), 115–118.
Konowalchuk, J., Spiers, J., and Stajric, S. (1977). *Infect. Immun.* **18,** 775–779.
Lee, W., Klock, J., and Macher, B. (1981). *Biochemistry* **20,** 3810–3814.
Lengyel, P. (1982). *Annu. Rev. Biochem.* **51,** 251–282.
Li, S.-C., Kundu, S. K., Degasperi, R., and Li, Y.-T. (1986). *Biochem. J.* **240,** 925–927.
Lindberg, A. A., Brown, J. E., Strömberg, N., Westling-Ryd, M., Schultz, J. E., and Karlsson, K.-A. (1987). *J. Biol. Chem.* **262**(4), 1779–1785.
Lindstedt, R., Baker, N., Falk, P., Hull, R., Hull, S., Karr, J., Leffler, H., Svanborg Edén, C., and Larson, G. (1989). *Infect. Immun.* **57**(11), 3389–3394.
Linggood, M. A., and Thompson, J. M. (1987). *J. Med. Microbiol.* **24,** 359–362.
Lingwood, C. A. (1993). *Nephron.* (in press).
Lingwood, C. A., and Yiu, S. C. K. (1992). *Biochem J.* **283**(1), 25–26.
Lingwood, C. A., Law, H., Richardson, S., Petrič, M., Brunton, J. L., DeGrandis, S., and Karmali, M. (1987). *J. Biol. Chem.* **262**(18), 8834–8839.
Lomberg, H., and Svanborg Edén, C. (1989). *FEMS Microbiol. Immunol.* **47,** 363–370.
Louise, C. B., and Obrig, T. G. (1991). *Infect. Immun.* **59**(11), 4173–4179.
Lund, B., Lindberg, F., Marklund, B.-I., and Normark, S. (1987). *Proc. Natl. Acad. Sci. U.S.A.* **84,** 5898–5902.
Lund, B., Marklund, B.-I., Strömberg, N., Lindberg, F., Karlsson, K.-A., and Normark, S. (1988). *Mol. Microbiol.* **2**(2), 255–263.
MacLeod, D. L., and Gyles, C. L. (1990). *Infect. Immun.* **58,** 1232–1239.
Madassery, J. V., Gillard, B., Marcus, D. M., and Nahm, M. H. (1991). *J. Immunol.* **147**(3), 823–829.
Magnani, J. L., Smith, D. F., and Ginsburg, V. (1980). *Anal. Biochem.* **109,** 399–402.
Makita, A. (1964). *J. Biochem. (Tokyo)* **55**(3), 269–276.
Mangeney, M., Richard, Y., Coulaud, D., Tursz, T., and Wiels, J. (1991). *Eur. J. Immunol.* **21,** 1131–1140.
Mangeney, M., Lingwood, C. A., Caillou, B., Taga, S., Tursz, T., and Wiels, J. (1993). Submitted for publication.

Mannori, G., Cecconi, O., Mugnai, G., and Ruggieri, S. (1990). *Int. J. Cancer* **45,** 984–988.
Marques, L. R. M., Peiris, J. S. M., Cryz, S. J., and O'Brien, A. D. (1987). *FEMS Microbiol. Lett.* **44,** 33–38.
Martensson, E. (1966). *Biochim. Biophys. Acta* **116,** 296–308.
Mobassaleh, M., Donohue-Rolfe, A., Jacewicz, M., Grand, R. J., and Keusch, G. T. (1988). *J. Infect. Dis.* **157,** 1023–1031.
Montesano, R., Roth, J., Robert, A., and Orci, L. (1982). *Nature (London)* **296,** 651–653.
Murray, L. J., Habeshaw, J. A., Wiels, J., and Greaves, M. F. (1985). *Int. J. Cancer* **36,** 561–565.
Newland, J. W., Strockbine, N. A., Miller, S. F., O'Brien, A. D., and Holmes, R. K. (1985). *Science* **230,** 179–181.
Newland, J. W., Strockbine, N. A., and Neill, R. J. (1987). *Infect. Immun.* **55**(11), 2675–2680.
Norbury, C. J., and Nurse, P. (1989). *Biochim. Biophys. Acta* **989,** 85–95.
O'Brien, A. D., and LaVeck, G. D. (1983). *Infect. Immun.* **40,** 675–683.
Obrig, T., Barley-Maloney, L., Louise, C., Boyd, B., Lingwood, C., and Daniel, T. (1993). *J. Biol. Chem.* (in press).
Obrig, T. G., Moran, T. P., and Brown, J. E. (1987). *Biochem. J.* **244,** 287–294.
Obrig, T. G., Vecchio, P. J. D., Brown, J. E., Moran, T. P., Rowland, B. M., Judge, T. K., and Rothman, S. W. (1988). *Infect. Immun.* **56,** 2373–2378.
Ohyama, C., Fukushi, Y., Satoh, M., Saitoh, S., Orikasa, S., Nudelman, E., Straud, M., and Hakomori, S.-I. (1990). *Int. J. Cancer* **45,** 1040-1044.
Oosterwijk, E., Kalisiak, A., Wakka, J., Scheinberg, D., and Old, L. J. (1991). *Int. J. Cancer* **48,** 848–854.
Pascher, I., and Sundell, S. (1977). *Chem. Phys. Lipids* **20,** 175–191.
Pellizzari, A., Pang, H., and Lingwood, C. A. (1992). *Biochemistry* **31,** 1363–1370.
Perera, L. P., Marques, L. R. M., and O'Brien, A. D. (1988). *J. Clin. Microbiol.* **26**(10), 2127–2131.
Perera, L. P., Samuel, J. E., Holmes, R. K., and O'Brien, A. D. (1991). *J. Bacteriol.* **173**(3), 1151–1160.
Pestka, S., Langer, J. A., Zoon, K. C., and Samuel, C. E. (1987). *Annu. Rev. Biochem.* **56,** 727–777.
Petrič, M., Karmali, M. A., Richardson, S., and Cheung, R. (1987). *FEMS Microbiol Lett.* **41,** 63–68.
Pudymaitis, A., and Lingwood, C. A. (1992). *J. Cell. Physiol.* **150,** 632–639.
Pudymaitis, A., Armstrong, G., and Lingwood, C. A. (1991). *Arch. Biochem. Biophys.* **286**(2), 448–452.
Ramotar, K., Boyd, B., Gariepy, J., Lingwood, C. A., and Brunton, J. (1990). *Biochem J.* **272,** 805–811.
Richardson, S. E., Karmali, M. A., Becker, L. E., and Smith, C. R. (1988). *Hum. Pathol.* **19**(9), 1102–1108.
Richardson, S. E., Rotman, T., Jay, V., Smith, C., Becker, L., Petric, M., Olivieri, N., and Karmali, M. (1992). *Infect. Immun.* **60,** 4154–4167.
Riley, L. W., Remis, R. S., Helgerson, S. D., McGee, H. B., Wells, J. G., Davis, B. R., Herbert, R. J., Olcott, E. S., Johnson, L. M., Hargrett, N. T., Blake, P. A., and Cohen, M. C. (1983). *N. Engl. J. Med.* **308,** 681–685.
Samuel, J. E., Perera, L. P., Ward, S., O'Brien, A. D., Ginsburg, V., and Krivan, H. C. (1990). *Infect. Immun.* **58**(3), 611–618.
Sandvig, K., Olnes, S., Brown, J., Peterson, O., and van Deurs, B. (1989). *J. Cell Biol.* **108,** 1331–1343.
Sandvig, K., Prydz, K., Ryd, M., and van Deurs, B. (1991). *J. Cell Biol.* **113**(3), 553–562.
Saxena, S. K., O'Brien, A. D., and Ackerman, E. J. (1989). *J. Biol. Chem.* **264**(1), 596–601.
Schmitt, C. K., McKee, M. L., and O'Brien, A. D. (1991). *Infect Immun.* **59**(3), 1065–1073.
Schwarting, G. A. (1980). *Biochem. J.* **189,** 407–412.
Schwartz-Albeiz, R., Dorken, B., Moller, P., Brodin, N. T., Monner, D. A., and Kniep, B. (1991). *Int. Immunol.* **2**(10), 929–936.

Scotland, S. M., Smith, H. R., and Rowe, B. (1985). *Lancet* **2,** 885–886.

Sehgal, P. B., May, L. T., Tamm, I., and Vilcek, J. (1987). *Science* **235,** 731–732.

Shivensky, L., Knowles, B., Damjanoy, I., and Solter, D. (1982). *Cell (Cambridge, Mass.)* **30,** 697–705.

Sixma, T. K., Pronk, S. E., Kalk, K. H., Wartna, E. S., van Zanten, B. A. M., Witholt, B., and Hol, W. G. J. (1991). *Nature (London)* **351,** 371–378.

Smythe, E., and Warren, G. (1991). *Eur. J. Biochem.* **202,** 689–699.

Stein, P. E., Boodhoo, A., Tyrell, G. J., Brunton, J. L., and Read, R. J. (1992). *Nature (London)* **355,** 748–750.

Strockbine, N. A., Marques, L. R. M., Holmes, R. K., and O'Brien, A. D. (1985). *Infect. Immun.* **50**(3), 695–700.

Strockbine, N. A., Jackson, M. P., Sung, L. M., Holmes, R. K., and O'Brien, A. D. (1988). *J. Bacteriol.* **170**(3), 1116–1122.

Strömberg, N., Marklund, B.-I., Lund, B., Ilver, D., Hamers, A., Gaastra, W., Karlsson, K.-A., and Normand, S. (1990). *EMBO J.* **9**(6), 2001–2010.

Strömberg, N., Nyholm, P.-G., Pascher, I., and Normark, S. (1991). *Proc. Natl. Acad. Sci. U.S.A.* **88,** 9340–9344.

Surewicz, W. K., Surewicz, K., Mants, H. H., and Auclair, F. (1989). *Biochem. Biophys. Res. Commun.* **160,** 126–132.

Takao, T., Tanabe, T., Hong, Y.-M., Shimonishi, Y., Kurazono, H., Yutsudo, T., Sasakawa, C., Yoshikawa, M., and Takeda, Y. (1988). *Microb. Pathog.* **5,** 357–369.

Taylor, C. M., Milford, D. V., Rose, P. E., Roy, T. C. F., and Rowe, B. (1990). *Pediatr. Nephrol.* **4,** 59–61.

Tesh, V. L., Samuel, J. E., Perera, L. P., Sharefkin, J. B., and O'Brien, A. D. (1991). *J. Infect. Dis.* **164,** 344–352.

Tyrrell, G. J., Ramotar, K., Lingwood, C. A., and Brunton, J. L. (1992). *Proc. Natl. Acad. Sci. U.S.A.* **89,** 524–528.

Uzé, G., Lutfalla, G., and Gresser, I. (1990). *Cell (Cambridge, Mass.)* **60,** 225–234.

van de Kar, N. C. A. J., Monnens, L. A. H., Karmali, M., and van Hinsbergh, V. W. M. (1992). *Blood* **80,** 2755–2764.

Waddell, T., Head, S., Petrič, M., Cohen, A., and Lingwood, C. A. (1988). *Biochem. Biophys. Res. Commun.* **152**(2), 674–679.

Waddell, T., Cohen, A., and Lingwood, C. A. (1990). *Proc. Natl. Acad. Sci. U.S.A.* **87,** 7898–7901.

Wadolkowski, E. A., Sung, L. M., Burris, J. A., Samuel, J. E., and O'Brien, A. D. (1990). *Infect. Immun.* **58**(12), 3959-3965.

Wiels, J., Fellous, M., and Tursz, T. (1981). *Proc. Natl. Acad. Sci. U.S.A.* **78,** 6485–6488.

Wiels, J., Holmes, E. H., Cochran, N., Tursz, T., and Hakomori, S.-I. (1984). *J. Biol. Chem.* **259**(23), 14783–14787.

Wiels, J., Mangeney, M., Tetaud, C., and Tursz, T. (1991). *Int. Immunol.* **3,** 1289–1300.

Wieruszeski, J.-M., Michalski, J.-C., Montreuil, J., and Strecker, G. (1987). *J. Biol. Chem.* **262**(14), 6650–6657.

Williams, B. R. G. (1991). *J. Interferon Res.* **11,** 207–213.

Yiu, S. C. K., and Lingwood, C. A. (1992). *Anal. Biochem.* **202,** 188–192.

Zoja, J. C., Corna, D., Farina, C., Sacchi, G., Lingwood, C. A., Doyles, M. P., Padhye, V. V., Abbate, M., and Remuzzi, G. (1992). *J. Lab. Clin. Med.* **120,** 229–238.

Glycosphingolipid Tumor Antigens

PAM FREDMAN

Department of Psychiatry and Neurochemistry
University of Göteborg
Mölndals sjukhus
S-43180 Mölndal, Sweden

I. Introduction

Glycosphingolipid alterations associated with virus-induced transformation of cells *in vitro* were reported during the late 1960s and a large number of tumor-associated glycosphingolipids in tumor cells and tissues, in cell lines established from tumors, and in transformed cells have since been reported. These tumor-associated glycosphingolipids have attracted a great deal of interest in tumor research, in particular with regard to the potential role of tumor-associated glycosphingolipid antigens as a target for monoclonal antibodies in the diagnosis and therapy of cancers. The role of glycosphingolipids has not been fully elucidated, but it has been suggested that they are involved in cellular events that may play an important biological role in tumor cell growth and in the invasive and metastatic properties of the tumor.

Established tumor cell lines are important in studies concerning the localization and treatment effects of monoclonal antibodies against tumor-associated antigens and in studies of the biological function of these antigens. However, the use of established cell lines has its limitations, as the glycosphingolipid expression *in vitro* often differs from that *in vivo*. The structures and expression of glycosphingolipid antigens in human tumors and cell lines established from them, their biological function in tumors, and their potential as targets for monoclonal antibodies in diagnosis and therapy are discussed here.

II. Glycosphingolipid Tumor Antigens

The glycosphingolipids are divided into the globo, lacto, neolacto, and ganglio series, depending on the composition of their terminal sugars (Fig. 1). The globo series is characterized by having a galactose (Gal) in an α1–4 linkage to galactose as the third sugar moiety. The lacto and neolacto series, also referred to as the type 1 and 2 lacto series, are both characterized by glucosamine linked β1–3 to the galactose as third sugar, but they differ in the linkage of the terminal galactose, linked β1–3 and β1–4, respectively. In the ganglio series, the third sugar moiety is galactosamine linked β1–4 to galactose instead. These are the core structures of the individual series of glycosphingolipids to which one or more sugar residues can be added (reviewed by Wiegandt, 1985; Alhadeff, 1989). Fucosylation and sialylation of glycosphingolipids in tumors are common and the glycosphingolipid tumor antigens in Tables I–IV are divided into nonsubstituted, fucosylated, sialylated and fucosylated, and sialylated only groups.

The expression of glycosphingolipids is related to the cell type and to cell differentiation. They contribute to the structural rigidity of the surface of the plasma cell membrane and are involved in transmembrane signaling, in cellular contacts, and in contacts between cells and matrices (for review, see Hakomori, 1981, 1990; Svennerholm, 1984; Fenderson *et al.*, 1990). Altered glycosylation patterns might thus be of importance in uncontrolled growth and metastatic and invasive properties of tumor cells. Changed glycosylation upon transformation was first demonstrated by Hakomori and Murakami (1968) and Mora *et al.*

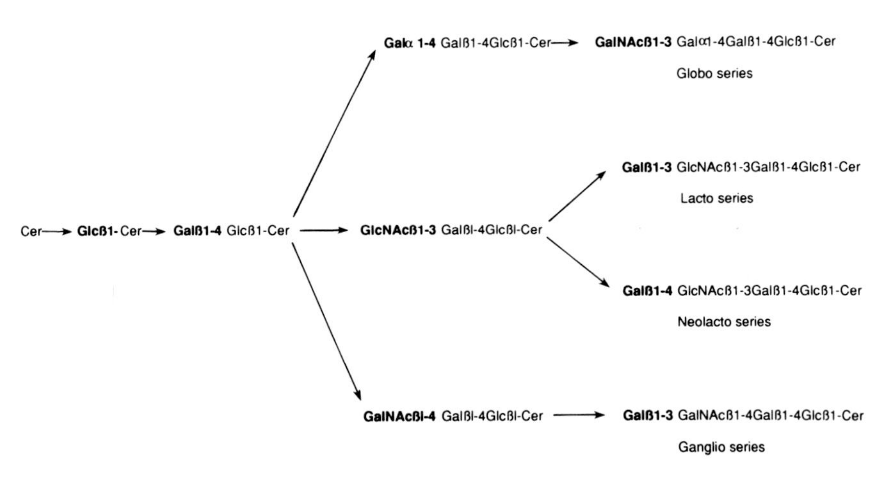

FIG. 1. Carbohydrate composition in the glycosphingolipid series: the globo, lacto, neolacto, and ganglio series. Glc, Glucose; Gal, galactose; GlcNAc, glucosamine; GalNAc, galactosamine; Cer, ceramide.

(1969), and was followed by several similar studies that have been reviewed previously (Hakomori, 1973; Brady and Fishman, 1974). The clear identification of tumor-associated glycosphingolipid antigens was based (1) initially on studies using rabbit antiserum against Gg3 (asialo-G_{M2}): specific accumulation of this antigen in tumors was demonstrated by immunochemical and immunohistological methods in murine sarcoma virus-induced tumors grown in Balb/c mice (Rosenfelder *et al.*, 1977); (2) subsequently on studies using monoclonal anti-Gg3 antibody with murine T cell lymphoma grown in C57BL mice (Young and Hakomori, 1981). Since then, a large number of glycosphingolipids associated with tumor tissues and tumor cell lines have been characterized (for review, see Hakomori, 1989; Alhadeff, 1989). None of the tumor-associated glycosphingolipid structures reported to date has been tumor specific, i.e., exclusively expressed in tumor cells. Glycosphingolipids considered to be tumor-associated antigens are (1) those absent or not detected with immunological methods in their progenitor cells or in the organ where the tumor is located, (2) those expressed in fetal but not adult tissue, or oncofetal antigens, and (3) those found in a significantly increased proportion and/or quantity in the tumors as compared to progenitor cells and/or surrounding tissue.

The glycosphingolipids show species specificity. This review therefore focuses on tumor-associated glycosphingolipids found in human tumors and in cell lines established from them, as only these glycosphingolipids are of interest in the biology of human malignancies. Human tumor-associated glycosphingolipids commonly described in the literature and their primary distributions are shown in Tables I–IV. In many cases, the occurrence of these antigens in other tumors and, in particular, in normal human tissues and cells cannot be excluded, as the tissue distribution has not been extensively investigated. Moreover, it should be noted that histopathological information in most studies is limited.

The majority of the tumor-associated glycosphingolipids are of the lacto or neolacto series (Tables I and II), which carry the epitopes for the blood group ABH system, the Lewis and the I-antigens. Le^a and Le^b of the lacto series (Table I) have been found to be accumulated in pancreatic and colon adenocarcinomas, and in addition they are often coexpressed (Hakomori and Andrews, 1970; Blaszczyk *et al.*, 1985; Tempero *et al.*, 1987). One of the most common antigens found to be detected by antibodies raised against various types of cancers is the Le^x determinant in the neolacto series (Table II) (Brockhaus *et al.*, 1982; Hansson *et al.*, 1983; Urdal *et al.*, 1983; Hakomori *et al.*, 1982). Le^x glycosphingolipid was initially found to be greatly accumulated in various human adenocarcinomas, and was suggested to be a tumor-associated antigen (Yang and Hakomori, 1971). The first antibody, defining Le^x, was produced by immunization with teratocarcinoma cells (Solter and Knowles, 1978) and the antigen, at that time unknown, was found to be developmentally regulated and therefore named "stage-specific embryonic antigen 1" (SSEA-1). Later, di- and trimeric

Table I
Glycosphingolipid Antigens of the Lacto Series (Type 1 Chain) in Human Tumors

Glycosphingolipid antigens	Designation	Tumor reactivity	References: Tumor reactivity	References: Production and specificity of monoclonal antibodies
Galβ1–3GlcNAcβ1–3Galβ1–R	L_{A1}	Embryonal carcinoma cells	Fukuda *et al.* (1986)	Rettig *et al.* (1985) Fukuda *et al.* (1986)
Fucosylated				
Galβ1–3GlcNAcβ1–3Galβ1–R (GlcNAc 4 — 1 Fucα)	Le^a Fuc-L_{A1}	 Colon adenocarcinoma* Pancreatic adenocarcinoma*	Hakomori and Andrews (1970) Blaszczyk *et al.* (1985) Tempero *et al.* (1987)	Blaszczyk *et al.* (1985) Mårtensson *et al.* (1988)
Galβ1–3GlcNAcβ1–3Galβ1–R (Gal 2 — 1 Fucα; GlcNAc 4 — 1 Fucα)	Le^b	 Colon adenocarcinoma* Pancreatic adenocarcinoma*	Hakomori and Andrews (1970) Blaszczyk *et al.* (1985) Tempero *et al.* (1987)	Brockhaus *et al.* (1981) Blaszczyk *et al.* (1985)
Sialylated				
Galβ1–3GlcNAcβ1–3Galβ1–R (Gal 3 — 2 NeuAcα)	3-isoL_{M1}	Small cell lung carcinoma Glioma Medulloblastoma Teratocarcinoma cells	Nilsson *et al.* (1985a) Fredman *et al.* (1988b) Gottfries *et al.* (1990) Fukuda *et al.* (1986)	Nilsson *et al.* (1985a) Rettig *et al.* (1985) Fredman *et al.* (1990a) Wikstrand *et al.* (1991)
Galβ1–3GlcNAcβ1–3Galβ1–R (Gal 3 — 2 NeuAcα; GlcNAc 6 — 2 NeuAcα)	3′,6′-isoL_{D1}	Liver metastasis of colon cancer Embryonal carcinoma cells	Fukushi *et al.* (1986) Fukuda *et al.* (1986)	Fukushi *et al.* (1986) Wikstrand *et al.* (1991)

Structure	Name	Tissue	Reference	Reference
Fucosylated and sialylated				
Galβ1–3GlcNAcβ1–3Galβ1–R (NeuAcα2–3 on Gal; Fucα1–4 on GlcNAc)	Fuc-3′-isoL$_{M1}$ Sialylated Le^a	Gastrointestinal cancer	Koprowski *et al*. (1979) Magnani *et al*. (1982) Chia *et al*. (1985)	Koprowski *et al*. (1979) Magnani *et al*. (1982) Chia *et al*. (1985)
		Pancreatic adenocarcinoma	Månsson *et al*. (1985)	
Galβ1–3GlcNAcβ1–3Galβ1–R (NeuAcα2–3 on Gal; NeuAcα2–6 and Fucα1–4 on GlcNAc)	Fuc-3′,6′-isoL$_{D1}$ Disialylated Le^a	Human colon adeno-carcinoma	Nudelman *et al*. (1986)	Nudelman *et al*. (1986)

*Le^a and Le^b are coexpressed. R, 4Glcβ1-ceramide.

Table II

GLYCOSPHINGOLIPID ANTIGENS OF THE NEOLACTO SERIES (TYPE 2 CHAIN) IN HUMAN TUMORS

Glycosphingolipid antigens*	Designation	Tumor reactivity	References: Tumor reactivity	References: Production and specificity of monoclonal antibodies
Galβ1–4GlcNAcβ1–3Galβ1–R	nL_{A1}	Colon cancer	Myoga *et al.* (1988)	Myoga *et al.* (1988)
Galβ1–4GlcNAcβ1–[3Galβ1–4GlcNAcβ1]$_n$–3Galβ1–R	i-Antigen	Lung cancer	Hirohashi *et al.* (1986)	Hirohashi *et al.* (1986)
Fucosylated				
Galβ1–4GlcNAcβ1–3Galβ1–R	SSEA-1	Teratocarcinoma	Solter and Knowles (1978)	Brockhaus *et al.* (1982)
3	Le^x	Colon cancer		Magnani *et al.* (1982)
| 1	Fuc-nL_{A1}			Fukushi *et al.* (1984a)
Fucα				Urdal *et al.* (1983)
Galβ1–4GlcNAcβ1–3Galβ1–4GlcNAcβ1–3Galβ1–R	Dimeric Le^x	Adenocarcinoma	Fukushi *et al.* (1984a,b)	Fukushi *et al.* (1984b)
3 3		Colon cancer		
| 1 | 1		Liver cancer	Hakomori *et al.* (1984)	
Fucα Fucα				
Galβ1–4GlcNAcβ1–3Galβ1–R	Le^y	Gastric cancer	Abe *et al.* (1983)	Abe *et al.* (1983)
2 4		Breast cancer	Brown *et al.* (1983)	
| 1 | 1		Colon cancer	Lloyd *et al.* (1983)	
Fucα Fucα				
Sialylated				
Galβ1–4GlcNAcβ1–3Galβ1–R	6′-L_{M1}	Colorectal carcinoma	Nilsson *et al.* (1985b)	Nilsson *et al.* (1985b)
6		Lung carcinomas		
| 2		Primary hepatoma	Taki *et al.* (1990)	Taki *et al.* (1990)
NeuAcα				
Fucosylated and sialylated				
Galβ1–4GlcNAcβ1–3Galβ1–R	Sialylated Le^x	Gastrointestinal cancer	Fukushima *et al.* (1984)	Fukushima *et al.* (1984)
3 3	or	Lung carcinoma		
| 2 | 1	Fuc-3′-L_{M1}			
NeuAcα Fucα				
Galβ1–4GlcNAcβ1–3Galβ1–4GlcNacβ1–3Galβ1–R		Gastric, colon, lung, breast, and renal cancers	Fukushi *et al.* (1985)	Fukushi *et al.* (1984b)
3 3 3				
| 2 | 1 | 1				
NeuAcα Fucα Fucα				

*R, 4Glcβ1-ceramide.

X-determinants were also found to be accumulated in human adenocarcinoma (Hakomori *et al.*, 1984).

Sialylated glycosphingolipids, gangliosides of the lacto, 3′-isoL_{M1}, 3′,6′-isoL_{D1} (Table I), and neolacto series, 6′-L_{M1}, (Table II) are found in various tumors. Ganglioside 3′-isoL_{M1} was first isolated and characterized from small cell lung carcinoma (Nilsson *et al.*, 1985a) and was later isolated from malignant gliomas (Fredman *et al.*, 1988b) and medulloblastomas (Gottfries *et al.*, 1990). Its disialylated form, 3′,6′-isoL_{D1}, has been isolated from colonic adenocarcinoma (Fukushi *et al.*, 1986), from embryonal carcinoma cells (Fukuda *et al.*, 1986), and has recently also been immunohistochemically detected in human gliomas (Wikstrand *et al.*, 1991). Ganglioside 6′-L_{M1} was detected in various carcinomas, but particularly accumulated in colorectal and lung carcinomas (Nilsson *et al.*, 1985b). Taki *et al.* (1990) found this antigen to be accumulated in human hepatoma.

Another group of tumor-associated glycosphingolipids in the lacto and neolacto series are both fucosylated and sialylated. Fuc-3′-isoL_{M1} and Fuc-3′,6′-isoL_{D1} (Table I), also referred to as mono- and disialylated Le^a, respectively, have been found to be highly expressed in adenocarcinomas of colon and pancreas and in adenocarcinomas of several other organs (Koprowski *et al.*, 1979; Magnani *et al.*, 1982; Månsson *et al.*, 1985). Within the neolacto series, Fuc-3′-L_{M1}, or sialylated Le^x, has been found in a variety of tumor tissues but in the highest frequency in stomach and colon cancers and in lung carcinoma (Fukushima *et al.*, 1984).

Fucosylated and/or sialylated forms of lacto and neolacto series glycosphingolipids are common among tumor antigens (Svennerholm, 1988; Hakomori, 1989; Alhadeff, 1989). However, there are only a few glycosphingolipid antigens that are neither fucosylated nor sialylated. These are L_{A1} in the lacto series (Table I) and nL_{A1} and the i-antigen in the neolacto series (Table II). The L_{A1} glycosphingolipid was isolated from human embryonal cancer cells (Fukuda *et al.*, 1986) and the two latter, nL_{A1} and the i-antigen, from colon and lung cancers, respectively (Myoga *et al.*, 1988; Hirohashi *et al.*, 1986).

The tumor-associated gangliosides in the ganglio series, listed in Table III, include G_{M3} and G_{D3}, reviewed by Hakomori (1989). These gangliosides do not contain the core structure that qualifies them as belonging to this series—they lack the galactosamine residue attached to the galactose—but they are included as they are precursors in the synthesis of the ganglio series gangliosides (Fig. 2). An alternative is to assign these gangliosides to the lactosyl series, as the core structure is lactosylceramide (Svennerholm, 1988). Human melanoma is the most extensively studied human tumor in regard to ganglioside antigens (reviewed by Tsuchida *et al.*, 1987a; Ravindrath and Irie, 1988; Ravindrath and Morton, 1991; Ritter and Livingston, 1991), and all of the ganglio series antigens in Table III, with the exception of Fuc-G_{M1}, have been described as melanoma-associated antigens. However, as for most glycolipid antigens, these are not restricted to melanoma. G_{M3} has been found in increased proportions in

Table III

GLYCOSPHINGOLIPID ANTIGENS OF THE GANGLIO SERIES IN HUMAN TUMORS

Glycosphingolipid antigens*	Designation	Tumor reactivity	References: Tumor reactivity	References: Production and specificity of monoclonal antibodies
GalNAcβ1–4Galβ1–R 3 | 2 NeuAcα	G_{M2}	Melanoma	Irie *et al.* (1982)	Tai *et al.* (1983)
			Tai *et al.* (1983)	Natoli *et al.* (1986)
			Tsuchida *et al.* (1987a,b)	Fredman *et al.* (1989)
		Glioma	Fredman *et al.* (1986b)	Vrionis *et al.* (1989)
		Germ cell tumors	Miyake *et al.* (1990)	
GalNAcβ1–4Galβ1–R 3 | 2 NeuAcα2–8NeuAcα	G_{D2}	Melanoma	Cahan *et al.* (1982)	Cahan *et al.* (1982)
		Neuroblastoma	Schulz *et al.* (1984)	Cheung *et al.* (1985)
			Wu *et al.* (1986)	Bosslet *et al.* (1989)
			Cheung *et al.* (1985)	Longee *et al.* (1991)
		Small cell lung carcinoma	Cheresh *et al.* (1986a)	
		Glioma	Fredman *et al.* (1986b)	
Galβ1–R 3 | 2 NeuAcα	G_{M3}	Melanoma	Tsuchida *et al.* (1987a,b)	Wakabayashi *et al.* (1984)
			Yamamato *et al.* (1990)	Hirabayashi *et al.* (1986)
			Furukawa *et al.* (1989)	Yamaguchi *et al.* (1987)
Galβ1–R 3 | 2 NeuAcα2–8NeuAcα	G_{D3}	Melanoma	Pukel *et al.* (1982)	Pukel *et al.* (1982)
		Melanoma	Nudelman *et al.* (1982)	Nudelman *et al.* (1982)
		Medulloblastoma	Gottfries *et al.* (1990)	He *et al.* (1989)
		Glioma	Fredman *et al.* (1986a, 1988b)	Brodin *et al.* (1985)
				Yamaguchi *et al.* (1987)
		Leukemia	Siddique *et al.* (1984)	Furukawa *et al.* (1989)
		Meningioma	Davidsson *et al.* (1989)	
Galβ1–R 3 | 2 9-*O*-Acetyl-NeuAcα2–8NeuAcα	9-*O*-Acetyl-G_{D3}	Melanoma	Cheresh *et al.* (1984b,c)	Cheresh *et al.* (1984b)
Fucosylated and sialylated				
Galβ1–3GalNAcβ1–4Galβ1–R 2 (Galβ1)	1 Fucα 3 (Galβ1–R)	2 NeuAcα	Fuc-G_{M1}	Small cell lung carcinoma	Nilsson *et al.* (1986)	Fredman *et al.* (1986a)

*R, 4Glcβ1-Cer.

human melanoma (Ravindrath and Irie, 1988) and in gliomas (reviewed by Yates, 1988). G_{M3} is a ganglioside common to most types of cells and tissues and is the major ganglioside in human red blood cells. G_{D3} is also found in normal tissue, but the expression is significantly increased in melanomas (Pukel *et al.*, 1982), gliomas (Fredman *et al.*, 1986b), medulloblastomas (Gottfries *et al.*, 1990), meningiomas (Davidsson *et al.*, 1989), and in leukemia (Siddique *et al.*, 1984). The 9-*O*-acetylated forms of G_{D3} have also been found in melanomas (Cheresh *et al.*, 1984b,c) and have appeared to be highly restricted to these tumors. However, in a review by Ritter and Livingston (1991), the authors claim to have shown that the *O*-acetylated forms of G_{D3} have the same distribution as non-*O*-aceylated G_{D3}.

Ganglioside G_{M2} is found in melanomas (Tsuchida *et al.*, 1987a), in human gliomas (Fredman *et al.*, 1986b; Yates, 1988), and in a variety of other tumors (Irie *et al.*, 1982; Tai *et al.*, 1983; Miyake *et al.*, 1990). An increase in its disialylated form, G_{D2}, has been noted in melanoma (Cahan *et al.*, 1982), neuroblastoma (Schulz *et al.*, 1984; Wu *et al.*, 1986), small cell lung carcinoma (Cheresh *et al.*, 1986a), and gliomas (Fredman *et al.*, 1988b; Longee *et al.*, 1991), but was not detected in non-neuroectodermal tumors (Cahan *et al.*, 1982; Schulz *et al.*,

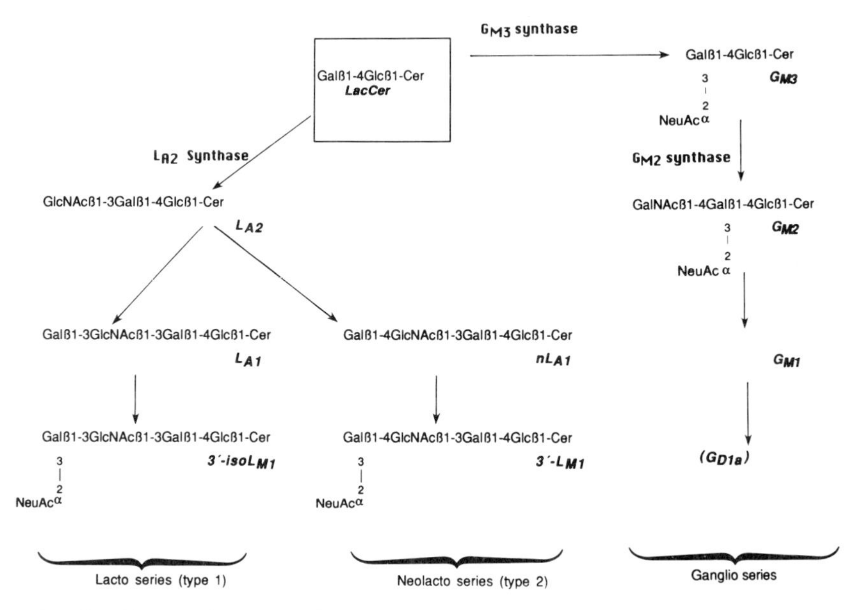

FIG. 2. Biosynthetic pathways for lacto, neolacto, and ganglio a and b series gangliosides. Enzymes: G_{M3} synthase (EC 2.4.99.9), G_{M2} synthase (EC 2.4.1.92). Glc, Glucose; Gal, galactose; GlcNAc, glucosamine; GalNAc, galactosamine; NeuAc, *N*-acetylneuraminic acid; Cer, ceramide.

1984). These results have led to the suggestion that G_{D2} is a marker for tumors of neuroectodermal origin. It has also been suggested that expression of G_{D2} is related to differentiation, as neuroblastoma contains a high proportion of G_{D2}, while the more differentiated forms of this tumor, ganglioneuroblastoma and ganglioneuronoma, contain little or no G_{D2} (Wu *et al.*, 1986).

The ganglioside antigen Fuc-G_{M1} is one of the more restricted tumor-associated glycosphingolipid antigens and has only been detected in small cell lung carcinoma (Nilsson *et al.*, 1986). It is not restricted to tumor tissue, as it is also found in small amounts in normal human brain tissue (Svennerholm *et al.*, 1989).

A few tumor-associated glycosphingolipids of the globo series have also been reported (Table IV). A relatively new group of tumor-associated glycosphingolipids are the sulfated ones. Recently, Kiguchi *et al.* (1992) reported that, among human ovarian tumors, the mucinous cystadenocarcinomas showed a unique high concentration of sulfated glycosphingolipids, the major one being sulfatide, I^3SO_3–GalCer. With the use of monoclonal antibody (SD1a) defining $GgOse_4CerII^3,IV^3$-disulfate and more complex sulfated glycosphingolipids, Hiraiwa *et al.* (1990) found these antigens to be increased in human hepatocellular carcinoma as compared to normal and cirrhotic liver.

N-Glycolylneuraminic acid-containing gangliosides, a "Hanganutziu-Deicher antigen," which has not been detected in normal human tissues or cells, have been reported to occur in various cancers (Higashi *et al.*, 1985; Kawai *et al.*, 1991). The detected quantity was low, comprising $<1\%$ of the total ganglioside sialic acid in colon cancer. Miyake *et al.* (1990) have also demonstrated that non-seminomatous differed from seminomatous germ cell tumors by the presence of *N*-glycolylneuraminic acid-containing G_{M2}. Investigation of human gliomas did not reveal any detectable *N*-glycolylneuraminic acid (Fredman, 1988), and most structural analyses of tumor-associated glycosphingolipids have not indicated the presence of this sialic acid. These results indicate that, except for the germ cell tumor, *N*-glycolylneuraminic acid-containing gangliosides, if present at all, are only minor components of human tumor tissue.

In summary, tumors of neuroectodermal origin mainly express ganglioside antigens of the ganglio series, while most tumor-associated glycosphingolipid antigens are of the lacto and neolacto series. There is, however, no definitive exclusion of one or the other antigens and not all tumors have been investigated regarding the expression of all tumor-associated glycosphingolipid antigens known today.

III. Monoclonal Antibodies to Tumor-Associated Glycosphingolipids

Antibodies against tumor antigens have been produced by immunization with tumor cells, tumor tissue extracts, and more or less purified antigens according

Table IV

OTHER GLYCOSPHINGOLIPID ANTIGENS, INCLUDING THE GLOBO SERIES, IN HUMAN TUMORS

Glycosphingolipid antigens*	Designation	Tumor reactivity	References	
			Tumor reactivity	Production and specificity of monoclonal antibodies
Globo series				
Galα1–4Galβ1–R	Gb3	Burkitt's lymphoma	Wiels *et al.* (1981)	Wiels *et al.* (1981)
Galβ1–3GalNAcβ1–3Galα1–4Galβ1–R (2 – 1 Fucα)	Globo-H	Breast cancer	Ménard *et al.* (1983)	Ménard *et al.* (1983) Bremer *et al.* (1984)
Sulfated				
SO_3–3Galβ1–1Cer	Sulfatide	Mucinous cyst-adenocarcinoma	Kiguchi *et al.* (1992)	Fredman *et al.* (1988a)
Galβ1–3GalNAcβ1–4Galβ1–R (3 – SO_3; 3 – SO_3)	Disulfated G_{A1}	Hepatocellular carcinoma	Hiraiwa *et al.* (1990)	Hiraiwa *et al.* (1990)
N-Glycolylneuraminic acid-containing gangliosides	*N*-Glycolylneuraminic acid	Colon cancer	Higashi *et al.* (1985) Kawai *et al.* (1991)	Higashi *et al.* (1985)
	N-Glycolyl-G_{M2}	*N*-Glycolyl-G_{M2}	Miyake *et al.* (1990)	Miyake *et al.* (1990)

*R, 4Glcβ1-ceramide.

to the procedure by Köhler and Milstein (1975). Many of these antibodies have not yet been characterized with regard to their epitope specificity, which often limits their usefulness. Antibodies to carbohydrate antigens have the advantage that they can be readily defined with regard to their binding epitope. Glycosphingolipids are often preferred as antigens for production and characterization of monoclonal antibodies to carbohydrates because they can be purified to carbohydrate homogeneity, in contrast to glycoproteins, which can show microheterogeneity. These antibodies are important tools for characterization of carbohydrate antigens, and are also invaluable for studies of the biodistribution and biological roles of glycosphingolipids in normal and neoplastic cells.

A large number of monoclonal antibodies to tumor-associated glycosphingolipids have been described, and a selection is presented in Tables I–IV. These antibodies define a specific sugar sequence epitope regardless of it being linked to a lipid, to a protein/peptide, or in a mucin (Hakomori, 1989; Alhadeff, 1989). Many tumor-associated carbohydrate epitopes of the lacto series, type 1 and 2, detected in serum of cancer patients have, for example, been shown to be associated with mucins and not glycosphingolipids (reviewed by Hakomori, 1989). The globo and the ganglio series epitopes, however, have so far been detected exclusively in glycosphingolipids.

The carbohydrate epitope defined by a monoclonal antibody is established by comparing its binding to various pure, structurally characterized glycosphingolipids. In general, the higher the number of sugar moieties included in the recognized epitope, the greater the specific reactivity. For example (Fredman *et al.*, 1986a), only 4 of 13 monoclonal antibodies reacting with ganglioside Fuc-G_{M1} (Table III) required both fucose, linked to the fourth sugar, and sialic acid, linked to the second sugar, for binding. The other 11 antibodies reacted with epitopes that were shared by structurally related glycosphingolipids. Several reacted with ganglioside G_{M1}, although with a lower affinity. Use of such a low-specificity monoclonal antibody might lead to incorrect conclusions when the antibody is used on tissues or cells where G_{M1} is a dominant antigen. It should be noted that, for use in diagnosis and therapy, antibodies with an unknown binding epitope, but with a high binding ratio to specific tumor cells as compared with other tumor cells and normal cells, might be useful (Levy and Miller, 1983; Wikstrand *et al.*, 1986), but in studies of the biological role of glycosphingolipids a high specificity to a defined carbohydrate epitope is a prerequisite for the use of monoclonal antibodies.

Some defined anti-ganglioside antibodies have been derived from human B cells. OFA-I-1 and OFA-I-2 were produced by transforming peripheral blood B lymphocytes of melanoma patients with Epstein-Barr virus (EBV). However, EBV-infected cells cannot be cloned, and a majority of them lose antibody production. By combining EBV transformation of human lymphocytes and fusion with mouse myeloma cells, Yamaguchi *et al.* (1987) have produced several human monoclonal antibodies to tumor antigens. Two of these (Furukawa *et al.*, 1989) reacted preferentially with gangliosides G_{M3} and G_{D3}.

Some antibodies have been shown to detect gangliosides with *N*-acetyl as well as *N*-glycolylneuraminic acid or both. In general, *N*-glycolylneuraminic acid-containing ganglioside antigens are, when detected, only minor components in human tumors (Higashi *et al.*, 1985; Miyake *et al.*, 1990). However, the expression of this sialic acid may be altered when tumors are cultivated *in vitro*. In gangliosides from D-54 MG human glioma cells, grown as solid tumors in nude mice, 90% of the sialic acid was the *N*-glycolyl form, although these cells grown *in vitro* only contained *N*-acetylneuraminic acid (Fredman, 1988). This may be species related, as the *in vitro* cultured cells were grown in media containing fetal calf serum. *N*-Glycolylneuraminic acid is the dominating sialic acid in mouse. The different growth state, two dimensional *in vitro* and three dimensional *in vivo*, has also to be considered.

The methods used for defining the carbohydrate epitope in glycosphingolipids are most commonly based on solid-phase immunoassays where the glycosphingolipids are adsorbed to wells of microtiter plates or chromatographed on thin-layer plates. The expression of the binding epitope for the monoclonal antibody might be very different from its expression in the cell membrane of the tumor cells. Preparation, fixation of cells, and, in particular, tissue specimens for immunohistochemical investigation, can have a strong influence on antigen expression. The affinity of a specific antibody to its antigen is thus related to the assay used. It has also been shown (Karlsson *et al.*, 1990) that the composition of the ceramide portion in G_{M2} strongly influences the affinity of specific monoclonal antibodies to the antigen in a given solid-phase assay. An aberrant ceramide composition of the tumor-associated antigen might thus increase the binding affinity of monoclonal antibodies to tumor cell antigen as compared to the same antigen in normal cells (Hakomori, 1975). The melanoma-associated ganglioside G_{D3} was found to have longer fatty acid chains than G_{D3} in normal brain (Nudelman *et al.*, 1982), and ganglioside G_{M2} in neuroblastoma was found to contain an increased proportion of α-hydroxyl fatty acids (Ladisch *et al.*, 1989). The density of antigen has also been implied to be an important factor for monoclonal antibody affinity (Nores *et al.*, 1987; Welt *et al.*, 1987).

IV. Expression of Tumor-Associated Glycosphingolipid in Tumor Cell Lines

Tumor biopsy or autopsy material is available only in very limited amounts, and the possibility of isolating glycosphingolipid or any other antigens is therefore restricted. Another factor limiting the use of tumor tissue specimens for biochemical analyses is that tumor tissue rarely consists of 100% tumor cells, but is mixed with various proportions of normal cells, including stroma, vessels, lymphocytes, and macrophages. This phenomenon is most pronounced in tumors like gliomas, that grow infiltratively without capsula or marked borders to the surrounding normal tissue. Thus, biochemical analyses of tumor tissues cannot in most cases be used to determine the tumor cell concentration or the cellular distribution of tumor-associated glycosphingolipids. Hersey and Jamal (1989) found that the melanoma-associated gangliosides G_{D3} and G_{D2} were also expressed on lymphocytes around melanoma cells.

These problems might be circumvented by the use of tumor cell lines established from human tumors. They provide a continuous source for antigen isolation and are also invaluable tools for studying the biological effect of monoclonal antibodies and the biodistribution and role of glycosphingolipid antigens in tumor cells. Many cell lines can also be grown as subcutaneous tumors in nude mice or rats. These tumors are also important experimental models for tumor-imaging studies with labeled antibodies. However, established tumor cell lines have several limitations, which have to be considered.

Tumors are heterogeneous regarding both genotypic and phenotypic characteristics both between and within individual tumors (Bigner *et al.,* 1981; Wikstrand *et al.,* 1985). The likelihood of finding one tumor-associated antigen common to all cells in a tumor is low. During establishment of a tumor cell line, one or a few of these clones, the most adaptable to *in vitro* culture, are selected (Bigner *et al.,* 1981), and several *in vivo* expressed antigens may not be found in the established cell lines. This heterogeneity has been shown to be reflected in ganglioside expression in various cell lines established from human glioma (Fredman, 1988) and melanoma (Tsuchida *et al.,* 1987a,b).

Many established cell lines can be grown as solid tumors in nude mice or rats, and the *in vivo* grown tumors demonstrate many characteristics of the original tumor (Schold *et al.,* 1982). The major ganglioside in the human glioma D-54 MG cell line grown in xenograft form in nude mice was 3′-isoL_{M1} of the neolacto series (Månsson *et al.,* 1986). However, in D-54 MG cells grown *in vitro,* the predominant ganglioside was G_{M2} of the ganglios series (Fredman, 1988), and the lacto series ganglioside 3′-isoL_{M1} could not be detected. Similar results were obtained with the human glioma cell line U-118MG (Fredman *et al.,* 1990a) and human medulloblastoma cell lines (Gottfries *et al.,* 1991), i.e., a switch from a high expression of the ganglio series *in vitro* to a high expression of the lacto series gangliosides, types 1 and 2, when transplanted to nude mice or rats (Fig. 2). As ganglioside 3′-isoL_{M1} had been detected in all malignant glioma (Fredman *et al.,* 1988b) and medulloblastoma (Gottfries *et al.,* 1990) biopsies analyzed, the ganglioside composition in the cell lines grown as solid tumors in nude mice or rats was more like the tumor tissue than that of the *in vitro* grown cells. In a recent study, 16 glioma cell lines were investigated with regard to the expression of 3′-isoL_{M1} and 3′,6′-isoL_{D1} antigens (Wikstrand *et al.*, 1991). Immunostaining of the ganglioside fractions from the cell lines showed that these gangliosides were increased in xenograft cells as compared to *in vitro* cultured cells.

Tsuchida *et al.* (1987b) studied the expression of melanoma-associated gangliosides of the cell line grown *in vitro* or as subcutaneous tumors in nude mice with that of the surgical specimen from which the cell line originated. The results in that study showed that the expression of gangliosides in the melanoma cells grown *in vitro* differed from those expressed in the tissue, a difference that was restored when the cells were grown as xenografts. In accordance with the studies on gliomas (Fredman *et al.*, 1990a) and medulloblastomas (Gottfries *et al.*, 1991), the melanoma cell lines grown *in vitro* showed higher expression of ganglioside G_{M2} than those grown as solid tumors.

Although the experimental design in those studies cannot exclude clonal selection, the results implied that the expression of tumor-associated glycosphingolipids is influenced by environmental factors. It had previously been shown that culture conditions might influence the composition of gangliosides in cells (Liepkalns *et al.*, 1981, 1983; Markwell *et al.*, 1984). Further support for the influence of environmental factors was given by the observation that the gangliosides from the human glioma D-54 MG cells grown as xenografts in nude mice consisted of 90% *N*-glycolylneuraminic acid, while the D-54 MG cells grown *in vitro* did not contain any detectable amounts of this sialic acid (Fredman, 1988).

The possible influence of genetic factors on the expression of tumor-associated glycosphingolipids is supported by several studies. In the first studies demonstrating altered glycosylation in cells transformed with simian virus 40 (SV-40) (Brady and Fishman, 1974; Hakomori, 1973, 1975), a lack of more complex gangliosides, including G_{M2}, was noted, and this was referred to as a blocked synthesis. These studies were performed on rodent cell lines. Recently, Hoffman *et al.* (1991) found that SV-40 transformation of rodent cells of both fibroblastic and neural origin resulted in an increased expression of G_{M3}, while transformation of corresponding human cells led to an increase in G_{M2}. These results (Hoffman *et al.*, 1991) also indicate that an increased content of ganglioside G_{M2} is a common finding in tumor cell lines of human origin grown *in vitro* (Fredman, 1988; Fredman *et al.*, 1990a; Tsuchida *et al.*, 1987a,b; Gottfries *et al.*, 1991).

The influence of genetic changes on ganglioside expression has also been demonstrated by transfection of oncogenes to the rat fibroblastic 3Y1 cell line (Sanai and Nagai, 1989). In c-*myc* transfected cells, where the proposed localization of the oncogene product is the nucleus, an increase of G_{D3} was noticed, while transfection with oncogenes with products localizing to the plasma cell membrane (v-*fes*, v-*ras*, v-*crs*, v-*fps*) led to an increase of ganglioside $3'L_{M1}$. In a recent study, we found that the expression of G_{D3} correlated to the loss of chromosome 22 in human meningiomas (Fredman *et al.*, 1990b).

In summary, these results show that the phenotypic expression of tumor-associated glycosphingolipid antigens in tumor cell lines depends not only on the genotype but is also strongly influenced by environmental factors. These alterations also emphasize the importance of investigating the expression of the tumor-associated glycosphingolipid antigen to be studied under the experimental conditions to be used.

Regardless of whether genetic and/or environmental factors are the cause, the aberrant glycosylation in tumor cells is the result of changes in the relative activity of the glycosyltransferases involved. Holmes and Hakomori (1983) showed that the accumulation of α-fucosyl-G_{M1} ($IV^2FucII^3NeuAcGgOse_4Cer$) and α-galactosyl-α-fucosyl-G_{M1} ($IV^3GalIV^2FucII^3NeuAcGgOse_4Cer$) in precancerous livers of rats fed chemical carcinogens was related to an increased activity of α-L-fucosyltransferase. More recently, they (Holmes *et al.*, 1987) demonstrated that the abundant expression of types 1 and 2 lacto series glycosphingolipids in human colonic adenocarcinoma was due to an activation of the normally unexpressed β1–3*N*-acetylglucosaminyltransferase (L_{A2} synthase in Fig. 2). In our study of human medulloblastoma cell lines (Gottfries *et al.*, 1991), the large proportion of the ganglio series ganglioside G_{M2} in the cells grown *in vitro* should have been the reflection of the high activity of G_{M3} synthase and/or G_{M2} synthase, and the increased proportion of lacto series gangliosides in the cells grown *in vivo* of a relatively higher activity of L_{A2} synthase (Fig. 2). Unexpectedly, no correlation between the ganglioside expression and the activity of the enzymes was detectable.

Although no glycosyltransferase so far has been shown to be the product of any oncogene, an increased activity of a transferase might be the result of genetic changes. Genetic alterations may produce products that inhibit or potentiate glycosylation. Other mechanisms that may be involved are altered intracellular transport, availability of substrates and enzymes within the Golgi, etc. (Schwartzmann and Sandhoff, 1990). These alterations may also be caused by environmental factors; factors that may be circumvented in the *in vitro* assays used.

V. Potential Biological Roles of Tumor-Associated Glycosphingolipid Antigens

The biological role of glycosphingolipids has not been completely elucidated, but there is evidence of their involvement in cell growth regulation and cell-to-cell and cell-to-matrix interactions (reviewed by Hakomori, 1981, 1990; Svennerholm, 1984; Feizi, 1991). The tumor-associated changes in glycosphingolipid composition in tumor cells may, therefore, play a role in tumor growth regulation and the invasive and metastatic properties of tumors. The potential of glycosphingolipids, and in particular of gangliosides, in diagnosis and therapy of tumors has been reviewed (Reisfeld and Cheresh, 1987; Herlyn and Koprowski, 1988; Ravindrath and Irie, 1988; Hakomori, 1989; Lloyd and Lloyd, 1989; Wong *et al.*, 1989; Ravindrath and Morton, 1991; Ritter and Livingston, 1991) and is not discussed here.

The role of glycosphingolipids in tumor cell growth *in vitro* is supported by the inhibitory effect of anti-glycosphingolipid antibodies on tumor cells. Monoclonal antibodies to the melanoma-associated ganglioside G_{D3} have been shown to cause lysis of melanoma cells *in vitro* (Dippold *et al.*, 1984; Cheresh *et al.*, 1985; Hellström *et al.*, 1985). G_{D3} expression varies between cell lines, and no growth inhibitory effect was observed in cell lines with little or no G_{D3} content (Dippold *et al.*, 1984; Welt *et al.*, 1987). Anti-G_{D3} monoclonal antibodies have also been shown to inhibit outgrowth of human melanoma xenografts in nude mice (Hellström *et al.*, 1985). Anti-G_{M2} monoclonal antibodies have been shown to cause necrosis in three-dimensional cultures (spheroids) of human glioma cells expressing this antigen (Bjerkvig *et al.*, 1991). Some cells survived the anti-G_{M2} treatment, and it was shown that these cells had a 50% lower proportion of G_{M2} than the original cell population, supporting the idea that a threshold number of antigens is required for antibody effect.

The involvement of specific tumor-associated glycosphingolipids in the attachment of melanoma cells to solid substrata was demonstrated by Cheresh *et al.* (1984a). They found that gangliosides G_{D2} and G_{D3} (Table III) attached to the cell surface of the melanoma cells and to their focal adhesion plaques, and that incubation of melanoma cells with anti-G_{D2} or anti-G_{D3} antibodies inhibited attachment to substratum (Cheresh *et al.*, 1986b). G_{D2} was also found to redistribute to the microprocesses during the cell attachment process (Cheresh and Klier, 1986).

VI. Summary and Conclusions

Tumor-associated glycosphingolipids are a common finding in human tumors, and monoclonal antibodies directed to defined carbohydrate epitopes of these structures have been developed. A majority of the glycosphingolipid antigens are of the lacto series, types 1 and 2, and these epitopes are often also found on glycoproteins. The tumor-associated gangliosides of the ganglio series, preferentially found in tumors of neuroectodermal origin, have binding epitopes that have not been detected in glycoproteins. A large number of monoclonal antibodies, mainly murine but some human, have been produced, and their epitope specificity defined. These antibodies have been used for investigation of the role of tumor-associated glycosphingolipids in tumor cells. Some antibodies have been shown to inhibit the growth of tumor cell lines *in vitro* and *in vivo* and to influence the adhesion properties of tumor cells, and some antibodies have been shown to give a clinical response. Tumor cell lines are commonly used in tumor biology studies, but the expression of tumor-associated glycosphingolipids in tumor cell lines does not always reflect their expression in the tumor tissue from which the cell line originates. The expression of tumor-associated glycosphingolipids has also been shown to be different in cells grown *in vitro* or as solid tumors in nude mice or rats.

In conclusion, monoclonal antibodies specific for tumor-associated glycosphingolipids may be valuable tools in diagnosis and, in addition, in therapy, glycosphingolipids may also be used in active immunotherapy. The availability of these antibodies also opens the possibility of studying the biological function of glycosphingolipids and may then lead to new approaches for the improved diagnosis and therapy of human tumors.

ACKNOWLEDGMENTS

I would like to express my gratitude to Professor Lars Svennerholm, Department of Psychiatry and Neurochemistry, University of Göteborg, Sweden, for introducing me to the field of gangliosides and for his encouragment throughout the years of our collaboration. I also wish to thank Professor Darell Bigner, M.D., Ph.D., and Dr. Carol Wikstrand for their review of the manuscript and Barbro Lundmark for invaluable expert secretarial work. The work cited herein from the author's laboratory was supported by grants from the Swedish Medical Research Council (03X-09909-01) and from NIH (303-8911).

References

Abe, K., McKibbin, J. M., and Hakomori, S.-I. (1983). *J. Biol. Chem.* **258,** 11793–11797.

Alhadeff, J. A. (1989). *CRC Crit. Rev. Oncol. Hematol.* **9,** 37–107.

Bigner, D. D., Bigner, S. H., Pontén, J., Westermark, B., Mahaley, M. S., Ruoslathi, E., Herschman, H., Eng, L. F., and Wikstrand, C. J. (1981). *J. Neuropathol. Exp. Neurol.* **40,** 201–229.

Bjerkvig, R., Engebraaten, O., Laerum, O. D., Fredman, P., Svennerholm, L., Vrionis, F. D., Wikstrand, C. J., and Bigner, D. D. (1991). *Cancer Res.* **51,** 4643–4648.

Blaszczyk, M., Pak, K. Y., Herlyn, M., Sears, H. F., and Steplewski, Z. (1985). *Proc. Natl. Acad. Sci. U.S.A.* **82,** 3552–3556.

Bosslet, K., Meenei, H. D., Rodden, F., Bauer, B. L., Wagner, F., Altmannsberger, A., Sedlacek, H. H., and Wiegandt, H. (1989). *Cancer Res.* **29,** 171–178.

Brady, R. O., and Fishman, P. (1974). *Biochim. Biophys. Acta* **355,** 121–148.

Bremer, E. G., Levery, S. B., Sonnino, S., Ghidoni, R., Canevari, S., Kannagi, R., and Hakomori, S.-I. (1984). *J. Biol. Chem.* **259,** 14773–14777.

Brockhaus, M., Magnani, J. L., Blaszczyk, M., Steplewski, Z., Koprowski, H., Karlsson, K.-A., Larson, G., and Ginsburg, V. (1981). *J. Biol. Chem.* **256,** 13223–13225.

Brockhaus, M., Magnani, J., and Herlyn, M. (1982). *Arch. Biochem. Biophys.* **217,** 647–651.

Brodin, T., Hellström, I., Hellström, K.-E., Karlsson, K. A., Sjögren, H.-O., Strömberg, N., and Thurin, J. (1985). *Biochim. Biophys. Acta* **837,** 349–353.

Brown, A., Feizi, T., Gooi, H. C., Embleton, M. J., Picard, J. K., and Baldwin, R. W. (1983). *Biosci. Rep.* **3,** 163–170.

Cahan, L. D., Irie, R. F., Singh, R., Cassidenti, A., and Paulson, J. C. (1982). *Proc. Natl. Acad. Sci. U.S.A.* **79,** 7629–7633.

Cheresh, D. A., and Klier, F. G. (1986). *J. Cell Biol.* **102,** 1887–1897.

Cheresh, D. A., Harper, J. R., Schulz, G., and Reisfeld, R. A. (1984a). *Proc. Natl. Acad. Sci. U.S.A.* **81,** 5767–5771.

Cheresh, D. A., Varki, A. P., Varki, N. M., Stallcup, W. B., Levine, J., and Reisfeld, R. A. (1984b). *J. Biol. Chem.* **259,** 7453–7459.

Cheresh, D. A., Reisfeld, R. A., and Varki, A. J. (1984c). *Science* **225,** 844–846.

Cheresh, D. A., Honsik, C. J., Staffileno, L. K., Jung, G., and Reisfeld, R. A. (1985). *Proc. Natl. Acad. Sci. U.S.A.* **82,** 5155–5159.

Cheresh, D. A., Rosenberg, J., Mujoo, K., Hirschowitz, L., and Reisfeld, R. (1986a). *Cancer Res.* **46,** 5112–5118.

Cheresh, D. A., Pierschbacher, M. D., Herzig, M. A., and Mujoo, K. (1986b). *J. Cell Biol.* **102,** 688–696.

Cheung, N.-KV., Saarinen, U. M., Neely, J., Landmeier, B., Donovan, C., and Coccia, P. F. (1985). *Cancer Res.* **45,** 2642–2649.

Chia, D., Terasaki, P. I., Suyama, N., Galton, J., Hirota, M., and Katz, D. (1985). *Cancer Res.* **45,** 435–437.

Davidsson, P., Fredman, P., Collins, V. P., von Holst, H., Månsson, J.-E., Granholm, L., and Svennerholm, L. (1989). *J. Neurochem.* **53,** 705–709.

Dippold, W. G., Knuth, A., and zum Buschenfelde, K. H. M. (1984). *Cancer Res.* **44,** 806–810.

Feizi, T. (1991). *TIBS* **16,** 84–86.

Fenderson, A. B., Eddy, E. M., and Hakomori, S.-I. (1990). *BioEssays* **12,** 173–179.

Fredman, P. (1988). *In* "New Trends in Ganglioside Research: Neurochemical and Neurogenerative Aspects" (R.W. Ledeen, E. L. Hogan, G. Tettamanti, A. J. Yates, and R. K. Yu, eds.), Fidia Res. Ser. Vol. 14, pp. 151–161. Liviana Press, Springer-Verlag, Padova/Berlin.

Fredman, P., Brezicka, F. T., Holmgren, J., Lindholm, L., Nilsson, O., and Svennerholm, L. (1986a). *Biochim. Biophys. Acta* **875,** 316–323.

Fredman, P., von Holst, H., Collins, V. P., Ammar, A., Dellheden, B., Wahren, B., Granholm, L., and Svennerholm, L. (1986b). *Neurol. Res.* **8,** 123–126.

Fredman, P., Mattsson, L., Andersson, K., Davidsson, P., Ishizuka, I., Jeansson, S., Månsson, J.-E., and Svennerholm, L. (1988a). *Biochem. J.* **251,** 17–22.

Fredman, P., von Holst, H., Collins, V. P., Granholm, L., and Svennerholm, L. (1988b). *J. Neurochem.* **50,** 912–919.

Fredman, P., Månsson, J.-E., Wikstrand, C. J., Vrionis, F. D., Rynmark, B.-M., Bigner, D. D., and Svennerholm, L. (1989). *J. Biol. Chem.* **264,** 12122–12125.

Fredman, P., Månsson, J.-E., Bigner, S. H., Wikstrand, C. J., Bigner, D. D., and Svennerholm, L. (1990a). *Biochim. Biophys. Acta* **1045,** 239–244.

Fredman, P., Dumanski, J., Davidsson, P., Svennerholm, L., and Collins, V. P. (1990b). *J. Neurochem.* **55,** 1838–1840.

Fukuda, M., Bothner, B., Lloyd, K. O., Rettig, W. J., Tiller, R. P., and Dell, A. (1986). *J. Biol. Chem.* **261,** 5145–5153.

Fukushi, Y., Hakomori, S., Nudelman, E., and Cochran, N. (1984a). *J. Biol. Chem.* **259,** 4681–4685.

Fukushi, Y., Nudelman, E., Levery, S. B., Rauvala, H., and Hakomori, S. (1984b). *J. Biol. Chem.* **259,** 10511–10517.

Fukushi, Y., Kannagi, R., Hakomori, S.-I., Shepard, T., Kulander, B. G., and Singer, J. W. (1985). *Cancer Res.* **45,** 3711–3717.

Fukushi, Y., Nudelman, E., Levery, S. B., Higuchi, T., and Hakomori, S.-I. (1986). *Biochemistry* **25,** 2859–2866.

Fukushima, K., Hirota, M., Terasaki, P., Wakisaka, A., Togashi, H., Chia, H., Suyama, N., Fukushi, Y., Nudelman, E., and Hakomori, S.-I. (1984). *Cancer Res.* **44,** 5279–5285.

Furukawa, K., Yamaguchi, H., Oettgen, H. F., Old, L. J., and Lloyd, K. O. (1989). *Cancer Res.* **49,** 191–196.

Gottfries, J., Fredman, P., Månsson, J.-E., Collins, V. P., von Holst, H., Armstrong, D. D., Percy, A. K., Wikstrand, C., Bigner, D. D., and Svennerholm, L. (1990). *J. Neurochem.* **55,** 1322–1326.

Gottfries, J., Percy, A. K., Månsson, J.-E., Fredman, P., Wikstrand, C. J., Friedman, H. S., Bigner, D. D., and Svennerholm, L. (1991). *Biochim. Biophys. Acta* **1081,** 105–116.

Hakomori, S.-I. (1973). *Adv. Cancer Res.* **18,** 265–315.

Hakomori, S.-I. (1975). *Biochim. Biophys. Acta* **417,** 55–89.

Hakomori, S.-I. (1981). *Annu. Rev. Biochem.* **50,** 733–764.

Hakomori, S.-I. (1989). *Adv. Cancer Res.* **52,** 257–331.

Hakomori, S.-I. (1990). *J. Biol. Chem.* **265,** 18713–18716.

Hakomori, S.-I., and Andrews, H. (1970). *Biochim. Biophys. Acta* **202,** 225–228.

Hakomori, S.-I., and Murakami, W. T. (1968). *Proc. Natl. Acad. Sci. U.S.A.* **59,** 254–261.

Hakomori, S., Nudelman, E., Kannagi, R., and Levery, S. B. (1982). *Biochem. Biophys. Res. Commun.* **109,** 36–44.

Hakomori, S.-I., Nudelman, E., Levery, S. B., and Kannagi, R. (1984). *J. Biol. Chem.* **252,** 4672–4680.

Hansson, G. C., Karlsson, K. A., Larson, G., McKibbin, J. M., Blaszczyk, M., Herlyn, M., Steplewski, Z., and Koprowski, H. (1983). *J. Biol. Chem.* **258,** 4091–4097.

He, X., Wikstrand, C. J., Fredman, P., Månsson, J.-E., and Svennerholm, L. (1989). *Acta Neuropathol.* **79,** 317–325.

Hellström, I., Brankovan, V., and Hellström, K. E. (1985). *Proc. Natl. Acad. Sci. U.S.A.* **82,** 1499–1502.

Herlyn, M., and Koprowski, H. (1988). *Annu. Rev. Immunol.* **6,** 283–308.

Hersey, P., and Jamal, O. (1989). *Pathology* **21,** 51–58.

Higashi, H., Hirabayashi, Y., Fukui, Y., Naiki, M., Matsumoto, M., Ueda, S., and Kato, S. (1985). *Cancer Res.* **45,** 3796–3802.

Hirabayashi, Y., Sugimoto, M., Ogawa, T., Matsumoto, M., Tagawa, M., and Taniguchi, M. (1986). *Biochim. Biophys. Acta* **875,** 126–128.

Hiraiwa, N., Fukkuda, Y., Imura, H., Tadano-Aritomi, K., Ishizuka, I., and Kannagi, R. (1990). *Cancer Res.* **50,** 2917–2928.

Hirohashi, S., Clausen, H., Nudelman, E., Inoue, H., Shimosato, Y., and Hakomori, S.-I. (1986). *J. Immunol.* **136,** 4163–4168.

Hoffman, L. M., Brooks, S. E., Stein, M. R., and Schneck, L. (1991). *Biochim. Biophys. Acta* **1084,** 94–100.

Holmes, E. H., and Hakomori, S.-I. (1983). *J. Biol. Chem.* **258,** 3706–3713.

Holmes, E. H., Hakomori, S.-I., and Ostrander, G. K. (1987). *J. Biol. Chem.* **262,** 15649–15658.

Irie, R. F., Sze, L. L., and Saxton, R. E. (1982). *Proc. Natl. Acad. Sci. U.S.A.* **79,** 5666–5670.

Karlsson, G., Månsson, J.-E., Wikstrand, C. J., Bigner, D. D., and Svennerholm, L. (1990). *Biochim. Biophys. Acta* **1043,** 267–272.

Kawai, T., Kato, A., Higashi, H., Kato, S., and Naiki, M. (1991). *Cancer Res.* **51,** 1242–1246.

Kiguchi, K., Takamutso, K., Tanaka, J., Nozawa, S., Iwamori, M., and Nagai, Y. (1992). *Cancer Res.* **52,** 416–421.

Köhler, G., and Milstein, C. (1975). *Nature (London)* **256,** 495–497.

Koprowski, H., Steplewski, Z., Mitchell, K., Herlyn, M., Herlyn, D., and Fuhrer, P. (1979). *Somatic Cell Genet.* **5,** 957–972.

Ladisch, S., Sweeley, C. C., Becker, H., and Gage, D. (1989). *J. Biol. Chem.* **264,** 12097–12105.

Levy, R., and Miller, R. A. (1983). *Fed. Proc. Fed. Am. Soc. Exp. Biol.* **42,** 2650–2656.

Liepkalns, V. A., Icard, C., Yates, A. J., Thompson, D. K., and Hart, R. W. (1981). *J. Neurochem.* **36,** 1959–1965.

Liepkalns, V. A., Icard-Liepkalns, C., Yates, A. J., Mattison, S., and Stephens, R.E. (1983). *J. Lipid Res.* **24,** 533–540.

Lloyd, K. O., and Lloyd, J. O. (1989). *Cancer Res.* **49,** 3445–3451.

Lloyd, K. O., Larson, G., Strömberg, N., Thurin, J., and Karlsson, K. A (1983). *Immunogenetics (N.Y.)* **17,** 537–541.

Longee, D. C., Wikstrand, C. J., Månsson, J.-E., He, X., Fuller, G. N., Bigner, S. H., Fredman, P., Svennerholm, L., and Bigner, D. D. (1991). *Acta Neuropathol.* **82,** 45–54.

Magnani, J., Nilsson, J., Brockhaus, B., Zopf, D., Steplewski, Z., Koprowski, H., and Ginsburg, V. (1982). *J. Biol. Chem.* **257,** 14365–14369.

Månsson, J.-E., Fredman, P., Nilsson, O., Lindholm, L., Holmgren, L., and Svennerholm, L. (1985). *Biochim. Biophys. Acta* **834,** 110–117.

Månsson, J.-E., Fredman, P., Bigner, D. D., Molin, K., Rosengren, B., Friedman, H. S., and Svennerholm, L. (1986). *FEBS Lett.* **201,** 109–113.

Markwell, M., Fredman, P., and Svennerholm, L. (1984). *Biochim. Biophys. Acta* **775,** 7–16.

Mårtensson, S., Due, C., Påhlsson, P., Nilsson, B., Eriksson, H., Zopf, D., Olsson, L., and Lundblad, A. (1988). *Cancer Res.* **48,** 2125–2131.

Ménard, S., Tagliabue, E., Canevari, S., Fossati, G., and Colnaghi, M. I. (1983). *Cancer Res.* **43,** 1295–1300.

Miyake, M., Hashimoto, K., Ito, M., Ogawa, O., Arai, E., Hitomi, S., and Kannagi, R. (1990). *Cancer (Philadelphia)* **65,** 499–505.

Mora, P. T., Brady, R. O., Bradley, R. M., and McFarland, V. W. (1969). *Proc. Natl. Acad. Sci. U.S.A.* **63,** 1290–1296.

Myoga, A., Taki, T., Arai, K., Sekiguchi, K., Ikeda, I., Kurata, K., and Matsumoto, M. (1988). *Cancer Res.* **48,** 1512–1516.

Natoli, E. J., Livingston, P. O., Pukel, C. S., Lloyd, K. O., Wiegandt, H., Szalay, J., Oettgen, H. F., and Old, L. J. (1986). *Cancer Res.* **46,** 4116–4120.

Nilsson, O., Månsson, J.-E., Lindholm, L., Holmgren, J., and Svennerholm, L. (1985a). *FEBS Lett.* **182,** 398–402.

Nilsson, O., Lindholm, L., Holmgren, J., and Svennerholm, L. (1985b). *Biochim. Biophys. Acta* **835,** 577–583.

Nilsson, O., Brezicka, F. T., Holmgren, J., Sörenson, S., Svennerholm, L., Yngvason, F., and Lindholm, L. (1986). *Cancer Res.* **46,** 1403–1407.

Nores, G. A., Ohi, T., Taniguchi, M., and Hakomori, S.-I. (1987). *J. Immunol.* **139,** 3171–3176.

Nudelman, E., Hakomori, S., Kannagi, R., Levery, S., Yeh, M. H., Hellström, K. E., and Hellström, I. (1982). *J. Biol. Chem.* **257,** 12752–12756.

Nudelman, E., Fukushi, Y., Levery, S. B., Higuchi, T., and Hakomori, S.-I. (1986). *J. Biol. Chem.* **261,** 5487–5495.

Pukel, C. S., Lloyd, K. O., Travassos, L. R., Dippold, W. G., Oettgen, H. F., and Old, L. J. (1982). *J. Exp. Med.* **155,** 1133–1147.

Ravindrath, M. H., and Irie, R. F. (1988). *In* "Malignant Melanoma: Biology, Diagnoses and Therapy" (L. Nathansson, ed.), pp. 17–43. Kluwer Academic Publishers, Boston.

Ravindrath, M. H., and Morton, D. L. (1991). *Int. Rev. Immunol.* **7,** 303–329.

Reisfeld, R. A., and Cheresh, D. A. (1987). *Adv. Immunol.* **40,** 323–377.

Rettig, W. J., Cordon-Cardo, C., Ng, J. S. C., Oettgen, H. F., Old, L. J., and Lloyd, K. O. (1985). *Cancer Res.* **45,** 815–821.

Ritter, G., and Livingston, P. O. (1991). *Cancer Biol.* **2,** 401–409.

Rosenfelder, G., Young, W. W., Jr., and Hakomori, S. (1977). *Cancer Res.* **37,** 1333–1339.

Sanai, Y., and Nagai, Y. (1989). *In* "Gangliosides and Cancer" (H. F. Oettgen, ed.), pp. 69–77. VCH Verlagsgesellschaft, Weinheim, Germany.

Schold, S. C., Bullard, D. E., Bigner, S. H., Jones, T. R., and Bigner, D. D. (1982). *J. Neurooncol.* **1,** 5–14.

Schulz, G., Cheresh, D. A., Varki, N. M., Yu, A., Staffileno, L. K., and Reisfeld, R. A. (1984). *Cancer Res.* **44,** 5914–5920.

Schwartzmann, G., and Sandhoff, K. (1990). *Biochemistry* **29,** 10866–10871.

Siddique, B., Buehler, J., Gregorio, M. W., and Macher, B. A. (1984). *Cancer Res.* **44,** 5262–5265.

Solter, D., and Knowles, B. B. (1978). *Proc. Natl. Acad. Sci. U.S.A.* **75,** 5565–5569.

Svennerholm, L. (1984). INSERM Symp. **126,** 21–44.

Svennerholm, L. (1984). *In* "Cellular and Pathological Aspects of Glycoconjugate Metabolism" (H. Dreyfus, R. Massarelli, L. Freysz, and G. Rebel, eds.), INSERM, Vol. 126, pp. 21–44. Strasbourg, France.

Svennerholm, L. (1988). *In* "New Trends in Ganglioside Research: Neurochemical and Neurogenerative Aspects" (R. W. Ledeen, E. L. Hogan, G. Tettamanti, A. J. Yates, and R. K. Yu, eds.), Fidia Res. Ser. Vol. 14, pp. 135–150. Liviana Press/Springer-Verlag, Padova/Berlin.

Svennerholm, L., Boström, K., Fredman, P., Månsson, J.-E., Rosengren, B., and Rynmark, B.-M. (1989). *Biochim. Biophys. Acta* **1005,** 109–117.

Tai, T., Paulsson, J., Cahan, L. D., and Irie, R. F. (1983). *Proc. Natl. Acad. Sci. U.S.A.* **80,** 5392–5396.

Taki, T., Yamamoto, K., Takamatsu, M., Ishii, K., Myoga, A., Sekiguchi, K., Ikeda, I., Kurata, K., Nakayama, J., and Handa, S. (1990). *Cancer Res.* **50,** 1284–1290.

Tempero, M. A., Uchida, E., Takasaki, H., Burnett, D. A., Steplewski, Z., and Pour, P. M. (1987). *Cancer Res.* **47,** 5501–5503.

Tsuchida, T., Saxton, R. E., Morton, D. L., and Irie, R. F. (1987a). *JNCI, J. Natl. Cancer Inst.* **78,** 45–54.

Tsuchida, T., Ravindrath, M. H., Saxton, R. E., and Irie, R. F. (1987b). *Cancer Res.* **47,** 1278–1281.

Urdal, D. L., Brentnall, T. A., Bernstein, I. D., and Hakomori, S.-I. (1983). *Blood* **62,** 1022–1026.

Vrionis, F. D., Wikstrand, C. J., Fredman, P., Månsson, J.-E., Svennerholm, L., and Bigner, D. D. (1989). *Cancer Res.* **49,** 6641–6649.

Wakabayashi, S. M., Saito, T., Shinohara, N., Okamoto, S., Tomioka, H., and Tanigushi, M. (1984). *J. Invest. Dermatol.* **83,** 128–133.

Welt, S. W., Carswell, E. A., Vogel, C.-V., Oettgen, H. F., and Old, L. J. (1987). *Clin. Immunol. Immunopathol.* **45,** 214–229.

Wiegandt, H. (1985). *In* "Glycolipids" (H. Wiegandt, ed.), pp. 199–260. Elsevier, Amsterdam.

Wiels, J., Fellous, M., and Tursz, T. (1981). *Proc. Natl. Acad. Sci. U.S.A.* **78,** 6485–6488.

Wikstrand, C. J., Grahmann, F. C., McComb, R. D., and Bigner, D. D. (1985). *J. Neuropathol. Exp. Neurol.* **44,** 229–241.

Wikstrand, C. J., McLendon, R. E., Bullard, D. E., Fredman, P., Svennerholm, L., and Bigner, D. D. (1986). *Cancer Res.* **46,** 5933–5940.

Wikstrand, C. J., He, X., Fuller, G. N., Bigner, S. H., Fredman, P., Svennerholm, L., and Bigner, D. D. (1991). *J. Neuropathol. Exp. Neurol.* **50,** 756–769.

Wong, J. H., Irie, R. F., and Morton, D. L. (1989). *Semin. Surg. Oncol.* **5,** 448–452.

Wu, Z.-I., Schwartz, E., Seeger, R., and Ladisch, S. (1986). *Cancer Res.* **46,** 440–443.

Yamaguchi, H., Furukawa, K., Fortunato, S. R., Livingston, P. O., Lloyd, K. O., Oettgen, H. F., and Old, L. J. (1987). *Proc. Natl. Acad. Sci. U.S.A.* **84,** 2416–2420.

Yamamoto, S., Yamamoto, T., Saxton, R. E., Hoon, D. S. B., and Irie, R. F. (1990). *J. Natl. Cancer Inst.* **82,** 1757–1760.

Yang, H.-J., and Hakomori, S. (1971). *J. Biol. Chem.* **246,** 1192–1200.

Yates, A. (1988). *Neurochem. Pathol.* **8,** 157–180.

Young, W. W., Jr., and Hakomori, S. (1981). *Science* **211,** 487–489.

ADVANCES IN LIPID RESEARCH, VOL. 25

Gangliosides and Modulation of the Function of Neural Cells

GUIDO TETTAMANTI AND LAURA RIBONI

Department of Medical Chemistry and Biochemistry
The Medical School
University of Milan
20133 Milan, Italy

I. Introduction: The Multifunctional Role of Gangliosides

Gangliosides, which are sialic acid-containing glycosphingolipids, are components of the plasma membrane of most vertebrate cells, and are particularly abundant in the nervous system (Wiegandt, 1985). They are asymmetrically located in the outer leaflet of the plasma membrane with the oligosaccharide portion exposed on the cell surface, in contact with extracellular substances, and the ceramide portion inserted into the membrane layer, in contact with various lipid and protein components of the membrane. Although present on the whole neuron surface, gangliosides are more concentrated in the synaptic region (Hansson *et al.*, 1977; Yu and Saito, 1989). In small amounts they are also located intracellularly, partly linked to the organelles responsible for their intracellular traffic and metabolism, and partly linked to soluble protein carriers (Sonnino *et al.*,

1979; Ledeen, 1989). Gangliosides are structurally heterogeneous in their oligosaccharide and ceramide portions (Yu and Saito, 1989), thus constituting a family of compounds, each with a potentially different capacity of interactions. These structural differences reflect coded expression trends, which are sensitive to environmental and functional influences (Caputto, 1988; Suzuki and Yamakawa, 1991).

Gangliosides represent recognition sites at the cell surface. Therefore, they can be instrumental to appropriate interactions between the plasma membrane and extracellular substances. This is the basis for the implication of gangliosides in receptor function and cell–cell or cell–substratum recognition (Brady and Fishman, 1979; Igarashi *et al.*, 1989; Hakomori, 1990). Moreover, gangliosides interact with membrane-bound functional proteins, modulating their activity, and influencing the membrane-mediated transfer of information (Fishman, 1988; Hakomori, 1990). Finally, the findings that metabolites deriving from sphingolipids (sphingosine, ceramide) act as regulators of protein phosphorylation (Hannun and Bell, 1989; Merrill and Jones, 1990) poses the possibility that gangliosides themselves act as precursors of intracellular metabolic regulators.

II. Chemical and Physicochemical Features of Gangliosides

A. The Chemical Heterogeneity of Gangliosides

The characteristic component of gangliosides is sialic acid (SA), the number of residues per ganglioside molecule varying from 1 to 7, with an average of 2–3 in the brain gangliosides of most vertebrates (Yu and Saito, 1989). The sialic acid residue(s) is (are) α-glycosidically attached to a neutral oligosaccharide core which may contain glucose (Glc), galactose (Gal), *N*-acetylgalactosamine (GalNAc), *N*-acetylglucosamine, and fucose. On the basis of their neutral oligosaccharide core, gangliosides are classified in different series (gala, hemato, ganglio, neolacto, globo, isoglobo series). The gangliosides that are predominant in the nervous tissue (see their structures in Fig. 1) belong to the ganglio series. In brain gangliosides, the most abundant sialic acid is *N*-acetylneuraminic acid (NeuAc), followed by *N*-glycolylneuraminic acid, and 4- or 9-*O*-acetyl-*N*-acetylneuraminic acid. The lactone form of ganglioside G_{D1b} has also been isolated from human and rat brain (Riboni *et al.*, 1986, 1989). The sialic acid-containing oligosaccharide is β-glycosidically linked to ceramide, formed by a long-chain fatty acid (commonly stearic acid) and a long-chain base (C_{18} or C_{20}), unsaturated (sphingosine) or saturated (sphinganine), linked together by an amide bond. Each ganglioside with a defined oligosaccharide structure is heterogeneous in the lipidic portion, and occurs in different molecular species carrying an individual long-chain base and fatty acid.

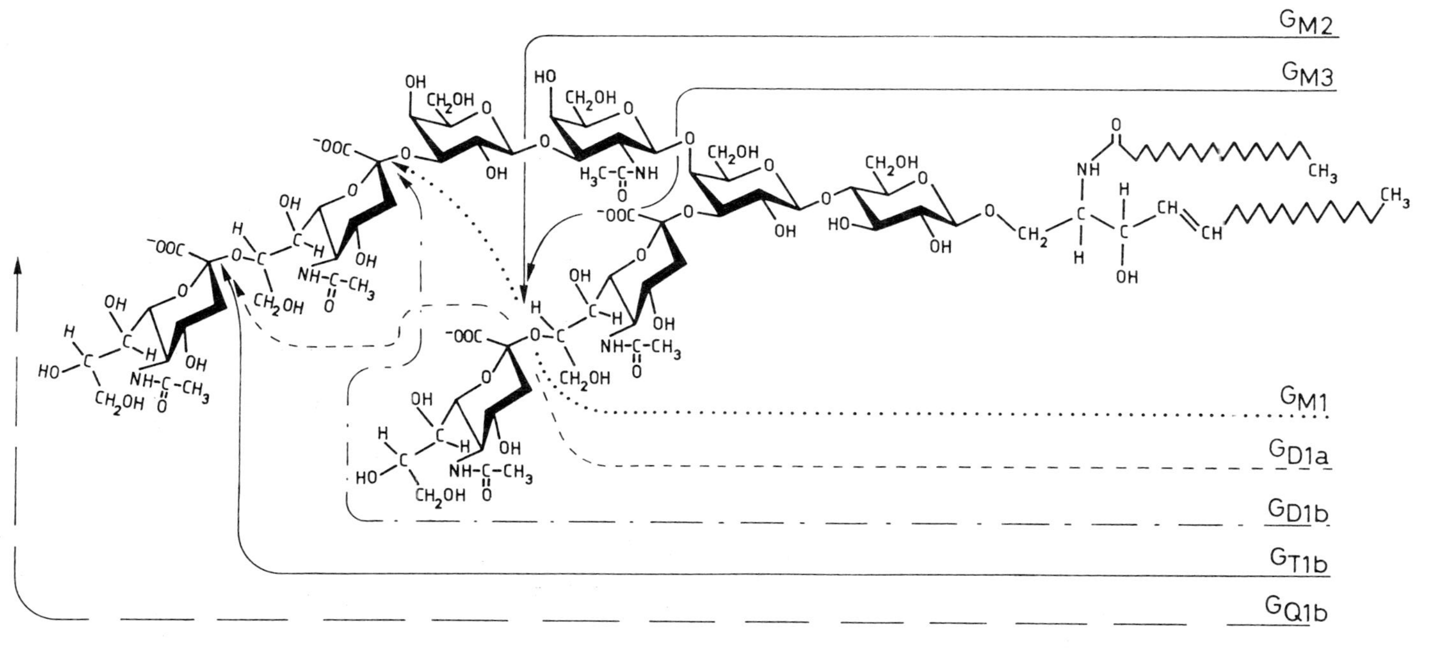

FIG. 1. Structure of some of the most abundant gangliosides in vertebrate brain. Ganglioside nomenclature is according to Svennerholm (1980).

The content and the chemical characteristics of brain gangliosides appear to be under genetic and epigenetic control (Caputto, 1988; Suzuki and Yamakawa, 1991; Rahmann, 1992). A genetic control correlates the ganglioside composition to animal species, regions or areas of the nervous system, development, and ageing. Among epigenetic influences, functional (learning, cycles of lightness and darkness) (Caputto, 1988), and environmental (thermal adaptation, hibernation, composition of the culture medium for cultured cells) factors have been identified (Rahmann, 1992; Staedel-Flaig *et al.,* 1987). Changes in ganglioside content and composition are one of the biochemical bases for neuronal plasticity. Each cell expresses the ganglioside pattern that is most convenient in a given condition and uses chemical diversity for selecting ligands and providing adequate responses.

B. Conformation of Ganglioside Molecules

The ceramide portion of gangliosides, containing both a hydrogen acceptor (amide carbonyl) and a hydrogen donor (hydroxyl group), adopts a rigid conformation, with parallel orientation of the axes of the two hydrocarbon chains, which are condensed in a closely packed arrangement (Pascher, 1976; Harris and Thornton, 1978; Hakomori, 1981).

The conformation of the oligosaccharide portion of gangliosides is dependent partly on the high stability of sialic acid (Czarniecki and Thornton, 1977) and partly on the number of sialic acid residues and on the type of vicinal sugars. In the case, for instance, of gangliosides G_{M1} and G_{D1a}, the plane of inner sialic acid is perpendicular to that of the neighboring *N*-acetylgalactosamine (Sillerud *et al.,* 1982). Moreover, the region of the inner galactose, linked to both *N*-acetylgalactosamine and sialic acid, defines an oxygen-rich surface particularly suitable for interaction with cations (Koerner *et al.,* 1983). This interaction, particularly with Ca^{2+}, is followed by a marked change of ganglioside conformation. Finally, the intramolecular hydrogen bond between the NH group of the long-chain base and the glycosidic oxygen may force the oligosaccharide chain to display a shovel position with respect to ceramide (Abrahamsson *et al.,* 1977). However, computational studies (Wynn and Robson, 1986) and measurements of surface potential in artificial membranes containing gangliosides (McDaniel *et al.,* 1984) showed that the oligosaccharide chain and the ceramide portion lie on the same axis. Possibly, the projected orientation of the oligosaccharide chain is favored by the ganglioside interactions with the other components of the membrane, and a reversible shovel/projected conformation transition constitutes a point of flexibility of potential functional significance.

C. Physicochemical Features of Gangliosides

Gangliosides have amphiphilic properties and undergo micellization above the critical micellar concentration, comprised in the 10^{-7}–10^{-9} *M* range (Tettamanti and Masserini, 1987; Corti *et al.*, 1987). Differently from phospholipids, gangliosides maintain a micellar structure in an extremely wide range of concentrations and generally do not form bilayer structures in the same way as phospholipids (Curatolo, 1987). This means that mixtures of gangliosides and phospholipids would tend to aggregate as micellar or lamellar structures depending on the preponderance of ganglioside or phospholipid in the mixture. It is well established that gangliosides decrease membrane fluidity (Bertoli *et al.*, 1981). In addition, they contribute to the electrostatic potential of the membrane (Thompson and Brown, 1988; Langner *et al.*, 1988) and influence the thermodynamic and geometric features of the membrane (Maggio, 1985). Their own conformation and charge can be modified by the surrounding environment of the membrane.

An interesting problem is whether gangliosides undergo lateral phase separation with formation of ganglioside-enriched microdomains in the membrane, as first suggested by Sharom and Grant (1978) and in analogy with the behavior of neutral glycosphingolipids (Thompson and Tillack, 1985). Differential scanning calorimetry studies showed that, under certain conditions, gangliosides do separate into enriched microdomains ("clusters") (Myers *et al.*, 1984). This ability (1) depends on the ganglioside concentration and is greater the higher is the content of sialic acid residues in the ganglioside molecule (Myers *et al.*, 1984); (2) is enhanced by the presence of Ca^{2+} ions (Masserini and Freire, 1986); (3) is dependent on the ceramide composition of ganglioside, phase separation being easier the higher is the compositional diversity between the hydrophobic tail of ganglioside and that of the other lipid components of the membrane (Masserini *et al.*, 1988). Interestingly, an enzyme (sialidase) is able to distinguish molecularly dispersed from cluster-associated gangliosides not only in artificial membranes (Masserini *et al.*, 1988), but also in neuronal membranes (Palestini *et al.*, 1991).

D. Binding Properties of Gangliosides

Gangliosides possess a high binding potential. The type of binding depends on the general surfactant character of gangliosides, or on the direct involvement of the ceramide or oligosaccharide portions of their molecule (Tettamanti and Masserini, 1987).

A number of agents interact primarily with the oligosaccharide portion of gangliosides. Among them are toxins of bacterial or nonbacterial origin (tetanus toxins, botulinum toxin, cholera toxin, *Escherichia coli* toxin, *Staphylococcus* toxin, *Streptococcus parahemolyticus* hemolysin, sea wasp hemolysin), viruses (Sendai virus, Newcastle disease virus), interferon, hormones (thyrotropin, chorionic gonadotropin, luteinizing hormone), serotonin, and fibronectin (Tettamanti and Masserini, 1987). The most specific (and apparently the only specific) interaction is that of cholera toxin with G_{M1} (Brady and Fishman, 1979) and Fuc-G_{M1} (Masserini *et al.*, 1992). Tetanus toxin is known to bind more specifically G_{D1b} and G_{T1b} (Van Heyningen, 1974), botulinum toxin binds G_{T1b} and G_{Q1b} (Kitamura *et al.*, 1980), thyrotropin binds G_{D1b}, G_{T1b}, and a not-yet-identified thyroid ganglioside (Kohn and Shifrin, 1982), and serotonin binds G_{D3} (Tamir *et al.*, 1980) and Sendai virus gangliosides with the terminal sequence NeuAcα2–8, NeuAcα–3Galβ1–3GalNac (Markwell *et al.*, 1981).

An example of interactions involving the ganglioside micelle is the binding of G_{M1} with bovine serum albumin (Tomasi *et al.*, 1980), with formation of two well-defined complexes. One of them is characterized by a stoichiometric G_{M1}/albumin ratio of one micelle per albumin molecule; the other complex derives from the former one by a slow process of dimerization. These two complexes, which result from hydrophobic interactions, are actually mixed ganglioside–albumin micelles. Similar complexes were described to occur between micellar gangliosides and enzymes such as neural cytosolic sialidase and an α-fucosidase from octopus hepatopancreas (Venerando *et al.*, 1985; Masserini *et al.*, 1985). The formation of the complex causes complete enzyme inactivation. However, treatment with detergents like Triton X-100 or lysolecithin restored, at least partially, the enzyme activity. The reversibility of the phenomenon suggests that interactions involving micellar gangliosides and active proteins may have a physiological meaning, since micelle-like assemblies of gangliosides (clusters) are present at the membrane level.

Evidence was provided for the occurrence of seemingly specific interactions of gangliosides with membrane proteins. These interactions are viewed as mediated mainly by the hydrophobic portion of ganglioside, it still being unclear whether individual ganglioside molecules or gangliosides micelles are implicated in the process. The cell surface component which is responsible for Paul-Bunnel antigenity was identified as a low-molecular-weight amphipathic polypeptide which is capable of binding specifically to, and is modulated by, ganglioside G_{M3} (Watanabe *et al.*, 1980). The occurrence in brain of plasma membrane proteins that specifically bind to gangliosides, particularly G_{M1}, has been demonstrated (Yasuda *et al.*, 1988; Tiemeyer *et al.*, 1990a,b; Fueshko and Schengrund, 1990; Sonnino *et al.*, 1992).

III. Ganglioside Implication in Neural Functions

Hundreds of papers in the past 15 years have dealt with ganglioside implication in events of biological significance. In these investigations, the following approaches were adopted: (1) description of the changes of ganglioside pattern and turnover accompanying biological events; (2) addition of exogenous gangliosides to a given system followed by observation of resulting effects on functional performances (assuming that exogenous gangliosides, inserted into the cell plasma membrane, mimic the action of endogenous gangliosides; Varon *et al.*, 1988); (3) use of gangliosides carrying special probes (paramagnetic, fluorescent, photoreactive, biotinylated) to inspect some properties of gangliosides (assuming that the used derivatives have the same behavior as natural gangliosides; Tettamanti, 1988); (4) addition of ligands capable of interacting specifically with individual gangliosides (anti-ganglioside antibodies; ganglioside-binding toxins, e.g., cholera toxin subunit B), on the assumption that the ligand binding would suppress ganglioside participation in a particular function (Fishman, 1988). A cautionary note on the use of B subunit of cholera toxin as a probe for ganglioside G_{M1} was published by Spiegel (1990), since commercial preparations of cholera toxin B subunit contain small amounts of the A subunit that can be responsible for the biological effects previously attributed to the simple binding of B subunit to G_{M1}.

A. Gangliosides in Neurodifferentiation, Neuritogenesis, and Synaptogenesis

Neuronal, as well as glial, differentiation is coupled to changes in the ganglioside content and pattern (Yates, 1986; Karpiak and Mahadik, 1990; Schengrund, 1990; Ledeen, 1991; Rahmann, 1992). The total ganglioside content increases until completion of differentiation and peculiar qualitative changes occur, in relation to the particular stage of development. Some gangliosides can be considered "markers" of differentiation stages (Table I). These changes were observed in full brains (or brain areas) and in primary cultures of neurons, with a remarkable parallelism between the two systems. A connection between ganglioside patterns and brain function seems likely. In fact, intracranial injection of purified anti-ganglioside antibodies into neonatal rats induces both morphological (reduced number of dendritic spines), and functional (greater uncorrect responses in learning tasks) abnormalities detectable in the adult stage. Moreover, normal rat pups treated with gangliosides displayed improved learning performances and increased cortex levels of acetylcholine esterase as compared to untreated animals (Karpiak and Mahadik, 1990). In parallel, neurite outgrowth from neural cells and tissue slices cultivated *in vitro* was inhibited by treatment with anti-G_{M1} antibodies (Schwartz and Spirman, 1982; Spirman *et al.*, 1982; Spoerri *et al.*, 1988).

Support for the concept that gangliosides are actively involved in neuronal differentiation was provided by the evidence that exogenously administered gangliosides have neuritogenic and synaptogenic effects (Obata *et al.*, 1977; Ledeen, 1984). All cultured neural cells that have been studied so far respond to the presence of gangliosides in the medium with outgrowth of processes and formation, under favorable conditions, of synaptic-like contacts (Spoerri, 1983).

Table I
EXPRESSION CHANGES OF GANGLIOSIDES IN THE DIFFERENT STAGES OF BRAIN DEVELOPMENT AND AGEING[a]

Marker ganglioside	Development stage
G_{D3}, G_{D2} (less)	Proliferation of neural and precursor cells
Highly sialylated gangliosides (G_{Q1b}, G_{T1c}, G_{Q1c}, G_{P1c})	Cell migration and differentiation (sprouting); dendritic arborization
G_{M1}, G_{D1a} (particularly in cerebrum), G_{D1b}, G_{T1b} (particularly in cerebellum)	Fiber growth; synapse formation
G_{M4}, G_{M1}, G_{D3} (less)	Myelination; oligodendrial cell proliferation
G_{M1}, G_{D1a}, G_{D1b}, G_{T1b}, G_{Q1b}	Adulthood
Similar pattern of adulthood; ganglioside species carrying prevalently C_{20} long-chain bases	Ageing

[a]The gangliosides that are predominant in each stage ("marker" gangliosides) are indicated. For more details, see the reviews of Ledeen (1989) and Rahmann (1992).

These effects have been observed in both neurotumoral cells, e.g., neuroblastoma and pheochromocytoma cells, and primary cultures of neurons (Dreyfus *et al.*, 1980, 1984; Ledeen *et al.*, 1990). The most important features of the neuritogenic effect of gangliosides in cultured cells, particularly neurotumoral cells, are listed in Table II.

After the first report by Ceccarelli *et al.* (1976), a large number of papers demonstrated that treatment with exogenous gangliosides markedly enhances neuritogenesis *in vivo* after traumatic lesions to the peripheral nervous system. This topic has been covered by extensive reviews (Ledeen *et al.*, 1990; Schengrund, 1990; Rodden *et al.*, 1991).

A tentative explanation (Ledeen *et al.*, 1990; Masco *et al.*, 1991) for the implication of gangliosides in neuritogenesis is that gangliosides, over certain concentrations, perturb the membrane structure. This "triggers" a biochemical response eventually yielding neurite outgrowth. This model may include the more specific suggestion (Nagai and Tsuji, 1989) that the neuritogenic effect exerted by G_{Q1b} on the human GOTO and NB-1 neuroblastoma cells is due to activation of an ectoprotein kinase. However, it can hardly be applied to the primary cultures of neurons and glial cells that are stimulated to differentiate by ganglioside treatment (Facci *et al.*, 1988; Levine and Goldman, 1988; Yim *et al.*, 1991). In this respect, it is worth considering that the neurodifferentiation effect of gangliosides is generally elicited when the cells are incubated in the presence of fetal calf serum or nerve growth factor. This means that factors are present in the medium which may be primarily responsible for the effect. However, cases are known where the neuritogenic effect of gangliosides is exhibited by cells cultivated in serum-free, though hormone-supplemented, media (Durand *et al.*, 1986;

Table II
COMMON FEATURES OF THE NEURITOGENIC EFFECT EXERTED BY GANGLIOSIDES (EXOGENOUS OR ENDOGENOUS) ON NEUROTUMORAL CELLS, PARTICULARLY NEUROBLASTOMA CELLS[a]

1. Different gangliosides may exhibit the effect at different optimal concentrations and with possible differential action regarding the number or length of the processes produced
2. The presence of sialic acid is necessary for the effect, but not that of the ceramide moiety of ganglioside
3. The carboxyl group of sialic acid is not necessary for the effect
4. The most responsive cells are those carrying (endogenous) gangliosides of the gangliotetraose family
5. The presence of (endogenous) G_{M1} seems essential for the effect
6. Sialidase treatment of cells, with a consequent increase of G_{M1} concentration, triggers the effect
7. Treatment of cells with anti-G_{M1} antibody, or cholera toxin B subunit blocks the effect
8. Insertion of exogenous ganglioside molecules into the cell plasma membrane is essential for the effect
9. The effect requires the presence of adequate concentrations of Ca^{2+} and is accompanied by a Ca^{2+} influx inside the cell
10. The effect requires the presence in the medium of promoting factors, possibly with the exception only of the human neuroblastoma cells of the GOTO and NB-1 lines
11. The effect is obtained generally at millimolar concentrations of exogenous gangliosides, with the exception of the GOTO and NB-1 lines, which require nanomolar concentrations of exogenous G_{Q1b}

[a]For detailed references, see the review by Ledeen (1989).

Nakajima *et al.,* 1986). The absence of added differentiating or trophic factors in the medium does not necessarily mean that such substances are not formed by the same cells during incubation. In fact, it was reported (Leon *et al.,* 1988) that dopaminergic neurons from embryonic mouse mesencephalon, cultured in a serum-free medium, undergo a ganglioside-stimulated differentiation that becomes evident only after a certain density of cells is reached in culture. As well, some clones of neuroblastoma cells are strongly induced by exogenous gangliosides to produce a microtubule-associated protein, MAP-2, in a way that is dependent on the cell density in culture (Ferreira *et al.,* 1990). MAP-2 is one of the proteins that regulate tubulin polymerization and microtubule stabilization, events that play a key role during neurite formation (Drubin *et al.,* 1985). Therefore, the hypothesis (Varon *et al.,* 1988) that gangliosides influence neurodifferentiation by modulating, or activating, receptors of exogenous differentiation factors might be accepted as a flexible model, which can apply to a wide range of different situations. Moreover, Yavin *et al.* (1988) showed that gangliosides modulate the expression of RNA-encoding cytoskeletal proteins in neurohybrid and pheochromocytoma cells, and that of tubulin RNA in substantia nigra and striatum of rats after unilateral hemitransection. This presumes a potential involvement of gangliosides in regulating the gene expression machinery.

B. Gangliosides in Neuronal Survival

It is generally accepted that survival of adult neurons depends on the presence of neurotrophic factors (an example of them is the nerve growth factor), that can be of neural or extraneural origin. In order to respond to these factors, neurons are assumed to possess the corresponding receptors. The interplay between functioning neurons and neurotrophic factors is the basis for the potential ability of damaged neurons to repair after injury.

Gangliosides appear to be involved in the response mechanisms of neurons to neurotrophic factors (Skaper *et al.,* 1988; Varon *et al.,* 1988; Cuello, 1990; Ledeen *et al.,* 1990). A first line of evidence concerns the capacity of exogenous gangliosides to increase the life span of neural cells cultivated *in vitro.* Leon *et al.* (1988) observed that primary cultures of fetal mesencephalic neurons prolonged their survival in serum-free media, in the presence of trophic influences (nerve growth factor included) and gangliosides, particularly G_{M1}. Chick spinal motoneurons, grown in a serum- and muscle extract-containing medium, displayed a 3-fold increase of survival in the presence of G_{M1} (Juurlink *et al.,* 1991). Rosner *et al.* (1992) confirmed the finding, adding that the ganglioside effect was more marked at a low (20 μM) Ca^{2+} concentration. A second line of evidence derives from studies where injuries of various origin (traumatic, ischemic, toxic) were produced in the central nervous system and recovery was evaluated by measuring some functional parameters. Intracerebroventricular administration of G_{M1}, following unilateral decortication of rats, protected forebrain cholinergic neurons from retrograde degeneration (Cuello *et al.,* 1989) in a manner comparable to that of β-nerve growth factor. In a number of experimental animals treated with MPTP (1-methyl-4-phenyl-1,2,3,6-tetrahydropyridine) (an experimental condition that is considered a model for Parkinson's disease; Burns *et al.,* 1983), chronic administration of gangliosides restored to near normal the activity of tyrosine hydroxylase and aromatic-L-amino-acid decarboxylase, and improved behavioral and neurological impairment (Hadjiconstantinou and Neff, 1990; Schneider *et al.,* 1992). In these conditions, exogenous gangliosides can either sensitize damaged neurons to locally released neurotrophic factors (Berg, 1984), and/or protect neurons against neurotoxic factors (Skaper *et al.,* 1988; Seren *et al.,* 1990).

An interesting study model for the protective action of gangliosides is constituted by cultured cerebellar granule cells that die after transient exposure to high doses (50 μM) of glutamate. This effect seems to be due to glutamate permanent activation of protein kinase C (following translocation from the cytosol to the neuronal plasma membrane) (Vaccarino *et al.,* 1987), with a consequent uncontrolled delayed and cytotoxic rise of free cytosolic Ca^{2+} (Manev *et al.,* 1989). Gangliosides protect these neurons from glutamate excitotoxicity by inhibiting protein kinase, preventing its translocation, and normalizing distorted intracellular free Ca^{2+} dynamics (Favaron *et al.*, 1988; De Erausquin *et al.*, 1990; Manev *et al.*, 1990). An excess of glutamate is generated in ischemic brain. Under these conditions, a persistent translocation of protein kinase C from the cytosol to the neuronal membrane occurs, the event being reversed by ganglioside administration (Louis *et al.*, 1988; Omodera *et al.*, 1989). These findings strengthen the hypothesis that gangliosides protect neurons from the cytotoxic action of glutamate excess. Whether protection depends on ganglioside action at the level of the glutamate receptor or the postreceptorial biochemical machinery has yet to be established.

IV. Cellular and Molecular Aspects of Ganglioside Action

A. Gangliosides and Ion Fluxes

1. *Sodium Flux*

Gangliosides were proven to influence Na^+ flux by modulating the activity of both (Na^+,K^+)-ATPase and Na^+ channel(s). Exogenous gangliosides, particularly G_{M1}, can both activate and inhibit (Na^+,K^+)-ATPase in preparations of neuronal membranes (Caputto *et al.*, 1977; Jeserich *et al.*, 1981; Leon *et al.*, 1981; Nagata *et al.*, 1987). The effect depends on the number of ganglioside molecules that are inserted into the neuronal membrane, a maximal activation being obtained at a certain degree of insertion (nanomolar range), an inhibition at higher concentrations. The activation does not imply any interference on the ouabain-binding site of the enzyme, or modifications of the apparent K_m for ATP. It is associated with an increase of the apparent V_{max} of the reaction, suggesting unmasking of cryptic molecules of the enzyme on addition of exogenous gangliosides. This implies the occurrence of strict interactions of gangliosides with the enzyme within the membrane (Esmann *et al.*, 1988). The activatory effect of ganglioside on (Na^+,K^+)-ATPase has also been demonstrated *in vitro*. It is known that peripheral nerves of animals with experimental diabetes have a markedly lower (Na^+,K^+)-ATPase activity than normal controls (Greene *et al.*, 1987). The enzyme levels are restored to the normal values after chronic treatment with G_{M1} (Bianchi *et al.*, 1988). A similar recovery of (Na^+,K^+)-ATPase levels by administration of exogenous gangliosides was observed in animals submitted to nigrostriatal and fimbria fibrosa transection or to ischemia (Li *et al.*, 1986; Fass *et al.*, 1987; Mahadik *et al.*, 1989), conditions known to cause a marked decrease of this enzyme activity. A modulatory effect of exogenous G_{M1} on Na^+ channels has been demonstrated by Spiegel *et al.* (1986) in cultured toad kidney epithelia.

2. Ca^{2+} *Flux*

The various aspects of the ganglioside interactions with Ca^{2+}, and the effects of gangliosides on both the Ca^{2+}-activated ATPase and Ca^{2+} channels have been comprehensively reviewed by Rahmann (1992). The same author presented a perfected version of the previously reported (Rahmann, 1983) hypothesis that calcium–ganglioside interactions act as modulators of neuronal function, particularly synaptic transmission and long-term adaptation (see also Thomas and Brewer, 1990). The experimental evidence supporting the hypothesis is the following: (1) gangliosides bind Ca^{2+} through the carboxyl groups and the cation-binding pocket, when present in their molecule; (2) binding with Ca^{2+} facilitates lateral phase separation of gangliosides, with formation of enriched microdomains (clusters); (3) gangliosides modulate the activity of Ca^{2+}-activated ATPase (Slenzka *et al.*, 1990) and Ca^{2+} channels (Spiegel, 1988); (4) gangliosides, especially those carrying disialosyl residues, appear to be located in a clustered organization on the surface of outgrowing nerve fibers and synaptic contact zones (Seybold *et al.*, 1989); (5) Ca^{2+} appears to be accumulated particularly within the synaptic cleft (Kortje *et al.*, 1990a), and Ca^{2+}-ATPase at the inner side of both presynaptic and postsynaptic membranes (Kortje *et al.*, 1990b). Rahmann's hypothesis, with some modifications (Thomas and Brewer, 1990), is presented in Table III.

B. Gangliosides and Protein Phosphorylation

1. *Protein Kinases*

Most of the signal transduction mechanisms operating in the cell are coupled with protein phosphorylation. In this process, specific protein kinases are involved that are located mostly intracellularly (endoprotein kinases), and a few are located on the cell surface (ectoprotein kinases) (Nagai and Tsuji, 1989). Many protein kinases, particularly those expressed in the nervous system, have been shown to be modulated by gangliosides (see Table IV). Modulation consists of activation or inactivation. One important point concerning the effect of gangliosides on protein kinases is the specificity of their action. In all cases, with the exception of tyrosine protein kinase associated with receptors of specific growth factors [epidermal growth factor (EGF), platelet-derived growth factor (PDGF), insulin] (Igarashi *et al.*, 1989), different gangliosides can affect the enzyme action, but at different concentrations. For example, G_{T1b} is a more potent inhibitor of brain protein kinase C and Ca^{2+}/calmodulin kinase II than G_{D1a}, and G_{D1a} is more potent than G_{M1} (Kreutter *et al.*, 1987; Fukunaga *et al.*, 1990). Similarly, G_{Q1b} is a much more potent activator of GOTO neuroblastoma cell ectoprotein kinase than G_{T1b}, G_{T1b} is more potent than G_{D1a}, and G_{D1a} is more potent than G_{M1} (Nagai *et al.*, 1986). A simple calculation shows that, using

Table III

THE ROLE OF CA^{2+}–GANGLIOSIDE INTERACTIONS IN SYNAPTIC TRANSMISSION: A HYPOTHESIS FOR GANGLIOSIDE INVOLVEMENT IN SYNAPSE FUNCTION[a]

1. Gangliosides act as modulators of the key proteins (Ca^{2+} channel, Ca^{2+}-ATPase, fusogenic proteins) involved in neurotransmission
2. In the resting state, the presynaptic membrane is refractory to ion permeation and membrane fusion due to the local rigidity conferred by clustering of gangliosides and (glyco)proteins through Ca^{2+} bridges. These bridges increase the actual concentration of Ca^{2+} (partly in a bound form) of the membrane surface and confer a closed conformation to Ca^{2+} channels
3. The action potential dissociates ganglioside Ca^{2+}–(glyco)protein complexes, with Na^+ substituting for Ca^{2+}. This causes Ca^{2+} channel facilitation to voltage-dependent opening and distribution of gangliosides on the membrane in a molecular (free) way. As a consequence, the membrane destabilizes and becomes more fluid, facilitating the interactions of fusogenic, or exocytosis-inducing, proteins, and leading to complete fusion of presynaptic membrane and synaptic vesicles. The higher availability of free negative charges provided by "free" gangliosides may facilitate exit of the positively charged neurotransmitter molecules
4. During repolarization, gangliosides enhance the transport capacity of the Ca^{2+}-ATPase, facilitating efflux of Ca^{2+} from the nerve terminal. The ganglioside Ca^{2+}–(glyco)protein complexes are reconstituted, and the membrane returns to the starting ridigity

[a]For details, see the reviews of Thomas and Brewer (1990) and Rahmann (1992).

other gangliosides at a concentration equal to that which is optimal for the most potent effector, no effect is produced. Therefore, specificity has to be related to active concentration. Moreover, in the above examples, the most potent gangliosides are multisialosylated, the least potent being G_{M1}, and G_{M1} can easily be produced from the multisialosylated gangliosides by the action of sialidase. This enzyme may thus transform an active ganglioside into another one that, at the actual concentration, has no activity. A second point is whether ganglioside interacts directly with the enzyme, or with the enzyme substrate. In general, gangliosides directly affect thc kinase. However, in the case of the kinase phosphorylating the different components of myelin basic protein (14, 18.5, and 62 kDa), gangliosides exert their effect (inhibitory on the lower molecular weight components, activatory on the higher molecular weight component) at the substrate level (Chan, 1989a,b). The formation of tight complexes between myelin basic protein(s) and gangliosides, namely G_{M1}, has been proven (Chan *et al.*, 1990).

It should be emphasized that administered gangliosides, particularly G_{M1}, were also shown to exert an influence on some protein kinases *in vivo* (Magal *et al.*, 1990). Magal *et al.* submitted fetal rats to ischemia and observed a significant shift of brain protein kinase C from the cytosolic to membrane-bound form with concurrent rise of Ca^{2+}/phosphatidylcholine-independent protein kinase (protein kinase M). Intraperitoneal administration of gangliosides almost entirely prevented both protein kinase C shift and rise of protein kinase M.

Table IV
EFFECT OF GANGLIOSIDES ON PROTEIN KINASES

Tissue	Type of protein kinase	Ganglioside involved	Effect	Reference
Brain	Protein kinase C	$G_{T1b} > G_{D1a} > G_{M1}$	Inhibition	Kreutter *et al.* (1987)
	Protein kinase C	G_{M3}	Activation	Momoi (1986)
Brain	Ganglioside-stimulated protein kinase (Ca^{2+}-independent)	$G_{T1b} > G_{D1a} > G_{D1b} > G_{M1}$	Activation	Chan (1987, 1989a)
Brain	Ganglioside-stimulated protein kinase II (Ca^{2+}/calmodulin-dependent)	$G_{T1b} > G_{D1a} > G_{M1}$	Activation	Goldenring *et al.* (1985); Fukunaga *et al.* (1990)
Brain	Ganglioside-inhibited protein kinase		Inhibition	Chan (1988)
GOTO, NB-1 cells	Ganglioside-stimulated ectoprotein kinase (Ca^{2+}-dependent)	$G_{Q1b} >> G_{T1b} > G_{D1a} > G_{M1}$	Activation	Tsuji *et al.* (1985)
Brain	Ca^{2+}/phosphatidylserine-independent protein kinase M	G_{M1}	Inhibition	Magal *et al.* (1990)
Brain	Ganglioside-dependent protein kinase (Ca^{2+}-independent) (acting on myelin basic protein)	G_{T1b}	Activation	Chan (1989b)
PC12 cells	Ca^{2+}/calmodulin-dependent protein kinase	G_{M1}	Activation	Hilbush and Levine (1991)
KB and A431 cells	Tyrosine protein kinase	G_{M3} > NeuAcnLc_4 > G_{M1} Lyso-G_{M3} de-*N*-acetyl-G_{M3}	Reduced phosphorylation of EGF receptor	Bremer *et al.* (1986) Igarashi *et al.* (1989)
Swiss 3T3 cells	Tyrosine protein kinase	G_{M1} > NeuAcnLc_4 > G_{M3}	Reduced phosphorylation of PDGF receptor	Igarashi *et al.* (1989)
HL-60 cells	Tyrosine protein kinase	NeuAcnLc_4	Reduced phosphorylation of insulin receptor	Nojri *et al.* (1991)

A relevant problem is whether endogenous gangliosides interact with protein kinases *in vivo*. The Ca^{2+}/calmodulin-dependent protein kinase II is localized at synaptic junctions (Quimet *et al.*, 1984) where gangliosides are located too, especially those carrying a disialosyl residue, which are the most active on the enzyme. With regard to the intracellularly located protein kinases, it cannot be excluded that they can interact with the gangliosides occurring in the cytosol (Sonnino *et al.*, 1979) or with ganglioside metabolites. The ectoprotein kinase described by Tsuji *et al.* (1985) in GOTO neuroblastoma cells is assumed to be located on the outer leaflet of the plasma membrane. It is suggested (Nagai and Tsuji, 1989) that the enzyme can be activated by interacting with a ganglioside molecule present either in the membrane of the same cell or of a vicinal cell.

2. *Cyclic Nucleotide Phosphodiesterase and Protein Dephosphorylation*

Brain phosphodiesterase acting on cyclic AMP was shown to be markedly stimulated by micromolar concentrations of gangliosides (Yates *et al.*, 1989). G_{T1b} exerted the highest effect, followed by G_{D1a} and G_{M1}. G_{T1b} stimulated the enzyme by increasing V_{max} and decreasing K_m (in a manner that is similar to that of calmodulin), whereas the effect of G_{D1a} and G_{M1} was only on V_{max}. These findings suggest that the process of protein dephosphorylation can be affected by gangliosides. Supporting this suggestion is the evidence (Bassi *et al.*, 1991) that the phosphorylation of five different proteins present in the crude mitochondrial fraction of rat brain is influenced by ganglioside G_{D1b} and its lactone form. G_{D1b} markedly enhanced the phosphorylation of some proteins and decreased that of others, whereas G_{D1b}-lactone exerted an activatory effect lower than that caused on the same proteins by G_{D1b}, or did not exhibit any effect at all.

C. Gangliosides and Neuroreceptors

The implication of gangliosides in receptor function has a lot of experimental support. Gangliosides themselves were suggested to act as receptors after the demonstration that G_{M1} (with the addition of Fuc-G_{M1}; Masserini *et al.*, 1992) specifically interacts with the B subunit of cholera toxin and is instrumental to the penetration of the active A subunit into the cell membrane (Brady and Fishman, 1979; Fishman, 1982). Other toxins (tetanus, botulinum, α-staphylococcal toxins, *E. coli* enterotoxin, *Vibrio parahemolyticus* hemolysin, and Sendai virus) might use gangliosides as membranes receptors (see Ledeen *et al.*, 1990). However, in all these cases, it is not yet clear whether gangliosides or glycoproteins constitute the physiological receptor.

Gangliosides exert a modulatory role on EGF, PDGF, and insulin receptors (Hakomori, 1990; Igarashi *et al.*, 1989; Weis and Davis, 1990). Generally, an individual ganglioside, or a ganglioside metabolic derivative, is involved. In the case of the EGF receptor, ganglioside G_{M3} inhibits receptor phosphorylation and dimerization, with consequent block of cell growth; in contrast, deNAc-G_{M3}, the derivative lacking the acetyl group on the sialic acid moiety, has an opposite effect on the same target.

As already anticipated, gangliosides are likely to act as modulators of the receptors for neurotrophic and neuritogenic factors (Varon *et al.,* 1988). Berry-Kravis and Dawson (1985) reported that exogenous gangliosides have a dramatic effect on the serotonin receptor-coupled adenylate cyclase in cultured NCB-20 cells, and Agnati *et al.* (1983) observed that peripherally administered gangliosides caused modulation of serotonin receptor function. A possible regulatory role of G_{M1} in the G_s-linked opioid receptor function has been suggested (Shen and Crain, 1990), with experiments showing that pretreatment of dorsal root ganglion neurons with the B subunit of cholera toxin or anti-G_{M1} antibodies blocked opioid-induced prolongation of the action potential duration.

The uptake of choline seems also to be affected by G_{M1}. In fact, treatment with micromolar concentrations of this ganglioside prevents the decrease of high-affinity choline uptake caused by the cholinergic neurotoxin hemicholinium[3*a*,*a*′]bis[di(2-chloroethyl)amino]-4, 4′,2-biacetophenone (Maysinger *et al.,* 1992). It was also observed that, after traumatic lesion of the rat nucleus basalis, the nerve growth factor-mediated recovery from the fall of high-affinity choline uptake was significantly potentiated by G_{M1} administration (Di Patre *et al.,* 1989).

Another aspect of the receptor-mediated interactions at the cell surface, where gangliosides are involved, is cell–cell adhesion processes. Cheresh *et al.* (1986, 1988) showed that G_{D2} accumulates near the vitronectin receptor and, in the presence of Ca^{2+}, facilitates adhesion of a number of cells (including neuroblastoma cells) to the pericellular supporting structure by formation of a bridge between the receptor and the pericellular structure. Supporting this line is the evidence (Mugnai *et al.,* 1984) that specific gangliosides accumulate in substratum adhesion sites of neuroblastoma cells, and the observation (Marchase, 1977) that a ganglioside, or related glycosphingolipid, is implicated in the preferential adhesion of chick retinal cells to the surfaces of intact optic tecta.

An interesting line of investigation concerns the presence in brain of proteins (probably glycoproteins) that act as membrane receptors for gangliosides (Yasuda *et al.,* 1988; Tiemeyer *et al.,* 1990a,b; Fueshko and Schengrund, 1990; Sonnino *et al.,* 1992). This evidence introduces two important concepts: (1) that gangliosides are instrumental to selective cell recognition processes based on carbohydrate–carbohydrate interaction (Hakomori, 1990), (2) that the binding of gangliosides to their receptors may trigger internalization of the complex via a receptor-mediated endocytosis process.

D. Gangliosides and Metabolic Second Messengers

1. Cyclic AMP

Partington and Daly (1979) observed that addition of gangliosides to rat cerebral membranes caused 50–95% activation of adenylate cyclase. The evidence was confirmed on cat cortical membranes by Claro *et al.* (1991). Cyclic nucleotide phosphodiesterase can also be activated by gangliosides, provided that proper conditions and critical concentrations of gangliosides are employed (Davis and Daly, 1980; Yates *et al.,* 1989).

2. Inositol Phosphates and Diacylglycerol

Treatment of primary neuron cultures and Neuro-2a neuroblastoma cells with gangliosides produced a substantial, although delayed, increase of phosphoinositide breakdown (Ferret *et al.,* 1987; Skaper *et al.,* 1987; Vaswani *et al.,* 1990). Treatment with gangliosides also resulted in increased formation of diacylglycerol (Leray *et al.,* 1988). In brain cortical membranes from cats affected by G_{M1}- and G_{M2}-gangliosidosis, the activation of phosphoinositide-specific phospholipase C by guanine nucleotide-binding proteins appears to be markedly impaired (Claro *et al.,* 1991). This suggests an action of gangliosides on the mechanism of activation of the phosphoinositide-specific phospholipase C. A further effect of gangliosides in modulating the concentration of diacylglycerol was discovered by Freysz *et al.* (1991), who observed in primary cultures of chicken neurons a pronounced time- and dose-dependent activation of mono- and diacylglycerol lipase by gangliosides.

3. Eicosanoids

The formation of eicosanoids seems to be influenced by gangliosides. In fact, the elevation of the levels of arachidonic acid cyclooxygenase and lipooxygenase metabolites, which takes place in rat brain after ischemic injury and reperfusion, is almost totally prevented by administration of G_{M1} or G_{M1}-lactone (Bertazzo *et al.,* 1988; Petroni *et al.,* 1989).

V. Ganglioside Turnover in Neural Cells and Formation of Metabolic Regulators

A. An Outline of Ganglioside Metabolism in Neural Cells

The following events characterize ganglioside metabolism in neural cells (Tettamanti *et al.,* 1988; Schwarzmann and Sandhoff, 1990): *de novo* biosynthesis, degradation, direct glycosylation following internalization, and recycling of degradation fragments for biosynthesis. These processes are likely to be concurrent and cooperate to determine the final pattern of neural gangliosides.

1. *De Novo Biosynthesis*

The oligosaccharide chain of gangliosides is formed by sequential addition of monosaccharide units to ceramide, which is biosynthesized in the endoplasmic reticulum (Walter *et al.*, 1983). The process takes place in the Golgi apparatus, with the involvement of specific membrane-bound glycosyltransferases and the corresponding sugar-nucleotides (Hirshberg and Snider, 1987). The reaction sequences for the biosynthesis of the ganglio series gangliosides and the involved enzymes (designated according to Basu *et al.*, 1987; Basu, 1991) are shown in Fig. 2. Initiation of the three distinct ganglioside lines (A, B, C) is dependent on the strict specificity of SAT-1, SAT-2, and SAT-3, which act on lactosylceramide, G_{M3}, and G_{D3}, respectively. Further glycosylations are catalyzed by enzymes

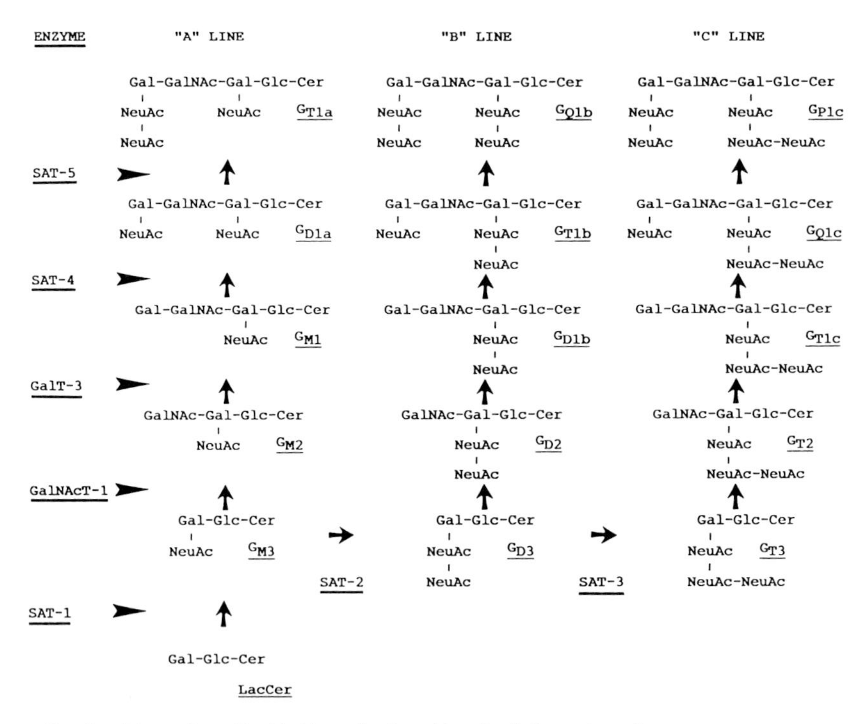

FIG. 2. Lines of ganglioside biosynthesis and involved glycosyltransferases.

having broader specificity. Particularly, and as first reported by Pohlentz *et al.* (1988), the single enzyme GalNacT-1 converts lactosylceramide to G_{A2} (the asialo derivative of G_{M2}), G_{M3} to G_{M2}, and G_{D3} to G_{D2}. Similarly, (1) GalT-3 catalyzes the transformation of G_{M2} to G_{M1}, and G_{D2} to G_{D1b}; (2) SAT-4 converts G_{M1} to G_{D1a}, G_{D1b} to G_{T1b}, and G_{T1c} to G_{Q1c} (SAT-4 is also capable of transforming lactosylceramide to G_{M3}); and (3) SAT-5 converts G_{D1a} to G_{T1a}, G_{T1b} to G_{Q1b}, and G_{Q1c} to G_{P1c} (SAT-5 also converts G_{D3} to G_{T3}).

As shown in liver and extraneural cells in culture, the glucosyltransferase forming glucosylceramide from ceramide shows an aspecific distribution along the different Golgi cisternae (Futerman and Pagano, 1991; Jeckel *et al.*, 1992) and has the active site oriented toward the cytosol (Coste *et al.*, 1986; Futerman and Pagano, 1991; Trinchera *et al.*, 1991b; Jeckel *et al.*, 1992). The galactosyltransferase forming lactosylceramide also seems to face the cytosolic side of the Golgi cisternae (Trinchera *et al.*, 1991c), whereas the subsequent glycosyltransferases show a luminal orientation (Schwarzmann and Sandhoff, 1990). Therefore, lactosylceramide should be translocated from the cytosolic to the luminal side of the Golgi membrane in order to be further glycosylated. The glycosyltransferases producing the different gangliosides from lactosylceramide show a gradient distribution on the Golgi system. Earlier sialosylations prevail in the cis/medial Golgi, and later glycosylations in the trans-Golgi/trans-Golgi network (Pohlentz *et al.*, 1988; Trinchera and Ghidoni, 1989; Trinchera *et al.*, 1990; Iber *et al.*, 1992). This implies that the growing glycolipid moves from the endoplasmic reticulum through the Golgi cisternae, presumably via a vesicle flow (Wattenberg, 1990), which is also assumed to transport the mature gangliosides to the plasma membrane of the neuron body. Newly synthesized gangliosides reach axons and nerve terminal membranes by fast axonal transport (Ledeen, 1989; Goodrum *et al.*, 1989). A scheme of the subcellular aspects of ganglioside metabolism in the neuron is presented in Fig. 3.

2. *Degradation*

Ganglioside catabolism consists of the sequential removal of individual sugar residues by exoglycohydrolases, with formation of ceramide (Sandhoff *et al.*, 1987). Ceramide is then degraded by ceramidase into sphingosine and fatty acid (Spence *et al.*, 1986). The glycohydrolases involved in neural ganglioside degradation reside in the lysosomes. In fact, gangliosides accumulate especially in the neurons after administration of the lysosomotropic drug chloroquine (Klinghardt *et al.*, 1981; Riboni *et al.*, 1991) and in the hereditary lysosomal diseases characterized by a defect of enzymes involved in ganglioside degradation (O'Brien, 1989; Sandhoff *et al.*, 1989). Moreover, highly purified lysosomes isolated from rat brain were proven to carry enzymes affecting gangliosides (Fiorilli *et al.*, 1989).

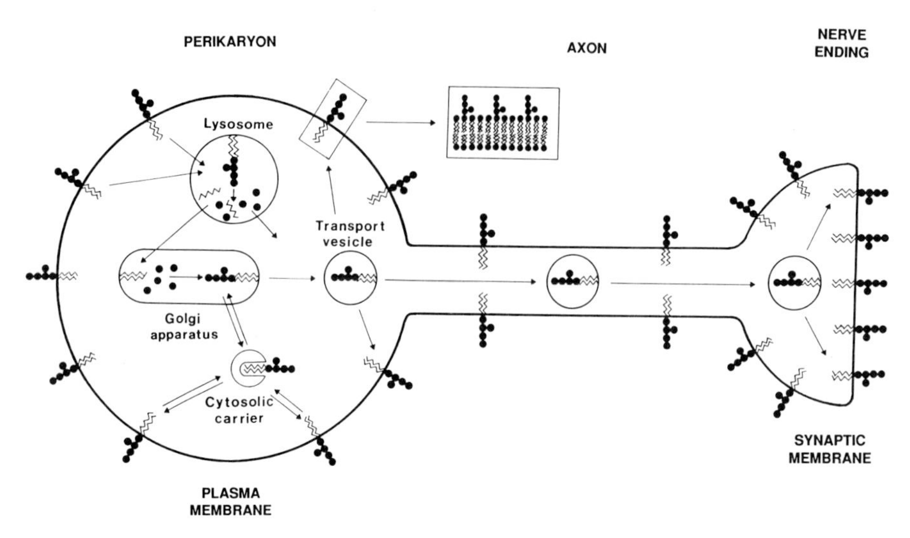

FIG. 3. Subcellular location and traffic of neuronal gangliosides.

3. *Recycling of Degradation Fragments for Biosynthesis (Salvage Pathway)*

The existence of recycling processes was demonstrated by experiments where animals or cultured cells were supplied with gangliosides carrying radioactivity in different portions of the molecule, and the metabolic fate of the radioactivity was followed. The assumption was that exogenous gangliosides, after being taken up by cells, enter the pool of endogenous gangliosides and undergo regular metabolic processing (Ghidoni *et al.,* 1989a; Schwarzmann *et al.,* 1987; Huang and Dietsch, 1991). Using this approach, it was shown that cerebellar granule cells cultivated *in vitro* take up and degrade gangliosides with liberation of galactose, *N*-acetylgalactosamine, sialic acid, fatty acid, and sphingosine, which are reused for the biosynthesis of novel gangliosides, glycoproteins, and phospholipids, particularly sphingomyelin (Ghidoni *et al.,* 1989b; Riboni *et al.,* 1990, 1991, 1992). Recycling of more complex fragments, like ceramide and glucosylceramide (the latter was shown to be recycled in the liver; Trinchera *et al.,* 1991a), cannot be excluded. The process of recycling is blocked by inhibiting lysosome function by chloroquine or by preventing endocytosis by low temperature treatment (Riboni *et al.,* 1991). This suggests (see Fig. 4) a primary role of lysosomes in the formation of compounds to be metabolically recycled and implies the exit of these metabolites from lysosomes, possibly by the action of specific carriers (Pisoni and Thoene, 1991).

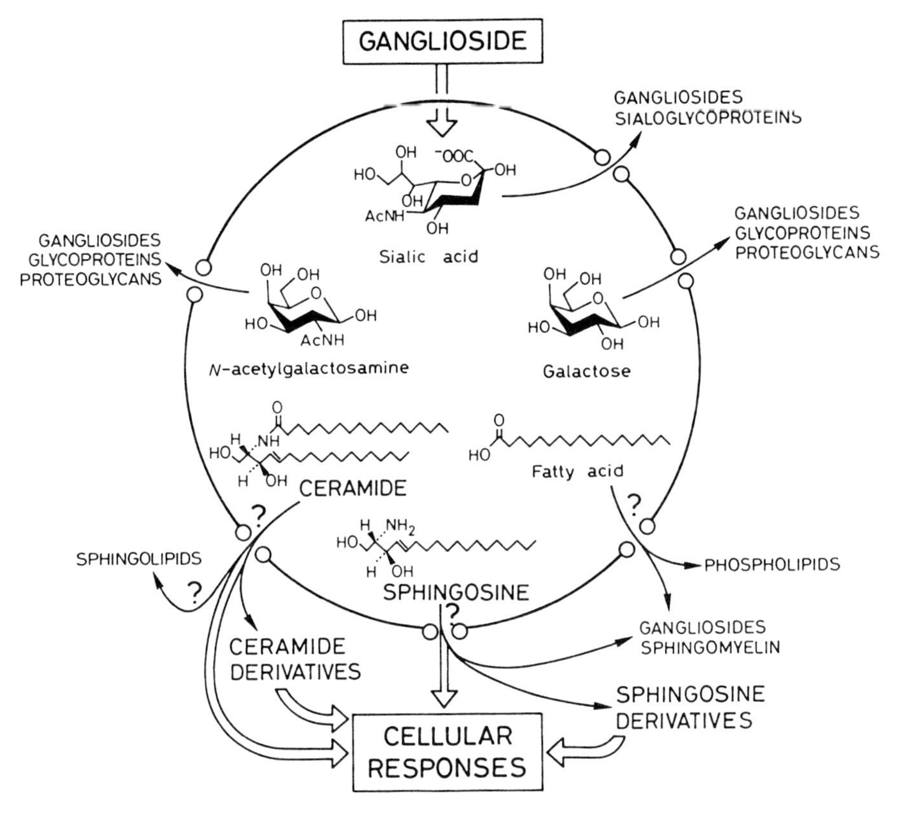

FIG. 4. Gangliosides and formation of metabolic second messengers of sphingoid nature: intralysosomal formation and exit from lysosomes of fragments of ganglioside degradation.

4. *Direct Glycosylation*

It consists of the formation of more complex gangliosides by glycosylation of membrane-bound gangliosides internalized in the cell. The occurrence of this process was demonstrated by the use of exogenously administered labeled gangliosides. For example, administration to rats (Ghidoni *et al.*, 1989a) or cultured neurons (Ghidoni *et al.*, 1989b) of G_{M1} radiolabeled in the terminal galactose moiety resulted in the formation of G_{D1a}, and that of G_{M2}, radiolabeled in the hexosamine moiety, in G_{M1} and G_{D1a}. In cultured neurons, inhibition of endocytosis (Riboni *et al.*, 1991) blocked direct glycosylation, implying that the exogenously added ganglioside had to be internalized to reach the intracellular site where glycosylations occur (presumably the Golgi apparatus). An important consequence of this evidence is the necessity to assume a specific sorting of the ganglioside-carrying vesicle from the early endosomal compartment to the Golgi apparatus in parallel to and independently from the sorting to lysosomes. The existence of this kind of lipid traffic is strongly supported by the evidence that, in cultured fibroblasts (Schwarzmann *et al.,* 1987), cultured kidney cells (Kok *et al.,* 1989), and rat liver (Trinchera *et al.,* 1991a), administration of a pulse of biotinylated, fluorescent, or radiolabeled glycolipids, respectively, is followed by appearance of the same compounds in the Golgi apparatus.

B. The Plasma Membrane in Ganglioside Turnover

The terminal sugars of gangliosides, particularly sialic acid, might undergo a local turnover at the plasma membrane level. It should be remembered that both sialidase and sialyltransferase activities are present in brain. Brain membrane-bound sialidase, first reported by Schengrund and Rosenberg (1970), was shown to be linked to synaptosomal plasma membranes (Tettamanti *et al.,* 1972) and to be enriched in the synaptic region (Schengrund and Nelson, 1976). This enzyme appears to be anchored to the membranes by a glycosylphosphatidylinositol bridge (Chiarini *et al.,* 1990). Moreover, a myelin-associated sialidase was described (Yohe *et al.,* 1983; Saito and Yu, 1986) that might play a role in myelin sheath maturation (Saito and Yu, 1992). On the other hand, sialyltransferases occur in the outer neural surface (Schengrund and Nelson, 1975; Preti *et al.,* 1980; Durrie *et al.,* 1987) with a specificity different from that of the main Golgi sialyltransferases (Durrie *et al.,* 1988).

The proof that desialosylation of gangliosides by membrane-bound sialidase is operative in living cells has been provided by Riboni *et al.* (1991), using primary cultures of cerebellar neurons. These authors observed that G_{D1a} and G_{D1b}, inserted into the cell plasma membrane, could be degraded to G_{M1} under conditions of complete block of lysosomal function by chloroquine, and of endocytosis by low temperature treatment. In parallel, studies performed on brain slices in the presence of CMP-NeuAc demonstrated the local labeling of membrane gangliosides by the plasma membrane-bound sialyltransferase (Durrie *et al.,* 1988). All together, these findings support the hypothesis that neural plasma membranes possess a sialosylation–desialosylation system that may modulate locally the degree of ganglioside sialosylation. It cannot be excluded that this system is instrumental to some functional performances of the neuron and/or to triggering ganglioside passage from the plasma membrane to the cell interior.

C. Role of Endocytosis in Ganglioside Turnover

Under culture conditions, cells are estimated to internalize via endocytosis about half their plasma membrane per hour (Steinman *et al.,* 1983). This event is followed by a parallel process of recycling or resynthesis of the membrane components (lipids and proteins), in order to maintain dynamically the cell surface area. The overall turnover rates of gangliosides, determined on whole brain, are quite variable, with half-lives ranging from a few days to several weeks (Ledeen, 1989). These lengthy turnover values, which contrast with the rapid turnover of membranes, might be interpreted considering that either (1) most of the ganglioside molecules are not involved in every endocytotic process, or (2) most of the same molecules are directly recycled to the membrane by transcytosis or via early endosomes, or (3) the saccharide and lipid fragments obtained from degradation of internalized gangliosides are rapidly used for biosynthesizing new molecules with a high degree of salvage.

Gangliosides, as well as glycosphingolipids in general, appear to undergo endocytosis (Schwarzmann *et al.*, 1987; Van Meer, 1989; Schwarzmann and Sandhoff, 1990; Pagano, 1990) in a number of cells, including neural cells (Van Echten *et al.*, 1990; Riboni and Tettamanti, 1991). A definite proof that endocytosis of membrane-bound gangliosides follows a receptor-mediated endocytic pathway is not available. However, the observation that, in both cultured fibroblasts and cerebellar granule cells (Sonnino *et al.*, 1989, 1992), exogenous G_{M1} (labeled with a photoreactive probe) rapidly binds to one (or few) membrane-bound protein(s) before being internalized and metabolized, strongly suggests this hypothesis. The intracellular destination of endocytosed gangliosides is only partly understood. Thus far, no approaches have been devised to inspect ganglioside transcytosis and direct return to the plasma membrane via the early endosome intracellular flow. Instead, convincing evidence was provided for vesicle transport to lysosomes (Schwarzmann *et al.*, 1987; Riboni and Tettamanti, 1991) and to the Golgi apparatus (Schwarzmann *et al.*, 1987; Tettamanti *et al.*, 1988). An evaluation of the percentages of distribution of endocytosed gangliosides to (1) direct recycling to the plasma membranes, (2) lysosomes, and (3) Golgi apparatus is presently impossible. However, a definite and substantial portion seems to be targeted to lysosomes.

Internalization of gangliosides by endocytosis is a rapid process. In cultured cerebellar granule cells fed with radiolabeled gangliosides, compounds of catabolic origin started being detectable 10–15 minutes after the pulse, and compounds of biosynthesis from recycled catabolites 15–30 minutes after the pulse (Riboni and Tettamanti, 1991). Both events appeared to be totally prevented by treating cells at 4°C, a condition known to stop endocytosis (Riboni *et al.*, 1991). Also, in a study where undifferentiated and differentiated cerebellar granule cells were employed, uptake and metabolic processing of exogenous ganglioside appeared to be dependent not on the cell unit but on the area of plasma membrane per cell, which is obviously higher in differentiated than undifferentiated cells (Riboni *et al.*, 1990). All this means that metabolic processing is strictly linked to endocytosis and is as rapid an event as endocytosis.

Pulse-chase experiments on cerebellar granule cells fed with G_{M1} carrying the radioactivity in the sphingosine moiety showed that only a small percentage of the sphingosine produced by ganglioside breakdown underwent complete degradation, as compared to the sphingosine that was metabolically recycled (Riboni *et al.*, 1990). Similar results (authors' unpublished observations) were obtained using exogenous gangliosides carrying the radioactivity in different sugar moieties. Therefore, it can be concluded that the ganglioside present in the membrane domain subjected to internalization via endocytosis is rapidly turned over, with a high incidence of salvage pathways.

D. Gangliosides as Precursors of Sphingosine and Ceramide

Studies have demonstrated that sphingosine and ceramide, together with sphingosine-1-phosphate, *N*-monomethylsphingosine, *N,N*-dimethylsphingosine, and ceramide-1-phosphate, exert a powerful regulatory effect on enzymes such as protein kinase C, which are fundamental in the control of cellular metabolism (Hannum and Bell, 1989; Igarashi *et al.*, 1989; Merrill and Jones, 1990; Dressler and Kolesnick, 1990). This evidence prompted the hypothesis that sphingosine and ceramide serve as metabolic second messengers (Hakomori, 1990; Merrill, 1991) and that sphingolipids may produce them under particular conditions of cell stimulation by external substances. Two questions appear to be crucial in assessing the validity of the above hypothesis: (1) by which pathway are sphingosine and ceramide formed from sphingolipids, and (2) which external stimuli are able to modify the rate of production of sphingosine and ceramide from sphingolipids?

Sphingosine and ceramide are present in free form in cultured cells and tissues (Van Veldhoven *et al.*, 1989; Dressler and Kolesnick, 1990; Goldkorn *et al.*, 1991), including brain (Merrill *et al.*, 1988). Both molecules are assumed to be formed during sphingolipid degradation. Quite recently it was demonstrated that ceramide and sphingosine are formed in cultured cerebellar granule cells during degradation of administered G_{M1} (radiolabeled at the sphingosine moiety) (Riboni *et al.*, 1992) (Fig. 5). Both substances appeared to be produced very rapidly (10 and 15 minutes after pulse for ceramide and sphingosine, respectively) and in higher amounts in differentiated than undifferentiated neurons. Since the block of endocytosis or lysosomal function inhibited ganglioside degradation as well as ceramide and sphingosine production, endocytosis mediates the process and lysosomes appear to be the compartments where these molecules are formed. In the above study, where a 1-hour pulse–4-hour chase with 2×10^{-6} *M* exogenous ganglioside was employed, it was calculated that 2 and 20 pmol/10^6 cells of sphingosine and ceramide were produced, respectively. Assuming that the levels of free sphingosine and ceramide in cerebellar granule

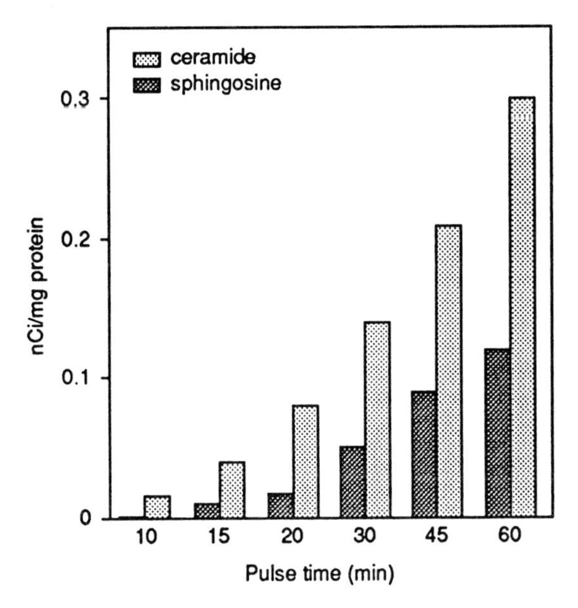

FIG. 5. Formation of sphingosine and ceramide by cerebellar granule cells in culture fed with 10^{-5} *M* G_{M1} ^{3}H-labeled at the level of the sphingosine moiety.

cells are in the range of those determined in a number of cells (3–20 and 130–230 pmol/10^6 cells, respectively (Van Veldhoven *et al.*, 1989; Dressler and Kolesnick, 1990; Goldkorn *et al.*, 1991), it should be inferred that gangliosides contribute to the maintenance of free sphingosine and ceramide levels under near-basal conditions.

Whether and how sphingosine and ceramide, produced by ganglioside degradation in the lysosomes, leave the lysosomal apparatus in order to be available at the subcellular sites where they exert their regulatory role are open to question. Sphingosine surely leaves the lysosomal apparatus, since it is recycled for the Golgi apparatus-assisted biosynthesis of both gangliosides and sphingomyelin (Tettamanti *et al.*, 1988; Riboni *et al.*, 1990; Riboni and Tettamanti, 1991).

VI. Conclusion and Perspectives

The direct implications of gangliosides in neural functions are suggested by a large crop of experimental evidence. In envisaging the molecular bases of these implications, two main questions arise: (1) the relationship between compositional diversity and functional involvement of gangliosides, and (2) the mechanism connecting gangliosides to different functional events. Regarding the first question, the oligosaccharide chains of gangliosides carry a high interaction potential that enables different and specific binding to a variety of ligands. Attachment of this chain to the hydrophobic portion transmits the conformation changes arising to the dynamic behavior of gangliosides in the membrane. With regards to the second question, the concept of the multifunctional potential of gangliosides has to be kept in mind. A variety of influences derive from specific interactions of gangliosides with proteins (receptors, enzymes, ion channels and carriers) at the membrane level, resulting in modulation of both sensitivity and response of cells to external ligands, particularly neurotrophic and neurodifferentiation factors. A challenging hypothesis is that gangliosides contribute to generate metabolic second messengers, like sphingosine and ceramide. According to this hypothesis, external ligand binding to gangliosides affects the rate of ganglioside endocytosis and metabolic processing, thus modifying the concentration of second messengers of sphingoid nature. Also, in this case, the final effect is a modification of cellular responses. Definition of the various aspects of this new perspective and assessment of its applicability to precise neural functions require novel experimental designs and much work.

References

Abrahamsson, S., Dahlén, B., Lofgren, H., Pascher, J., and Sandell, S. (1977). *In* "Structure of Biological Membranes" (S. Abrahamsson and J. Pascher, eds.), pp. 1–23. Plenum, New York.

Agnati, L. F., Benfenati, F., Battistini, N., Cavicchioli, L., Fuxe, K., and Toffano, G. (1983). *Acta Physiol. Scand.* **117,** 311–314.

Bassi, R., Chigorno, V., Fiorilli, A., Sonnino, S., and Tettamanti, G. (1991). *J. Neurochem.* **57,** 1207–1211.

Basu, M., De, T., Das, K. K., Kyle, J. W., Chon, H.-C., Schaper, R. J., and Basu, S. C. (1987). *In* "Methods in Enzymology" (V. Ginsburg, ed.), Vol. 138, pp. 575–607. Academic Press, Orlando, FL.

Basu, S. C. (1991). *Glycobiology* **1,** 469–475.

Berg, D. K. (1984). *Annu. Rev. Neurosci.* **7,** 149–170.

Berry-Kravis, E., and Dawson, G. G. (1985). *J. Neurochem.* **45,** 1739–1747.

Bertazzo, A., Petroni, A., Sarti, S., Colombo, C., and Galli, C. (1988). *Pharmacol. Res. Commun.* **20,** 95–98.

Bertoli, E., Masserini, M., Sonnino, S., Ghidoni, R., Cestaro, B., and Tettamanti, G. (1981). *Biochim. Biophys. Acta* **467,** 196–202.

Bianchi, R., Marini, P., Merlini, S., Fabris, M., Triban, C., Mussini, E., and Fiori, M. G. (1988). *Diabetes* **37,** 1340–1345.

Brady, R. O., and Fishman, P. H. (1979). *Adv. Enzymol.* **50,** 303–324.

Bremer, F., Schlessinger, J., and Hakomori, S. I. (1986). *J. Biol. Chem.* **261,** 2434–2440.

Burns, S. P., Chiueh, C. C., Markey, S. P., Ebert, M. H., Jacobowitz, D. M., and Kopin, I. J. (1983). *Proc. Natl. Acad. Sci. U.S.A.* **80,** 4546–4550.

Caputto, R. (1988). *In* "New Trends in Ganglioside Research: Neurochemical and Neurogenerative Aspects" (R. W. Ledeen, E. L. Hogan, G. Tettamanti, A. J. Yates, and R. K. Yu, eds.), Fidia Res. Ser., Vol. 14, pp. 3–14. Liviana Press/Springer-Verlag, Padova/Berlin.

Caputto, R., Maccioni, H., and Caputto, B. L. (1977). *Biochem. Biophys. Res. Commun.* **74,** 1046–1052.

Ceccarelli, B., Aporti, F., and Finesso, M. (1976). *In* "Ganglioside Function" (G. Porcellati, B. Ceccarelli, and G. Tettamanti, eds.), pp. 275–293. Plenum, New York.

Chan, K. F. J. (1987). *J. Biol. Chem.* **262,** 5248–5255.

Chan, K. F. J. (1988). *J. Biol. Chem.* **263,** 568–574.

Chan, K. F. J. (1989a). *Neurosci. Res. Commun.* **5,** 95–104.

Chan, K. F. J. (1989b). *Biochem. Biophys. Res. Commun.* **165,** 93–100.

Chan, K. F. J., Robb, N. D., and Chen, W. H. (1990). *J. Neurosci. Res.* **25,** 535–544.

Cheresh, D. A., Pierschbacher, M. D., Herzig, M. A., and Mujoo, K. (1986). *J. Cell Biol.* **102,** 688–696.

Cheresh, D. A., Pytela, R., Pierschbacher, M. D., Ruoslahti, E., and Reisfeld, R. A. (1988). *In* "New Trends in Ganglioside Research: Neurochemical and Neuroregenerative Aspects" (R. W. Ledeen, E. L. Hogan, G. Tettamanti, A. J. Yates, and R. K. Yu, eds.), Fidia Res. Ser., Vol. 14, pp. 203–217. Liviana Press/Springer-Verlag, Padova/Berlin.

Chiarini, A., Fiorilli, A., Siniscalco, C., Tettamanti, G., and Venerando, B. (1990). *J. Neurochem.* **55,** 1576–1584.

Claro, E., Wallace, M. A., Fain, J. N., Nair, B. G., Patel, T. B., Shanker, G., and Baker, H. J. (1991). *Mol. Brain Res.* **11,** 265–271.

Corti, M., Cantù, L., and Sonnino, S. (1987). *NATO ASI Ser., Ser. H* **7,** 101–118.

Coste, H., Martel, M. B., and Got, R. (1986). *Biochim. Biophys. Acta* **858,** 6–12.

Cuello, A. C. (1990). *Adv. Pharmacol.* **21,** 1–50.

Cuello, A. C., Garofalo, L., Kenisberg, R. L., and Maysinger, D. (1989). *Proc. Natl. Acad. Sci. U.S.A.* **86,** 2056–2060.

Curatolo, W. (1987). *Biochim. Biophys. Acta* **906,** 111–136.

Czarniecki, M. F., and Thornton, E. R. (1977). *J. Am. Chem. Soc.* **99,** 8273–8278.

Davis, C. W., and Daly, J. W. (1980). *Mol. Pharmacol.* **17,** 206–211.

De Erausquin, G. A., Manev, H., Guidotti, A., Costa, E., and Brooker, G. (1990). *Proc. Natl. Acad. Sci. U.S.A.* **87,** 8017–8021.

Di Patre, P. L., Casamenti, F., Cenni, A., and Pepey, G. (1989). *Brain Res.* **480,** 219–224.

Dressler, K. A., and Kolesnick, R. N. (1990). *J. Biol. Chem.* **265,** 14917–14929.

Dreyfus, H., Louis, J. C., Harth, S., and Mandel, P. (1980). *Neuroscience* **5,** 1647–1655.

Dreyfus, H., Ferret, B., Harth, S., Gorio, A., Freysz, L., and Massarelli, R. (1984). *In* "Ganglioside Structure, Function and Biomedical Potential" (R. W. Ledeen, R. K. Yu, M. M. Rapport, and K. Suzuki, eds.), pp. 513–524. Plenum, New York.

Drubin, D., Feinstein, S. C., Shooter, E. M., and Kirschner, M. W. (1985). *J. Cell Biol.* **101,** 1790–1807.

Durand, M., Guerold, B., Lombard-Golly, D., and Dreyfus, H. (1986). *In* "Gangliosides and Neuronal Plasticity" (G. Tettamanti, R. W. Ledeen, K. Sandhoff, Y. Nagai, and G. Toffano, eds.), Fidia Res. Ser., Vol. 6, pp. 295–308. Liviana Press/Springer-Verlag, Padova/Berlin.

Durrie, R., Saito, M., and Rosenberg, A. (1987). *J. Neurosci. Res.* **18,** 456–465.

Durrie, R., Saito, M., and Rosenberg, A. (1988). *Biochemistry* **27,** 3759–3764.

Esmann, M., Marsch, D., Schwarzmann, G., and Sandhoff, K. (1988). *Biochemistry* **27,** 2398–2403.

Facci, L., Skaper, S. D., Favaron, M., and Leon, A. (1988). *J. Cell Biol.* **106,** 821–828.

Fass, B., Ramirez, J. J., Stein, D. G., Mahadik, S. P., and Karpiak, S. E. (1987). *J. Neurosci. Res.* **17,** 45–50.

Favaron, M., Manev, H., Alho, H., Bertolino, M., Ferret, B., Guidotti, A., and Costa, E. (1988). *Proc. Natl. Acad. Sci. U.S.A.* **85,** 7351–7355.

Ferreira, A., Busciglio, J., Landa, C., and Caceres, A. (1990). *J. Neurosci.* **10,** 293–302.

Ferret, B., Massarelli, R., Freysz, L., and Dreyfus, H. (1987). *C.R. Hebd. Seances Acad. Sci.* **304,** 97–99.

Fiorilli, A., Venerando, B., Siniscalco, C., Monti, E., Bresciani, R., Caimi, L., Preti, A., and Tettamanti, G. (1989). *J. Neurochem.* **53,** 672–680.

Fishman, P. H. (1982). *J. Membr. Biol.* **69,** 85–97.

Fishman, P. H. (1988). *In* "New Trends in Ganglioside Research: Neurochemical and Neuroregenerative Aspects" (R. W. Ledeen, E. L. Hogan, G. Tettamanti, A. J. Yates, and R. K. Yu, eds.), Fidia Res. Ser., Vol. 14, pp. 183–201. Liviana Press/Springer-Verlag, Padova/Berlin.

Freysz, L., Farooqui, A. A., Horrocks, L. A., Massarelli, R., and Dreyfus, H. (1991). *Neurochem. Res.* **16,** 1241–1244.

Fueshko, S. M., and Schengrund, C. L. (1990). *J. Neurochem.* **54,** 1791–1797.

Fukunaga, K., Miyamoto, E., and Soderling, T. R. (1990). *J. Neurochem.* **54,** 102–109.

Futerman, H., and Pagano, R. E. (1991). *Biochem. J.* **280,** 295–302.

Ghidoni, R., Fiorilli, A., Trinchera, M., Venerando, B., Chigorno, V., and Tettamanti, G. (1989a). *Neurochem. Int.* **15,** 455–465.

Ghidoni, R., Riboni, L., and Tettamanti, G. (1989b). *J. Neurochem.* **53,** 1567–1574.

Goldenring, J. R., Otis, L. C., Yu, R. K., and DeLorenzo, R. J. (1985). *J. Neurochem.* **44,** 1229–1234.

Goldkorn, T., Dressler, K. A., Muindi, J., Radin, N. S., Mendelsohn, J., Menaldino, D., Liotta, D., and Kolesnick, R. N. (1991). *J. Biol. Chem.* **266,** 16092–16097.

Goodrum, J. F., Stone, G. C., and Morell, P. (1989). *In* "Neurobiology of Glyconjugates" (R. U. Margolis and R. K. Margolis, eds.), pp. 277–308. Pergamon, New York.

Greene, D. A., Lattiner, S. A., and Sima, A. A. F. (1987). *N. Engl. J. Med.* **316,** 599–607.

Hadjiconstantinou, M., and Neff, N. H. (1990). *Eur. J. Pharmacol.* **181,** 137–139.

Hakomori, S. I. (1981). *Annu. Rev. Biochem.* **50,** 733–764.

Hakomori, S. I. (1990). *J. Biol. Chem.* **265,** 18713–18716.

Hannun, Y. A., and Bell, R. M. (1989). *Science* **243,** 500–507.

Hansson, H. A., Holmgren, L., and Svennerholm, L. (1977). *Proc. Natl. Acad. Sci. U.S.A.* **74,** 3782–3786.

Harris, P. L., and Thornton, E. R. (1978). *J. Am. Chem. Soc.* **100,** 6738–6745.

Hilbush, B. S., and Levine, J. M. (1991). *Proc. Natl. Acad. Sci. U.S.A.* **88,** 5616–5620.

Hirshberg, C. B., and Snider, M. D. (1987). *Annu. Rev. Biochem.* **56,** 63–88.

Huang, T. C., and Dietsch, E. (1991). *FEBS Lett.* **281,** 39–42.

Iber, H., Van Echten, G., and Sandhoff, K. (1992). *J. Neurochem.* **58,** 1533–1537.

Igarashi, Y., Nojiri, H., Hanai, N., and Hakomori, S. I. (1989). *In* "Methods in Enzymology" (V. Ginsburg, ed.), Vol. 179, pp. 521–541. Academic Press, San Diego.

Jeckel, D., Karrenbauer, A., Burger, K. N. J., Van Meer, G., and Wieland, F. (1992). *J. Cell Biol.* **117,** 259–267.

Jeserich, G., Breer, H., and Duvel, M. (1981). *Neurochem. Res.* **6,** 465–472.

Juurlink, B. H. J., Munoz, D. G., and Ang, L. C. (1991). *Neurosci. Lett.* **133,** 25–28.

Karpiak, S. E., and Mahadik, S. P. (1990). *Prog. Brain Res.* **85,** 299–309.

Kitamura, M., Iwamori, M., and Nagai, Y. (1980). *Biochim. Biophys. Acta* **628,** 1980–1985.

Klinghardt, G. W., Fredman, P., and Svennerholm, L. (1981). *J. Neurochem.* **37,** 897–908.

Koerner, T. A. W., Prestegard, J. H., Demou, P. C., and Yu, R. K. (1983). *Biochemistry* **22,** 2676–2684.

Kohn, L. D., and Shifrin, S. (1982). *In* "Hormone Receptors" (L. D. Kohn, ed.), pp. 1–23. Academic Press, New York.

Kok, J., Eskelinen, S., Hoekstra, K., and Hoekstra, D. (1989). *Proc. Natl. Acad. Sci. U.S.A.* **86,** 9896–9900.

Kortje, K. H., Freihofer, D., and Rahmann, H. (1990a). *J. Histochem. Cytochem.* **38,** 895–900.

Kortje, K. H., Freihofer, D., and Rahmann, H. (1990b). *Ultramicroscopy* **32,** 12–17.

Kreutter, D., Kim, J. Y. H., Goldenring, J. R., Rasmussen, H., Ukomadu, Y., DeLorenzo, R. J., and Yu, R. K. (1987). *J. Biol. Chem.* **262,** 1633–1637.

Langner, M., Winiski, A., Eisenberg, M., and McLaughlin, S. (1988). *In* "New Trends in Ganglioside Research: Neurochemical and Neuroregenerative Aspects" (R. W. Ledeen, E. L. Hogan, Y. Nagai, G. Tettamanti, A. J. Yates, and R. K. Yu, eds.), Fidia Res. Ser., Vol. 14, pp. 121–131. Liviana Press/Springer-Verlag, Padova/Berlin.

Ledeen, R. W. (1984). *J. Neurosci. Res.* **12,** 147–159.

Ledeen, R. W. (1989). *In* "Neurobiology of Glycoconjugates" (R. U. Margolis and R. K. Margolis, eds.), pp. 43–83. Plenum, New York.

Ledeen, R. W. (1991). *Encycl. Hum. Biol.* **3,** 737–742.

Ledeen, R. W., Wu, G., Vaswani, K. K., and Cannella, M. S. (1990). *In* "Trophic Factors and the Nervous System" (L. A. Horrocks, N. H. Neff, A. J. Yates, and M. Hadjicostantinon, eds.), Fidia Res. Found. Symp. Ser., Vol. 3, pp. 17–33. Raven Press, New York.

Leon, A., Facci, L., Toffano, G., Sonnino, S., and Tettamanti, G. (1981). *J. Neurochem.* **37,** 350–357.

Leon, A., Dal Toso, R., Presti, D., Benvegni, D., Facci, L., Kirschner, G., Tettamanti, G., and Toffano, G. (1988). *J. Neurosci.* **8,** 746–753.

Leray, C., Ferret, B., Freysz, L., Dreyfus, H., and Massarelli, R. (1988). *Biochim. Biophys. Acta* **944,** 79–84.

Levine, S. M., and Goldman, J. E. (1988). *J. Comp. Neurol.* **277,** 456–464.

Li, Y. S., Mahadik, S. P., and Rapport, M. M., and Karpiak, S. E. (1986). *Brain Res.* **377,** 292–297.

Louis, J. C., Magal, E., and Yavin, E. (1988). *J. Biol. Chem.* **263,** 19282–19285.

Magal, E., Louis, J. C., Aguilera, J., and Yavin, E. (1990). *J. Neurochem.* **55,** 2126–2131.

Maggio, B. (1985). *Biochim. Biophys. Acta* **815,** 245–258.

Mahadik, S. P., Hawver, D. B., Hungund, B. L., Li, Y. S., and Karpiak, S. E. (1989). *J. Neurosci. Res.* **24,** 402–412.

Manev, H., Favaron, M., Guidotti, A., and Costa, E. (1989). *Mol. Pharmacol.* **36,** 106–112.

Manev, H., Favaron, M., Vicini, S., and Guidotti, A. (1990). *Acta Neurobiol. Exp.* **50,** 475–488.

Marchase, R. B. (1977). *J. Cell Biol.* **75,** 237–257.

Markwell, M. A., Svennerholm, L., and Paulson, J. C. (1981). *Proc. Natl. Acad. Sci. U.S.A.* **78,** 5406–5410.

Masco, D., Van de Walle, M., and Spiegel, S. (1991). *J. Neurosci.* **11,** 2443–2552.

Masserini, M., and Freire, E. (1986). *Biochemistry* **25,** 1043–1049.

Masserini, M., Giuliani, A., Venerando, B., Fiorilli, A., D'Aniello, A., and Tettamanti, G. (1985). *Biochem. J.* **229,** 595–603.

Masserini, M., Palestini, P., Venerando, B., Fiorilli, A., Acquotti, D., and Tettamanti, G. (1988). *Biochemistry* **27,** 7973–7978.

Masserini, M., Freire, E., Palestini, P., Calappi, E., and Tettamanti, G. (1992). *Biochemistry* **31,** 2422–2426.

Maysinger, D., Leavitt, B. R., Zorc, B., Butula, I., Fernandes, L. G., and Ribeiro-da-Silva, A. (1992). *Neurochem. Int.* **20,** 289–297.

McDaniel, R. V., McLaughlin, A., Winiski, A. P., Eisenberg, M., and McLaughlin, S. (1984). *Biochemistry* **23,** 4618–4624.

Merrill, A. H., Jr. (1991). *J. Bioenerg. Biomembr.* **23,** 83–104.

Merrill, A. H., Jr., and Jones, D. D. (1990). *Biochim. Biophys. Acta* **1044,** 1–12.

Merrill, A. H., Jr., Wang, E., Mullins, R. E., Jamison, W. C. L., Nimkar, S., and Liotta, D. C. (1988). *Anal. Biochem.* **171,** 373–381.

Momoi, T. (1986). *Biochem. Biophys. Res. Commun.* **138,** 865–871.

Mugnai, G., Tombaccini, D., and Ruggieri, S. (1984). *Biochem. Biophys. Res. Commun.* **125,** 142–148.

Myers, M., Wortman, C., and Freire, E. (1984). *Biochemistry* **23,** 1442–1448.

Nagai, Y., and Tsuji, S. (1989). *In* "Gangliosides and Cancer" (H. F. Oettgen, ed.), pp. 33–39. VCH Verlaggesellschaft, Weinheim, Germany.

Nagai, Y., Tsuji, S., and Nakajima, J. (1986). *In* "Metabolism and Development of the Nervous System" (S. Tucek, ed.), pp. 113–118. Wiley, Chichester.

Nagata, Y., Ando, M., Iwata, H., Hara, A., and Taketomi, T. (1987). *J. Neurochem.* **49,** 201–207.

Nakajima, J., Tsuji, S., and Nagai, Y. (1986). *Biochim. Biophys. Acta* **876,** 65–71.

Nojri, H., Stroud, M., and Hakomori, S.-I. (1991). *J. Biol. Chem.* **266,** 4531–4537.

Obata, K., Oide, M., and Handa, S. (1977). *Nature (London)* **266,** 369–371.

O'Brien, J. S. (1989). *In* "The Metabolic Basis of Inherited Disease" (C. R. Scriver, A. L. Beaudet, W. S. Sly, and D. Valle, eds.), 6th ed., pp. 1797–1806. McGraw-Hill, New York.

Omodera, H., Araki, T., and Kogure, K. (1989). *Brain Res.* **481,** 1–7.

Pagano, R. E. (1990). *Curr. Opin. Cell Biol.* **2,** 652–663.

Palestini, P., Masserini, M., Fiorilli, A., Calappi, E., and Tettamanti, G. (1991). *J. Neurochem.* **57,** 748–753.

Partington, C. R., and Daly, J. W. (1979). *Mol. Pharmacol.* **15,** 484–491.

Pascher, J. (1976). *Biochim. Biophys. Acta* **455,** 433–451.

Petroni, A., Bertazzo, A., Sarti, S., and Galli, C. (1989). *J. Neurochem.* **53,** 747–752.

Pisoni, R. L., and Thoene, J. G. (1991). *Biochim. Biophys. Acta* **1071,** 351–373.

Pohlentz, G., Klein, D., Schwarzmann, G., Schmitz, D., and Sandhoff, K. (1988). *Proc. Natl. Acad. Sci. U.S.A.* **85,** 7044–7048.

Preti, A. S., Fiorilli, A., Lombardo, A., Caimi, L., and Tettamanti, G. (1980). *J. Neurochem.* **35,** 281–296.

Quimet, C. C., McGuiness, T. L., and Greengard, P. (1984). *Proc. Natl. Acad. Sci. U.S.A.* **81,** 5604–5608.

Rahmann, H. (1983). *Neurochem. Int.* **5,** 539–547.

Rahmann, H. (1992). *In* "Current Aspects of the Neurosciences" (N. N. Osborne, ed.), Vol. 4, pp. 87–125. Macmillan, London.

Riboni, L., and Tettamanti, G. (1991). *J. Neurochem.* **57,** 1931–1939.

Riboni, L., Sonnino, S., Acquotti, D., Malesci, A., Ghidoni, R., Egge, H., Mingrino, S., and Tettamanti, G. (1986). *J. Biol. Chem.* **261,** 8514–8519.

Riboni, L., Ghidoni, R., and Tettamanti, G. (1989). *J. Neurochem.* **52,** 1401–1406.

Riboni, L., Prinetti, A., Pitto, M., and Tettamanti, G. (1990). *Neurochem. Res.* **15,** 1175–1183.

Riboni, L., Prinetti, A., Bassi, R., and Tettamanti, G. (1991). *FEBS Lett.* **287,** 42–46.

Riboni, L., Bassi, R., Sonnino, S., and Tettamanti, G. (1992). *FEBS Lett.* **300,** 188–192.

Rodden, F. A., Wiegandt, H., and Bauer, B. L. (1991). *J. Neurosurg.* **74,** 606–619.

Rosner, H., Al-Aqtum, M., Sonnentag, U., Wurster, A., and Rahmann, H. (1992). *Neurochem. Int.* **20,** 409–419.

Saito, M., and Yu, R. K. (1986). *J. Neurochem.* **47,** 632–641.

Saito, M., and Yu, R. K. (1992). *J. Neurochem.* **58,** 83–87.

Sandhoff, K., Schwarzmann, G., Sarmientos, F., and Conzelmann, E. (1987). *NATO ASI Ser., Ser. H* **7,** 231–250.

Sandhoff, K., Conzelmann, E., Neufeld, E. F., Kaback, M. M., and Suzuki, K. (1989). *In* "The Metabolic Basis of Inherited Disease" (C. R. Scriver, A. L. Beudet, W. S. Sly, and D. Valle, eds.), 6th ed., pp. 1807–1839. McGraw-Hill, New York.

Schengrund, C. L. (1990). *Brain Res. Bull.* **24,** 131–141.
Schengrund, C. L., and Nelson, J. T. (1975). *Biochem. Biophys. Res. Commun.* **63,** 217–223.
Schengrund, C. L., and Nelson, J. T. (1976). *Neurochem. Res.* **1,** 171–180.
Schengrund, C. L., and Rosenberg, A. (1970). *J. Biol. Chem.* **245,** 6196–6200.
Schneider, J. S., Pope, A., Simpson, K., Taggart, J., Smith, M. G., and DiStefano, L. (1992). *Science* **256,** 843–846.
Schwartz, M. A., and Spirman, N. (1982). *Neurobiology* **79,** 6080–6083.
Schwarzmann, G., and Sandhoff, K. (1990). *Biochemistry* **29,** 10865–10871.
Schwarzmann, G., Hinrichs, U., Sonderfeld, S., Marsh, D., and Sandhoff, K. (1986). *In* "Enzymes of Lipid Metabolism II" (L. Freysz, H. Dreyfus, R. Massarelli, and S. Gatt, eds.), pp. 553–562. Plenum, New York.
Schwarzmann, G., Marsh, D., Herzog, V., and Sandhoff, K. (1987). *NATO ASI Ser., Ser. H* **7,** 217–229.
Seren, M. S., Lipartiti, M., Lazzaro, A., Rubini, R., Mazzari, S., Facci, L., Vantini, G., Zanoni, R., Zanotti, A., Bonvento, G., and Leon, A. (1990). *In* "Trophic Factors in the Nervous System" (L. A. Horrocks, N. H. Neff, A. J. Yates, and M. Hadjiconstantinou, eds.), Fidia Res. Found. Symp. Ser., Vol. 3, pp. 339–348. Raven Press, New York.
Seybold, V., Rösner, H., Greis, C., Beck, E., and Rahmann, H. (1989). *J. Neurochem.* **52,** 1958–1961.
Sharom, F. J., and Grant, C. W. M. (1978). *Biochim. Biophys. Acta* **507,** 280–293.
Shen, K. F., and Crain, S. M. (1990). *Brain Res.* **531,** 1–7.
Sillerud, L. O., Yu, R. K., and Schafer, D. E. (1982). *Biochemistry* **21,** 1260–1271.
Skaper, S. D., Favaron, M., Facci, L., and Leon, A. (1987). *Soc.Neurosci. Abstr.* **13,** 1119.
Skaper, S. D., Favaron, M., Facci, L., Vantini, G., Fusco, M., Ferrari, G., and Leon, A. (1988). *In* "New Trends in Ganglioside Research: Neurochemical and Neuroregenerative Aspects" (R. W. Ledeen, E. L. Hogan, G. Tettamanti, A. J. Yates, and R. K. Yu, eds.), Fidia Res. Ser., Vol. 14, pp. 351–360. Liviana Press/Springer-Verlag, Padova/Berlin.
Slenzka, K., Appel, R., and Rahmann, H. (1990). *Neurochem. Int.* **17,** 609–614.
Sonnino, S., Ghidoni, R., Galli, C., and Tettamanti, G. (1979). *J. Neurochem.* **33,** 117–121.
Sonnino, S., Chigorno, V., Acquotti, D., Pitto, M., Kirschner, G., and Tettamanti, G. (1989). *Biochemistry* **28,** 77–84.
Sonnino, S., Chigorno, V., Valsecchi, M., Pitto, M., and Tettamanti, G. (1992). *Neurochem. Int.* **20,** 315–321.
Spence, M. W., Reed, S., and Cook, H. W. (1986). *Biochem. Cell Biol.* **64,** 400–404.
Spiegel, S. (1988). *In* "New Trends in Ganglioside Research: Neurochemical and Neuroregenerative Aspects" (R. W. Ledeen, E. L. Hogan, G. Tettamanti, A. J. Yates, and R. K. Yu, eds.), Fidia Res. Ser., Vol. 14, pp. 405–421. Liviana Press/Springer-Verlag, Padova/Berlin.
Spiegel, S. (1990). *J. Cell Biochem.* **42,** 143–152.
Spiegel, S., Handler, J. S., and Fishman, P. H. (1986). *J. Biol. Chem.* **261,** 15755–15760.
Spirman, N., Sela, B. A., and Schwartz, M. A. (1982). *J. Neurochem.* **39,** 874–877.
Spoerri, P. E. (1983). *Int. J. Rev. Neurosci.* **1,** 383–391.
Spoerri, P. E., Rapport, M. M., Mahadik, S. P., and Roisen, F. J. (1988). *Brain Res.* **469,** 71–77.
Staedel-Flaig, C., Beck, J. P., Gabellec, M., and Rebel, G. (1987). *Cancer Biochem. Biophys.* **9,** 233–244.
Steinman, R. M., Mellman, I. S., Muller, W. A., and Cohn, Z. A. (1983). *J. Cell Biol.* **16,** 1–27.
Suzuki, A., and Yamakawa, T. (1991). *Encycl. Hum. Biol.* **3,** 725–735.
Svennerholm, L. (1980). *Adv. Exp. Med. Biol.* **125,** 11–21.
Tamir, H., Brunner, W., Casper, D., and Rapport, M. (1980). *J. Neurochem.* **34,** 1719–1724.
Tettamanti, G. (1988). *In* "New Trends in Ganglioside Research: Neurochemical and Neuroregenerative Aspects" (R. W. Ledeen, E. L. Hogan, G. Tettamanti, A. J. Yates, and R. K. Yu, eds.), Fidia Res. Ser., Vol. 14, pp. 625–646. Liviana Press/Springer-Verlag, Padova/Berlin.

Tettamanti, G., and Masserini, M. (1987). *In* "Biomembrane and Receptor Mechanism" (E. Bertoli, D. Chapman, A. Cambria, and U. Scapagnini, eds.), Fidia Res. Ser., Vol. 7, pp. 223–260. Liviana Press/Springer-Verlag, Padova/Berlin.

Tettamanti, G., Morgan, I. G., Gombos, G., Vincendon, G., and Mandel, P. (1972). *Brain Res.* **47,** 515–518.

Tettamanti, G., Ghidoni, R., and Trinchera, M. (1988). *Indian J. Biochem. Biophys.* **25,** 106–111.

Thomas, P. P., and Brewer, G. J. (1990). *Biochim. Biophys. Acta* **1031,** 277–289.

Thompson, T. E., and Brown, R. E. (1988). *In* "New Trends in Ganglioside Research: Neurochemical and Neuroregenerative Aspects" (R. W. Ledeen, E. L. Hogan, Y. Nagai, G. Tettamanti, A. J. Yates, and R. K. Yu, eds.), Fidia Res. Ser., Vol. 14, pp. 65–78. Liviana Press/Springer-Verlag, Padova/Berlin.

Thompson, T. E., and Tillack, T. W. (1985). *Annu. Rev. Biophys. Chem.* **14,** 361–386.

Tiemeyer, M., Yasuda, Y., and Schnaar, R. L. (1990a). *J. Biol. Chem.* **264,** 1671–1681.

Tiemeyer, M., Swank-Hill, R., and Schnaar, R. L. (1990b). *J. Biol. Chem.* **265,** 11990–11999.

Tomasi, M., Roda, C., Ausiello, C., D'Agnolo, G., Venerando, B., Ghidoni, R., Sonnino, S., and Tettamanti, G. (1980). *Eur. J. Biochem.* **111,** 315–324.

Trinchera, M., and Ghidoni, R. (1989). *J. Biol. Chem.* **264,** 15766–15769.

Trinchera, M., Pirovano, B., and Ghidoni, R. (1990). *J. Biol. Chem.* **265,** 18242–18247.

Trinchera, M., Carrettoni, D., and Ghidoni, R. (1991a). *J. Biol. Chem.* **266,** 9093–9099.

Trinchera, M., Fabbri, M., and Ghidoni, R. (1991b). *J. Biol. Chem.* **266,** 20907–20912.

Trinchera, M., Fiorilli, A., and Ghidoni, R. (1991c). *Biochemistry* **30,** 2719–2724.

Tsuji, S., Nakajima, J., Sasaki, T., and Nagai, Y. (1985). *J. Biochem. (Tokyo)* **97,** 969–972.

Vaccarino, F., Guidotti, A., and Costa, E. (1987). *Proc. Natl. Acad. Sci. U.S.A.* **84,** 8707–8711.

Van Echten, G., Iber, H., Stotz, H., Takatsuki, A., and Sandhoff, K. (1990). *Eur. J. Cell Biol.* **51,** 135–139.

Van Heyningen, W. E. (1974). *Nature (London)* **249,** 415–416.

Van Meer, G. (1989). *Annu. Rev. Cell Biol.* **5,** 247–275.

Van Veldhoven, P. P., Bishop, W. R., and Bell, R. M. (1989). *Anal. Biochem.* **183,** 177–189.

Varon, S., Pettman, B., and Manthorpe, M. (1988). *In* "New Trends in Ganglioside Research: Neurochemical and Neuroregenerative Aspects" (R. W. Ledeen, E. L. Hogan, G. Tettamanti, A. J. Yates, and R. K. Yu, eds.), Fidia Res. Ser., Vol. 14, pp. 607–623. Liviana Press/Springer-Verlag, Padova/Berlin.

Vaswani, K. K., Wu, G., and Ledeen, R. W. (1990). *J. Neurochem.* **55,** 492–499.

Venerando, B., Fiorilli, A., Masserini, M., Giuliani, A., and Tettamanti, G. (1985). *Biochim. Biophys. Acta* **833,** 82–92.

Walter, V. P., Sweeney, K., and Morré, D. J. (1983). *Biochim. Biophys. Acta* **750,** 346–352.

Watanabe, K., Hakomori, S. I., Powell, M. E., and Yokota, M. (1980). *Biochem. Biophys. Res. Commun.* **92,** 638–641.

Wattenberg, B. W. (1990). *J. Cell Biol.* **111,** 421–428.

Weis, F. M. B., and Davis, R. J. (1990). *J. Biol. Chem.* **265,** 12059–12066.

Wiegandt, H. (1985). *In* "Glycolipids, New Comprehensive Biochemistry" (A. Neuberger and L. L. M. van Deenen, eds.), Vol. 10, pp. 199–260. Elsevier, Amsterdam.

Wynn, C. G., and Robson, B. (1986). *J. Theor. Biol.* **123,** 221–230.

Yasuda, Y., Tiemeyer, M., Blackburn, C. C., and Schnaar, R. (1988). *In* "New Trends in Ganglioside Research: Neurochemical and Neuroregenerative Aspects" (R. W. Ledeen, E. L. Hogan, G. Tettamanti, A. J. Yates, and R. K. Yu, eds.), Fidia Res. Ser., Vol. 14, pp. 229–243. Liviana Press/Springer-Verlag, Padova/Berlin.

Yates, A. J. (1986). *Neurochem. Pathol.* **5,** 309–329.

Yates, A. J., Walters, J. D., Wood, C. L., and Johnson, J. D. (1989). *J. Neurochem.* **53,** 162–167.

Yavin, E., Consolazione, A., Gil, S., Ginzburg, I., Leon, A., Del Toso, R., and Rybak, S. (1988). *In* "New Trends in Ganglioside Research: Neurochemical and Neuroregenerative Aspects" (R. W. Ledeen, E. L. Hogan, G. Tettamanti, A. J. Yates, and R. K. Yu, eds.), Fidia Res. Ser., Vol. 14, pp. 579–593. Liviana Press/Springer-Verlag, Padova/Berlin.

Yim, S. H., Yavin, E., Hammer, J. A., and Quarles, R. H. (1991). *J. Neurochem.* **57,** 2144–2147.

Yohe, H. C., Jacobson, R. I., and Yu, R. K. (1983). *J. Neurosci. Res.* **9,** 401–412.

Yu, R. K., and Saito, M. (1989). *In* "Neurobiology of Glycoconjugates" (R. U. Margolis and R. K. Margolis, eds.), pp. 1–41. Plenum, New York.

ADVANCES IN LIPID RESEARCH, VOL. 25

Protection by Gangliosides against Glutamate Excitotoxicity

H. MANEV,* A. GUIDOTTI, AND E. COSTA

Fidia-Georgetown Institute for the Neurosciences
Georgetown University Medical School
Washington, D.C. 20007

I. Introduction

A. Gangliosides as Drugs

Purified and well-characterized gangliosides (Fig. 1) are currently being used clinically for the treatment of the symptoms of several acute and chronic neurodegenerative disorders. The first report on the experimental use of gangliosides for the treatment of a peripheral nervous system dystrophy appeared in 1976, when Ceccarelli *et al.* showed that the parenteral administration of a ganglioside mixture to cats improves the time course of the functional recovery of the denervated nictating membrane.

This discovery opened a new avenue for the clinical treatment of acute degenerative diseases of peripheral nerves, brain, and spinal cord. Thus, gangliosides, in particular the monosialoganglioside G_{M1} [nomenclature according to Svennerholm (1980)] proved to be clinically efficacious in the treatment of stroke (Bassi *et al.*, 1984; Battistin *et al.*, 1985; Argentino *et al.*, 1989; Carolei *et al.*, 1991; Rocca *et al.*, 1992; Angeleri *et al.*, 1992), subarachnoid hemorrhage (Papo *et al.*, 1991), and head (Hoermann, 1988) and spinal cord (Geisler *et al.*, 1991; Walker, 1991) traumatic injuries.

These clinical studies were preceded by a vast body of experimental work showing the neuroprotective action of gangliosides against glutamate excitotox-

*Present address: Fidia Research Laboratories, FIDIA S.p.A., Abano Terme, Italy.

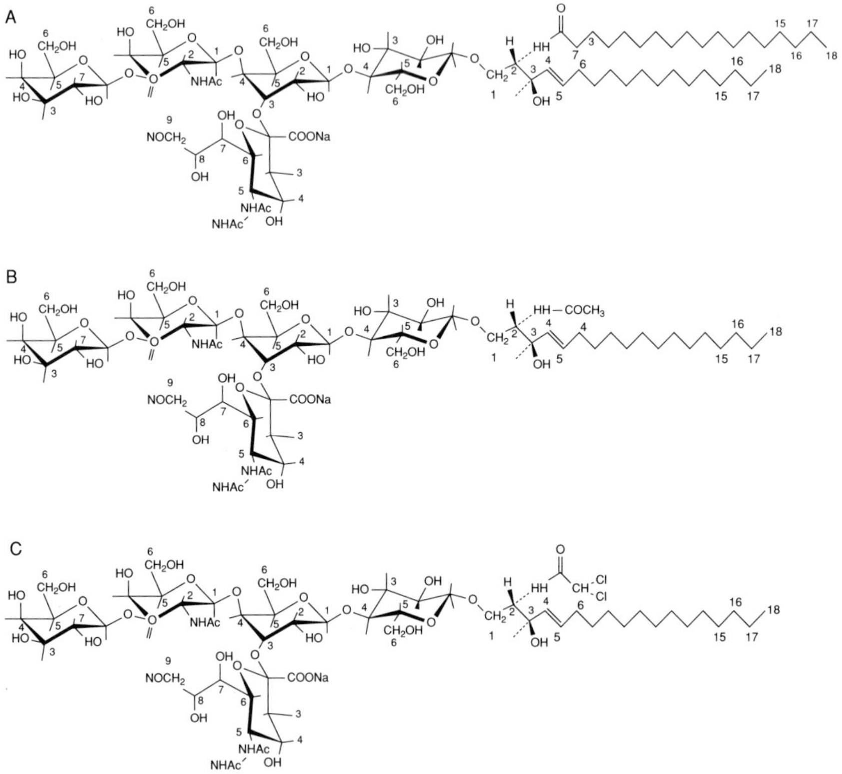

FIG. 1. Structures of (A) G_{M1}, (B) LIGA 4, (C) LIGA 20.

icity (Carolei *et al.,* 1991; Lipartiti *et al.,* 1991). Moreover, *in vitro* and *in vivo* pharmacological research with gangliosides helped to improve our understanding of the mechanism of glutamate excitotoxicity (Manev *et al.,* 1990b; Costa *et al.,* 1993; Comelli *et al.,* 1992). Finally, the use of gangliosides enabled us to elucidate new possible targets in the sequelae of events leading to excitotoxicity that can be used for the development of new drugs acting on acute neurodegenerative disorders (i.e., brain traumas and ischemias).

B. GLUTAMATE, A NEUROTRANSMITTER CAPABLE OF CAUSING EXCITOTOXICITY

Glutamate, a dicarboxylic amino acid (Fig. 2) that is stored in neuronal synaptic vesicles and is released following neuronal stimulation, fulfills all the criteria

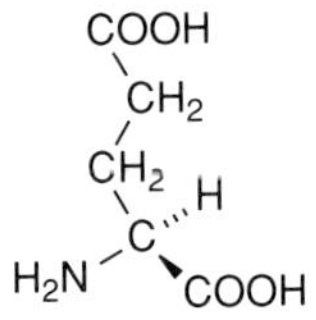

FIG. 2. Structure of glutamate.

of a classical neurotransmitter (Fonnum, 1991). Probably, it is the most abundant neurotransmitter present in human and other mammalian brains. Neurons that are the target of this neurotransmitter express specific membrane proteins which bind glutamate with high affinity and transduce this chemical message into a variety of regulatory events important for the function and plasticity of the receiving cells. Recently, numerous protein subunits specific for the structure of each glutamate receptor subtype were cloned (Boulter *et al.*, 1992; Meguro *et al.*, 1992; H. Monyer *et al.*, 1992; M. D. Monyer *et al.*, 1992; Sakimura *et al.*, 1992). Several of these subunits form heterooligomeric membrane channels acting as ligand-gated conveyors of ion fluxes for the receiving cells. In addition, a class of glutamate receptor subunits is linked (via guanyl nucleotide-binding proteins) to enzymes and they function as metabotropic receptors (Masu *et al.*, 1991; Nakanishi, 1992; Tanabe *et al.*, 1992).

Several reports have delineated the pharmacological profile of glutamate receptors. To summarize: on the basis of the ability of the receptor to be activated selectively and specifically by synthetic agonists, ionotropic glutamate receptors are divided into a class of *N*-methyl-D-aspartate (NMDA)-selective and another class of non-NMDA-selective receptors (for review, see Fagg and Massieu, 1991). The latter class is subdivided into the kainate (KA)- and α-amino-3-hydroxy-5-methyl-4-isoaxolepropionate (AMPA)-sensitive receptors. The specific agonist for metabotropic glutamate receptors is (1*S*,3*R*)-aminocyclo-pentane-1,3-decarboxylate (1*S*,3*R*-ACPD). Quisqualate can stimulate both metabotropic and non-NMDA-selective ionotropic glutamate receptors.

The stimulation of ionotropic glutamate receptors results in depolarization caused by Na^+ influx. Depending on the structure of the transmembrane domain, ionotropic glutamate receptors may allow a Na^+ influx associated with a Ca^{2+} influx (Boulter *et al.*, 1992; M. D. Monyer *et al.*, 1992). The latter, acting as second messenger, brings about the stimulation of Ca^{2+}-dependent enzymes in the target cells. A similar Ca^{2+}-dependent enzyme activation also occurs as a result of the stimulation of some metabotropic glutamate receptors, justifying giving the name of "excitatory amino acid" (EAA) to glutamate (Olney, 1990).

Functionally, ionotropic glutamate receptors account for the majority of the neuronal responses to glutamate; however, the study of metabotropic receptors is just beginning. In general, glutamate participates in the short- and long-term processing of information at excitatory synapses and modulates the expression of nuclear information by activating the transcription of specific genetic programs via the expression of immediate early responsive genes encoding for proteins acting as third nuclear messengers (Szekely *et al.,* 1990). These genetic programs mediate changes in synaptic plasticity by suppressing or enhancing the transcription of mRNAs encoding for various neuronal regulatory proteins, including ionotropic receptor subunits (Szekely *et al.*, 1990). Such glutamatergic regulation of neuronal plasticity may be important in learning, adaptation to environmental changes, and in the structural repair and compensation for functional deficits that are determined by ischemic and traumatic neuronal damage.

Physiologically, quanta of glutamate are released rapidly and intermittently from the nerve terminals during neuronal firing (Fonnum, 1991). This neurally released glutamate interacts with its receptors for only a millisecond or less before being cleared from the synaptic cleft by reuptake and/or enzymatic degradation. If the homeostasis of these processes is overwhelmed by excessive glutamate release, due to long-lasting depolarization or uncontrolled transmitter release caused by neuronal damage, or by a combination of these events, glutamate then reveals its pathological nature. Namely, the accumulation of glutamate at the synaptic cleft, in persistently high concentrations, causes a continuous and protracted binding of glutamate to its receptors, leading to neuronal damage or death (Olney, 1990). *In vitro* experiments have shown that, if the glutamate receptor stimulation is protracted for 30 minutes or longer, it irreversibly endangers neuronal viability and can elicit Ca^{2+}-dependent neuronal death (Favaron *et al.,* 1988). This type of stimulation has been referred to as abusive receptor stimulation which usually is followed by a delayed neuronal death, a phenomenon termed "excitotoxicity" (Olney, 1990). Several acute and chronic neurodegenerative diseases originate from such excitotoxicity. For instance, excitotoxicity is currently believed to be operative in brain and spinal cord trauma and ischemia, as well as in several chronic degenerative diseases (Alzheimer's disease, Parkinson's disease, Huntington's disease, lateral sclerosis, and epilepsy) (Rothstein *et al.,* 1992; Beal *et al.,* 1986; Heafield *et al.,* Chapman, 1992).

C. Glutamate Receptors: Heterogeneity and Excitotoxicity

The neurotoxic action of glutamate is triggered by the stimulation of the same ionotropic receptors that mediate the physiological action of this neurotransmitter (Choi, 1991). Excitotoxicity follows the abusive stimulation of NMDA or non-NMDA ionotropic glutamate receptors that are permeable to Ca^{2+}. Although the "abuse" (an excessive and prolonged stimulation) of the receptors is the critical departing point in distinguishing between receptor physiology and pathology, the neurons of every brain region are not equally susceptible to glutamate excitotoxicity.

Histochemical studies employing specific labeled ligands of glutamate ionotropic and metabotropic receptors (Fagg and Massieu, 1991), specific antisera (antibodies) directed against peptides included in the structure of ionotropic receptors, or protein subunits (Wenthold *et al.,* 1992) or oligonucleotide probes hybridizing the mRNAs encoding for different subunits of ionotropic or metabotropic glutamate receptors (Boulter *et al.,* 1992; Meguro *et al.,* 1992; H. Monyer *et al.,* 1992; M. D. Monyer *et al.,* 1992; Sakimura *et al.,* 1992), revealed the uneven brain distribution of the glutamate receptors and helped to monitor the location and density of the various receptor subtypes operative in glutamate signal transduction. In line with the variable distribution of glutamate receptor subtypes in different structures, pharmacotherapeutic interventions aimed at preventing or protecting from glutamate excitotoxicity have different efficacy in different brain regions. For example, an allosteric antagonist of NMDA-selective glutamate receptors such as dizocilpine (MK-801) provides partial, albeit consistent, protection against cortical infarction after middle cerebral artery (MCA) occlusion, while it exerts only minimal protection against hippocampal and striatal damage after global ischemia (Meldrum *et al.,* 1992). At the same time, it must be noted that MK-801 can per se cause neuronal damage in limbic brain structures (Olney *et al.,* 1989). In contrast, hippocampus and striatum are protected by non-NMDA receptor antagonists [for example, 2,3-dihydroxy-6-nitro-7-sulfamoyl-benzo(*f*) quinoxaline (NBQX) (Sheardown *et al.,* 1990; Meldrum *et al.,* 1992)]. These observations suggest the importance of the structural characteristics of glutamate receptors, expressed in a given brain area, for the development of excitotoxicity.

Only recently, after the metabotropic glutamate receptors were cloned (Masu *et al.,* 1991; Tanabe *et al.,* 1992), did another aspect of the complexity in the glutamate–receptor function in neurotoxicity and neuroplasticity of the CNS begin to attract wide attention. In fact, it is now agreed that the metabotropic glutamate receptor does not mediate excitotoxicity (Koh *et al.,* 1991a), since its pharmacological stimulation might actually be operative in the process of neuroplasticity (Conn and Desai, 1991). Recent reports had in fact indicated that quisqualate could reduce kainic acid excitotoxicity (Nicoletti *et al.,* 1989), and *in vitro* (Koh *et al.,* 1991b) and *in vivo* (Siliprandi *et al.,* 1992) experiments, using the specific metabotropic receptor agonist 1*S*,3*R*-ACPD, have suggested that an interplay between metabotropic and ionotropic glutamate receptor subtypes becomes important in reducing glutamate excitotoxicity.

This interplay can be dynamically regulated by external stimuli. For example, in primary neuronal cultures the expression of metabotropic glutamate receptors is influenced by the depolarizing levels of the culture media (Favaron *et al.,* 1992). Similarly, the plasticity of the ionotropic non-NMDA receptors is also dependent on external signals. The expression of these receptors was shown to be affected by kainate-induced seizures (Gall *et al.,* 1990). Moreover, the expression of different non-NMDA glutamate receptor mRNAs (GluR1 to GluR6) that are developmentally regulated (Zukin *et al.,* 1992) appears to change in pathological conditions such as brain ischemia (A. Pellegrini, personal communication). The latter finding might indicate that the ionotropic receptor structure might change brain susceptibility to ischemic damage, and such change may in fact result in a protection against excitotoxicity. In this regard, the expression of GluR2 subunits appears to be crucial. The presence or absence of this subunit in the structure of a given ionotropic glutamate receptor controls the channel permeability to Ca^{2+} (Boulter *et al.,* 1992; M. D. Monyer *et al.,* 1992). Hence, when the glutamate receptor includes the GluR2 subunit, the receptor permeability to Ca^{2+} is virtually absent. Since neuronal $[Ca^{2+}]_i$ homeostasis plays a crucial role in excitotoxicity (see below), a regulated inclusion of a GluR2 subunit in the ionotropic receptor structure might hinder the progression of excitotoxicity. It can be envisioned that, in the future, attempts will be made to find pharmacological tools for the treatment of excitotoxicity (connected to brain ischemias) by targeting drugs at the mechanisms that modify the assembly of glutamate receptor subunits and facilitate the recruiting of GluR2 subunits that decrease Ca^{2+} permeability.

D. Strategies of Protection from Excitotoxicity: The Concept of Receptor Abuse-Dependent Antagonism

Currently, pharmacological research directed to reduce excitotoxicity by glutamate follows two major trends: (1) the blockade of ionotropic glutamate receptors utilizing not only the isosteric receptor blockers but also compounds that affect the regulatory sites on the NMDA-sensitive receptor complex (inhibitors of recognition sites operative in the glycine-positive modulation of glutamate action at NMDA receptors or ligands of the polyamine regulatory site of this receptor) (McCulloch *et al.,* 1991; Sheardown *et al.,* 1990; Kemp *et al.,* 1988; Moroni and Pelliciari, 1992; McGeer and Zhu, 1990; Scatton *et al.,* Simon *et al.,* 1992), and (2) the inhibition of critical steps in the cascade of events that are selectively triggered by the continuous pathological stimulation of glutamate receptors while leaving the primary physiological receptor responses intact, as happens with the receptor abuse-dependent antagonism (RADA) observed with gangliosides (Manev *et al.,* 1990a).

Although differing substantially, both approaches are targeted at counteracting the essential mechanisms operative in excitoxicity, that is, the destabilization of the neuronal homeostatic mechanisms that maintain intraneuronal levels of $[Ca^{2+}]_i$ within the limits of an oscillation compatible with neuronal function

(Connor, 1991). In fact, the destabilization of $[Ca^{2+}]$ homeostasis is a common denominator of the events causing excitotoxicity. Neuronal $[Ca^{2+}]$ overload induced by glutamate receptor activation may accelerate several molecular events, including enhanced lipolysis and proteolysis, activation and translocation of protein kinases, and changes in gene expression (Manev *et al.*, 1990a). Nitric oxide (NO) formation is also stimulated by $[Ca^{2+}]_i$ (Kiedrowski *et al.*, 1991); it depends on protein kinase C (PKC) activation (Marin *et al.*, 1992) and may contribute to hydroxyl radical formation and to activation of specific enzymes such as guanylate cyclase and protein NAD ribosylation (Wroblewski *et al.*, 1991).

The complex balance between Ca^{2+} release from endoplasmic reticulum stores and Ca^{2+} influx from extraneuronal sources is a crucial target for the pathological destabilization of $[Ca^{2+}]_i$ homeostasis elicited by glutamate receptor abusive stimulation (de Erausquin *et al.*, 1990; Randall and Thayer, 1992). The Ca^{2+} influx mechanism is regulated by the opening time and frequency of voltage- or inositol trisphosphate (IP_3)-regulated Ca^{2+} channels that increase free $[Ca^{2+}]_i$. The Ca^{2+} efflux includes the calmodulin-sensitive Ca^{2+}-ATPase and probably other regulatory mechanisms that are not readily understood at this time.

These processes are important for the regulation of $[Ca^{2+}]_i$ homeostasis and can be affected in a positive or a negative manner by phosphorylation. In particular, the phosphorylation mediated by PKC appears to play an important role in the process. The translocation of cytosolic PKC to neuronal membranes is a dynamic process believed to be prompted in a slowly reversible manner by the continuous and persistent stimulation of glutamate receptors and might be pivotal in the destabilization of $[Ca^{2+}]_i$ homeostasis (Manev *et al.*, 1989; Favaron *et al.*, 1990; Candeo *et al.*, 1992). The physicochemical characteristics of such PKC interactions with neuronal membranes appear to depend on the mode of glutamate receptor stimulation. The PKC translocation resulting from abusive glutamate receptor stimulation has a dissociation kinetic that is much slower than that of the translocation triggered by physiological receptor stimulation (Vaccarino *et al.*, 1987; Manev *et al.*, 1990c). An understanding of the molecular mechanisms mediating this difference in PKC dissociation from neuronal membranes may open an avenue to an efficacious treatment of the sequelae of events that cause excitotoxicity in the surrounding area of a primary focus of brain ischemia.

Additional support for the view that PKC plays a role in excitotoxicity comes from experiments in which an activation of protein kinases was shown to cause the degeneration and death of cultured neurons (Mattson, 1991; Candeo *et al.*, 1992) and to play a pivotal role in the cascade of events destabilizing neuronal viability in the area penumbra surrounding brain infarcts (Nabeshima *et al.*, 1991; Joo *et al.*, 1989; Domanska-Janik and Zalewska, 1992). In primary cultures of cerebellar granule cells the addition of okadaic acid (OKA), the murine toxin which increases protein phosphorylation by inhibiting the serine/threonine protein phosphatases 1 and 2A, causes neuronal death (Fig. 3). Protection

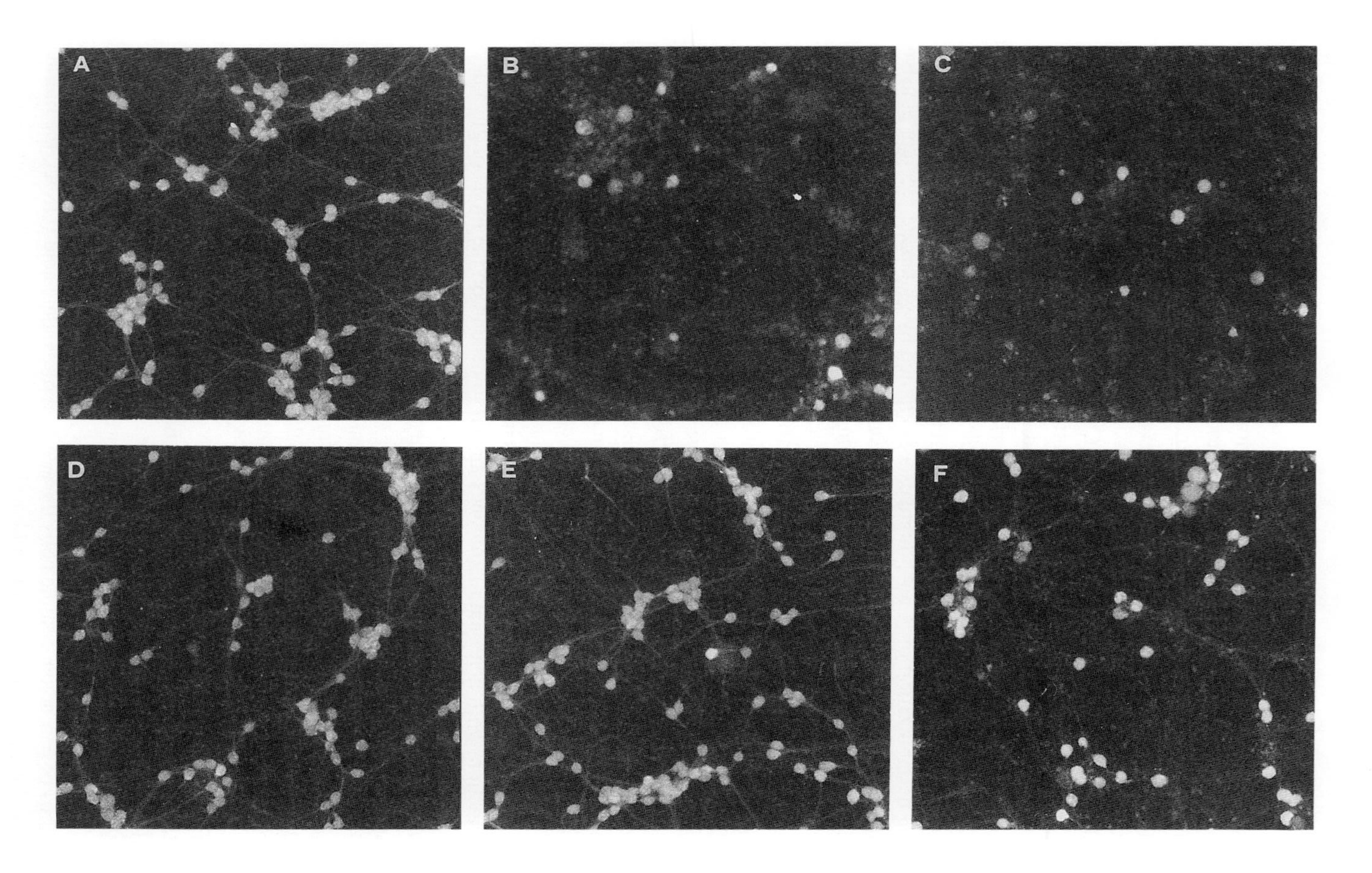
A
B
C
D
E
F

Table I
PATHOLOGICAL PHOSPHORYLATION MAY CAUSE CELL DEATH IN NEURONAL CULTURES: ANTAGONISM BY GANGLIOSIDES

Treatment	Effect on PKC	Effect on neurons	Antagonism of the effect	Reference
Glutamate	Translocation –activation	Not determined	Gangliosides	Vaccarino *et al.* (1987)
Glutamate	Translocation	Neurotoxicity	Gangliosides	Favaron *et al.* (1988)
Glutamate	Translocation	Neurotoxicity	Gangliosides	Manev *et al.* (1989)
Glutamate`	Translocation	Neurotoxicity	Ganglioside derivatives	Manev *et al.* (1990b)
Glutamate	Translocation	Neurotoxicity	PKC downregulation	Favaron *et al.* (1990)
Phorbol ester	Activation	Neurotoxicity	H7	Mattson (1991)
Okadaic acid	Activation	Neurotoxicity	None	Fernandez *et al.* (1991)
Okadaic acid	Activation	Neurotoxicity	H7; PKC downregulation	Candeo *et al.* (1992)

against OKA neurotoxicity was obtained either by the concomitant treatment of the cultures with the protein kinase inhibitor 1-(5-isoquinolinsulfonyl)-2-methylpiperazine (H7), or by downregulating PKC prior to OKA administration (Table I). The PKC downregulation can be induced by an overnight treatment with phorbol 12-myristate 13-acetate (TPA). A similar downregulation of PKC also protects the cultures of cerebellar granule neurons from glutamate excitotoxicity (Table I). Importantly, natural gangliosides and their semisynthetic lyso derivatives can inhibit PKC translocation induced by neurotoxic doses of glutamate and provide protection from the associated excitotoxicity induced by this neurotransmitter (Table I). Natural gangliosides and semisynthetic lysogangliosides reduce the PKC activation elicited by photochemically induced cortical lesions in rats and decrease the severity of these lesions (Costa *et al.,* 1993). Moreover, the PKC inhibitors staurosporine and H7 attenuate the amnesia associated with basal forebrain lesions induced by ibotenic acid, or with lesions of

FIG. 3. Pathological phosphorylation induces neurotoxicity. Cerebellar granule neurons in culture were exposed to (B) glutamate (50 μM, no Mg^{2+}, 15 min) or (C) okadaic acid (20 n*M;* 24 hours). These treatments resulted in neurotoxicity, shown as the loss of fluorescein diacetate fluorescence when compared to (A) controls (for methods, see Manev *et al.,* 1990b). The toxicity of glutamate was decreased by (D) G_{M1} treatment or by (E) downregulation of PKC (for details, see Favaron *et al.,* 1990). PKC downregulation also reduced (F) the toxicity of okadaic acid (Candeo *et al.,* 1992).

brain cholinergic neurons, or with ischemia-induced formation of brain edema (Nabeshima *et al.*, 1991; Joo *et al.*, 1989).

The exact target for PKC-modulated phosphorylation that is triggered by the glutamate and is associated with neuronal degeneration has not been identified as yet. One can surmise that it may be a protein operative in the regulation of Ca^{2+} extrusion from neurons or in the regulation of neuroplasticity induced by NMDA-selective glutamate receptor stimulation (Catania *et al.*, 1991; Favaron *et al.*, 1992), or in some other function of NMDA receptors (Chen and Huang, 1992). Recently, using isolated trigeminal neurons, it was demonstrated that intracellularly applied PKC reduces the Mg^{2+} blockade of NMDA-receptor channels (Chen and Huang, 1992).

The potential clinical use of isosteric or allosteric glutamate receptor antagonists appears for the present to be outweighed by the disadvantages. This is not surprising, because glutamatergic receptors represent more than 40% of the transmitter receptors present in mammalian brain and therefore are operative in several neuronally regulated vital functions. Also, *in vivo* allosteric glutamate receptor blockers are known to be endowed with a significant intrinsic neurotoxic action. For these reasons, great hopes arise from the finding that *in vitro* in primary brain neuronal cultures and *in vivo* in models of focal brain ischemia, protection against glutamate-mediated excitotoxicity can be achieved without blocking the channel gating mechanisms of NMDA and non-NMDA receptors with gangliosides (Favaron *et al.*, 1988; Carolei *et al.*, 1991). Based on the finding that the pathological sequelae triggered by continuous and prolonged glutamate receptor stimulation differ from the physiological sequelae, which are triggered by an intermittent physiological stimulation of the same receptor (where each stimulus lasts 1 msec or less), it became apparent that it was possible to act on the selective processes of the pathological sequelae while sparing the chain of events associated with the physiological mode of receptor operation. As previously mentioned, this strategy of selective pharmacological action on the events deriving from abusive and continuous stimulation of glutamate receptors was termed RADA, and the gangliosides and their semisynthetic lysogangliosides derivatives, which protect neurons from excitotoxicity without blocking the channel gating mechanism of NMDA and non-NMDA receptors, can be considered as the prototypes of such RADA drugs (Manev *et al.*, 1990a,b,c; Favaron *et al.*, 1988).

The target of such pharmacological strategy is the downregulation of the processes that are pathologically amplified in the signal transduction at NMDA-selective glutamate receptors. The protracted translocation of PKC and the prolonged disturbance of $[Ca^{2+}]_i$ homeostasis are those processes that can be downregulated by gangliosides and appear to possess a relevance in the physiopathology of excitotoxicity (Manev *et al.*, 1990a, 1991).

II. Gangliosides and Excitotoxicity

A. Mechanism of Neuroprotection by Gangliosides: RADA Drugs

Gangliosides are natural constituents of cell membranes and are particularly abundant in neurons (Kracun *et al.*, 1992). The functional importance of these molecules can be inferred on the basis that the relative composition profile of gangliosides in the CNS changes during brain development and maturation, and that an hereditary abnormality in the qualitative profile of the gangliosides content results in serious neurological disorders (Purpura, 1978; Baumann *et al.*, 1980). Moreover, in newborn infants who died of hypoxia it was found that brain ganglioside content decreases, and this decrease correlates with the severity of brain damage (Qi and Xue, 1991).

Gangliosides added to neurons in culture are incorporated into neuronal membranes (Favaron *et al.*, 1988; Ghidoni *et al.*, 1989), and thereafter they may participate in the physiological metabolism and recycling of brain glycosphingolipids, modifying in a predictable manner the complex series of their metabolic pathways (Ghidoni *et al.*, 1989). The incorporation of exogenous gangliosides into neurons does not appear to affect the basal function or the physiological sequelae following the stimulation of several transmitter receptors, including ionotropic glutamate receptors (Manev *et al.*, 1990a). This absence of perturbations of transmitter receptor functions was studied by voltage-clamp technology in the whole-cell or membrane-patch mode (Favaron *et al.*, 1988; Manev *et al.*, 1990b), and by Ca^{2+} imaging technology with fura-2 (de Erausquin *et al.*, 1990; Milani *et al.*, 1991). Hence, gangliosides added to cellular or cortical neuronal cultures did not affect intracellular $[Ca^{2+}]_i$ homeostasis per se, but normalized the glutamate-induced prolonged destabilization of $[Ca^{2+}]_i$ homeostasis. Whereas in the absence of a ganglioside pretreatment a neurotoxic dose of glutamate elicited a prolonged $[Ca^{2+}]_i$ increment (lasting longer than 1 hour following glutamate removal), a pretreatment of the cultures with gangliosides abated this $[Ca^{2+}]_i$ increase that outlasts glutamate withdrawal and protected from ensuing neuronal death (Favaron *et al.*, 1988; Manev *et al.*, 1989, 1990b; de Erausquin *et al.*, 1990). The fact that gangliosides failed to affect the basal $[Ca^{2+}]_i$ or the increase of $[Ca^{2+}]_i$ following physiological stimuli of NMDA-selective glutamate receptors suggests that the action of gangliosides follows the expectations of attaining a specific reduction of the crucial pathological events leading to neuronal death that ensue after the abusive receptor stimulation. Therefore, gangliosides fulfill the expectations of drugs having a modus operandi that follows the concept of RADA action.

Ganglioside protection against glutamate-induced destabilization of $[Ca^{2+}]_i$ homeostasis depends on the reduction of the pathological phosphorylation of membrane proteins operative in $[Ca^{2+}]_i$ homeostasis regulation, which are presumably mediated by the ganglioside's ability to inhibit the exaggerated and prolonged PKC-mediated phosphorylation and to reduce the prolonged slowly reversible translocation of PKC elicited by pathological stimulation of glutamate receptors (Vaccarino *et al.,* 1987; Manev *et al.,* 1990a,b,c).

The major advantage of the RADA approach is the lack of serious side effects. The ganglioside G_{M1}, when used in experimental animals and later in human therapy, proved to be a safe drug. The delayed and limited bioavailability of this compound after its systemic administration, and its inability to reach therapeutically valid brain concentrations when administered orally, prompted the synthesis of new ganglioside derivatives with suitable physicochemical properties to assure an appropriate oral absorption and better membrane permeability (Polo *et al.*, 1992).

B. Natural Gangliosides and Excitotoxicity

In primary cultures of cerebellar granule neurons, stimulation of ionotropic receptors by glutamate or NMDA in the absence of Mg^{2+}, or by kainate in the presence of Mg^{2+}, induces a dose-dependent neuronal death. A pulse of excitotoxin (about 15 minutes for NMDA or 30–60 minutes for kainate) induces delayed neuronal death that begins 1–2 hours following the end of the pulse and reaches completion in about the next 10 or more hours (Favaron *et al.,* 1988). Natural gangliosides inhibit the delayed neuronal death. The number of the sialic acid molecules in the ganglioside structure is crucial to obtain the RADA action. Hence, the asialo-G_{M1} failed to exert any protective action, while the efficacy of different naturally occurring gangliosides was directly proportional to the increasing number of sialic acid residues present: $G_{T1b} > G_{D1b} > G_{M1}$ (Favaron *et al.,* 1988). A similar protective action was also reproduced by using models of anoxia- or hypoglycemia-induced neuronal death (Facci *et al.,* 1990).

The physicochemical properties of gangliosides are such that, when dissolved in water-based solvents (buffers, biological fluids) even in concentrations as low as 1 μ*M*, a very small proportion of the compound is dissolved as a monomer, while the bulk forms large micelles (Sonnino *et al.,* 1990). Since the monomeric form is the one that crosses the blood–brain membrane and favors the membrane insertion of the gangliosides, it takes G_{M1} longer than 30 minutes to reach an equilibrium for optimizing the condition necessary for neuronal insertion. Both processes are of key importance because it was shown that the intensity of ganglioside protection linearly depends on the amount of ganglioside that is inserted into the neuronal membrane (Favaron *et al.,* 1988).

C. Ganglioside Derivatives and Excitotoxicity

Because natural gangliosides cannot be administered by mouth, and because they insert and concentrate in neuronal membranes relatively slowly, the synthesis of ganglioside derivatives having physicochemical properties suitable for oral administration and possessing facilitated rates of insertion into neuronal membranes was initiated.

One might predict two major strategies to achieve this goal, one directed toward the lipidic moiety of the molecule, the other toward the carbohydrate moiety. Up to now, positive results have been obtained by following the first strategy. Among the different glycosphingolipids tested for RADA action *in vitro*, two semisynthetic G_{M1} derivatives, namely, the single-chain compounds having in the lipidic moiety *N*-acetylsphingosine, LIGA 4 (II^3Nen-5-Ac-GgOse_4-2D-*erythro*-1,3-dihydroxy-2-acetylamide-4-*trans*-octadecene), or *N*-dichloroacetylsphingosine, LIGA 20 (II^3Nen-5-Ac-GgOse_4-2D-*erythro*-1,3-dihydroxy-2-dichloro-acetyl-amide-4-*trans*-octadecene), were found to be the most effective (Manev *et al.*, 1990b) (Fig. 1).

The structural differences between LIGAs and G_{M1} are manifested in the different aggregation of properties in aqueous solutions. The micellar concentration of LIGA 4 is four orders of magnitude higher than that of G_{M1} (Sonnino *et al.*, 1990). It is possible that the formation of micelles might also be hindering the oral absorption of natural gangliosides. That is to say, ganglioside derivatives which tend to form a lower proportion of micelles when they are in an aqueous medium, such as lysoganglioside derivatives, tend to insert more rapidly and by a greater extent into the membranes. Moreover, after oral administration, these single-chain ganglioside derivatives can be found in greater amounts both in the plasma and in various organs, including the brain (Fig. 4).

In neuronal cultures, protection against excitotoxicity by LIGAs was obtained with doses that were one order of magnitude lower than those of G_{M1}. Moreover, the occurrence of the neuroprotective action was instantaneous (as opposed to the pretreatment needed for G_{M1}) and, in the case of LIGA 20, the protection persisted longer than 24 hours (Manev *et al.*, 1990b). The latter effect might be attributed to the formation of an active metabolite. That is, in cerebellar granule neurons in culture, the LIGAs undergo metabolic processing with the formation of ceramide-like derivatives [PKS1 (D-*erythro*-1,3-dihydroxy-2-acetylamide-4-*trans*-octadecene) from LIGA 4 and PKS3 (D-*erythro*-1,3-dihydroxy-2-dichloroacetylamide-4-*trans*-octadecene) from LIGA 20] (Pitto *et al.*, 1991). Among these derivatives, PKS3, the metabolite of LIGA 20, possesses an anti-excitotoxic action (Manev *et al.*, 1990b).

As shown for G_{M1} (Favaron *et al.*, 1988), the neuroprotective effect of LIGAs is not achieved via the inhibition of glutamate receptors (Manev *et al.*, 1990b); rather, it is related to the ability of these molecules to prevent the glutamate-triggered destabilization of the $[Ca^{2+}]_i$ dynamics, at least in neuronal cultures (de Erausquin *et al.*, 1990).

Recently, a Ca^{2+}-dependent ganglioside-binding protein (gangliomodulin) was isolated from a soluble fraction of mouse brain and found to be calmodulin

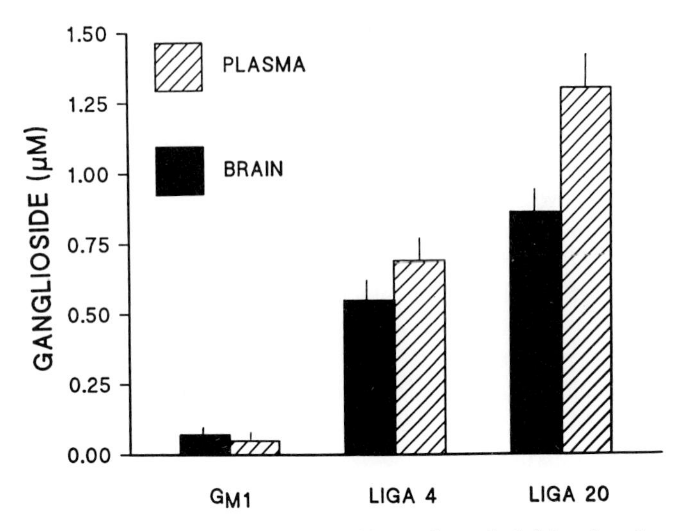

FIG. 4. Brain and plasma ganglioside content 6 hours after oral administration of natural (G_{M1}) and semisynthetic (LIGA 4 and LIGA 20) gangliosides. Rats were injected by oral gavage with 66 µmol/kg of each ganglioside. Each bar represents the mean ± SE of 5 rats (A. Polo, unpublished data).

(Higashi and Yamagata, 1992). It has also been shown that natural gangliosides, by binding to calmodulin, modulate the ability of calmodulin-dependent enzymes to catalyze their reaction (Higashi *et al.*, 1992). Based on these observations, it can be inferred that the function of the calmodulin-dependent Ca^{2+}-ATPase pump on the plasma neuronal membranes is modulated by the binding of gangliosides to calmodulin or to the pump itself. This interaction may be the basis for the ganglioside prevention of the destabilization of $[Ca^{2+}]_i$ homeostasis elicited by excitotoxic doses of glutamate.

If oxygen is withdrawn from rat hippocampal slices, a spreading depression-like response occurs, and, after reoxygenation, recovery of synaptic transmission correlates inversely with the time spent in spreading depression. The ganglioside G_{M1} accelerated the recovery time from hypoxic depression in the synaptic transmission of the hippocampal region CA1 (Somjen *et al.*, 1990). Hence, the delay of anoxic depolarization (AD) protected neurons from subsequent death (Somjen *et al.*, 1990). The treatment of rat hippocampal slices with LIGA 4 or PKS3 was also shown to prolong the latency to AD (Balestrino *et al.*, 1992).

Taken together, the results of *in vitro* experiments are indicative of a superior RADA action by single-chain derivatives of G_{M1}.

D. *In Vivo* Excitotoxicity: Protection via RADA Mechanisms

The best proof that gangliosides also operate with a RADA mechanism in living animals is given by the *in vivo* protection afforded by G_{M1} against pathologies in which a role for excitotoxicity has been implied [(stroke (Angeleri *et al.*, 1992; Argentino *et al.*, 1989; Battistin *et al.*, 1985; Ceccarelli *et al.*, 1976; Facci *et al.*, 1990; Papo *et al.*, 1991; Skaper *et al.*, 1989); brain ischemia, brain and spinal cord trauma (Somjen *et al.*, 1990; Geisler *et al.*, 1991; Hoermann, 1988; Walker, 1991); and Parkinson's disease models (Hadjiconstantinou *et al.*, 1986; Schneider *et al.*, 1992)]. In all these situations, no serious side effects of the G_{M1} treatment were observed, this being in line with the finding that the neuroprotective action of the compound is achieved without the blockade of specific neurotransmitter receptors, including glutamate receptors.

In the neonatal rat model of NMDA neurotoxicity, the subcutaneous administration of G_{M1} (20 mg/kg; 1 hour before and immediately after i.c.v. injection of NMDA) reduced (1) loss in hemispheric weight, (2) loss in tissue choline acetyltransferase activity, and (3) morphological damage in various brain regions (Lipartiti *et al.*, 1991). Similarly, in the established model of NMDA neurotoxicity in adult rat retina (Siliprandi *et al.*, 1993), the intraocular administration of G_{M1} reduced significantly the loss in choline acetyltranferase activity (Siliprandi *et al.*, 1992).

In vivo, the neuroprotective action of G_{M1} (or that of the inner ester of G_{M1}) was most extensively studied in various models of cerebral lesions of an anoxic–ischemic nature (Bharucha *et al.*, 1991; Seren *et al.*, 1990). Recently, a detailed review of literature describing these experiments was published by Carolei *et al.* (1991). More importantly, in the therapy of human stroke and spinal cord injury, G_{M1} administration showed a statistically significant beneficial effect (Geisler *et al.*, 1991). This effect is currently being evaluated with a more extensive approach, for example, the designing of the ambitious Early Stroke Trial (Rocca *et al.*, 1992).

Encouraging *in vivo* animal data were obtained with the new single-chain ganglioside derivatives (LIGAs). As expected from studies on the physicochemical properties of these molecules and on the kinetics of their membrane insertion *in vitro,* the bioavailability of LIGAs given orally to rats was significantly better than that of natural gangliosides (Fig. 4). While natural gangliosides given orally fail to generate appreciable blood or brain levels, the brain concentrations of orally administered LIGAs were close to the concentration range needed for their *in vitro* efficacy. This quality of LIGAs casts a new light on the pharmacological potential of these molecules in the therapy of chronic neurodegenerative diseases such as Alzheimer's and Parkinson's disease.

When tested *in vivo,* LIGAs confirmed the neuroprotective action previously found *in vitro*. For example, in the model of NMDA neurotoxicity in rat pups, subcutaneously administered LIGA 20 protected the lesioned hemisphere in doses 10 times lower than the effective dose of G_{M1} (M. Lipartiti, unpublished observation). In the model in which neuronal death in the area penumbra is due to an excess of glutamate in the interstitial fluid of MCA occlusion, the intravenous administration of LIGA 4, initiated immediately after the focal ischemia of the brain (10 mg/kg, given immediately after the stroke and then daily for 7 days), resulted in better recovery of the stroke-induced behavioral deficit (Fig. 5).

Similarly, in the model of photochemical-thrombotic stroke (Boquillon *et al.,* 1992), where an increase of glutamate in the penumbra region is linked to neurotoxicity, the same LIGA 4 administration schedule resulted in behavioral improvement of lesioned animals (Manev *et al.,* 1990a). Moreover, a detailed histological examination of perifocal brain areas after LIGA 4 or LIGA 20 administration to rats revealed a reduction in the number of degenerating neurons in the area perifocal to the infarct (Table II). The stroke-induced increase in PKC translocation was also diminished by intravenous LIGA 4 or LIGA 20 pretreatment (50 mg/kg i.v.; 1 hour before the lesion) (Table II).

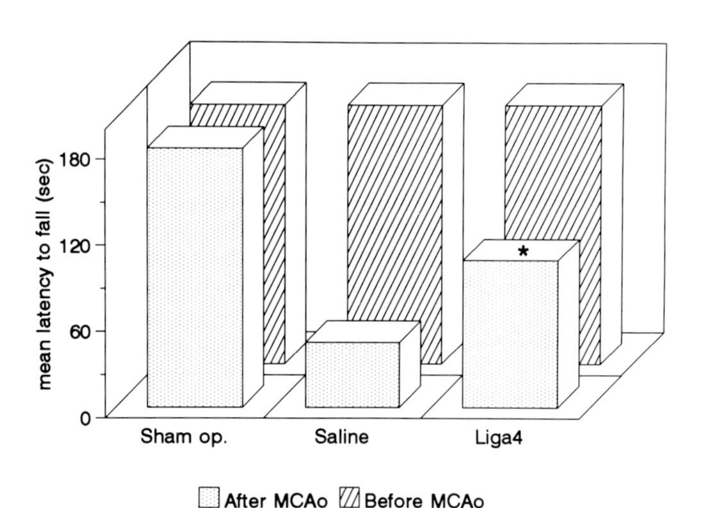

FIG. 5. LIGA 4 treatment reduced the motor deficit caused by the middle cerebral artery occlusion (MCAo). Eight animals were sham-operated, 16 were MCAo animals treated with saline, and 14 were MCAo animals treated with LIGA 4 (6.6 μmol/kg; i.v.; immediately after the stroke, 3 hours later, and then daily). The results shown were obtained with a rota-rod test (12 rpm) measured before and 3 days following the stroke. *, $p < 0.05$ versus corresponding saline (A. Lazzaro and S. Seren, unpublished).

Table II
HISTOLOGICAL AND BIOCHEMICAL EFFECTS OF LIGA 4 AND LIGA 20 IN THE *in Vivo* MODEL OF PHOTOCHEMICAL STROKE

	Number of neurons[a]	PKC activity[b] ([^{3}H]PDBU binding)
Control	25 + 1.2	5.8 + 0.8
LIGA 4	27 + 1.2	6.8 + 0.9
LIGA 20	26 + 1.5	6.5 + 1.1
Stroke	14 + 1.0	21 + 1.6
Stroke + LIGA 4	22 + 1.6[c]	12 + 0.9[c]
Stroke + LIGA 20	20 + 20[c]	13 + 1.9[c]

[a]Nissl-positive neurons were counted in a 0.125 × 0.125 mm area adjacent to the infarcted zone.
[b][^{3}H]Phorbol 12,13-dibutyrate binding is evaluated in optical density in the areas surrounding the infarcted area.
[c]$p < 0.01$ when compared with the stroke group.

As already mentioned, pharmacokinetic studies revealed that a significant portion of LIGA 4 can be found in blood and brain following its oral administration (Fig. 4). Thus, with the appropriate oral administration schedule, similar brain levels to those obtained after intravenous administration (which was neuroprotective) can be obtained. Experiments in progress are aimed at exploring the potential neuroprotective action of orally administered LIGAs in models of chronic neurodegenerative diseases (i.e., epilepsy, Parkinson's, Alzheimer's) in which glutamate could be the neurotoxic agent.

References

Angeleri, F., Scarpino, O., Martinazzo, C., Mauro, A., Magi, M., Pelliccioni, G., Rapex, G., and Bruno, R. (1992). *Cereb. Dis.* **2**(3), 163–168.

Argentino, C., Sacchetti, M. L., Toni, D., Savoini, G., D'Arcangelo, E., Erminio, F., Federico, F., Ferro-Milone, F., Gallai, V., Gambi, D., Mamoli, A., Ottonello, G. A., Ponari, O., Rebucci, G., Senin, U., and Fieschi, C. (1989). *Stroke* **20**(9), 1143–1149.

Balestrino, M., Cogliolo, I., Lunardi, G., Leon, A., and Mazzari, S. (1992). *Soc. Neurosci. Abstr* **522**(14), 1256.

Bassi, S., Albizzati, M.G., Sbacchi, M., Frattola, L., and Massarotti, M. (1984). *J. Neurosci. Res.* **12,** 493–498.

Battistin, L., Cesari, A., Galligioni, F., Marin, G., Massarotti, M., Paccagnella, D., Pellegrini, A., Testa, G., and Tonin, P. (1985). *Eur. Neurol.* **24**(5), 343–351.

Baumann, N., Harpin, M. L., and Jacque, C. (1980). *INSERM Symp.* **14,** 257–262.

Beal, M. F., Kowall, N. W., Ellison, D. W., Mazurek, M. F., Swartz, K. J., and Martin, J. B. (1986). *Nature (London)* **321,** 169–171.

Bharucha, V. A., Wakade, C. G., Mahadik, S. P., and Karpiak, S. E. (1991). *Exp. Neurol.* **114,** 136–139.

Boquillon, M., Boquillon, J. P., and Bralet, J. (1992). *J. Phys. Med.* **27**(1), 1–6.

Boulter, J., Bettler, B., Dingledine, R., Duvoisin, R., Egebjerg, J., Gasic, G., Hartley, M., Hermans-Borgmeyer, I., Hollmann, M., Hughes, T. E., Hume, I. R., Moll, C., Rogers, S., and Heinemann, S. (1992). *In* "Excitatory Amino Acids" (R. P. Simon, ed.), Fidia Res. Found. Symp. Ser., Vol. 9, pp. 9–13. Thieme, New York.

Candeo, P., Favaron, M., Lengyel, I., Manev, R. M., Rimland, J. M., and Manev, H. (1992). *J. Neurochem.* **59,** 1558–1561.

Carolei, A., Fieschi, C., Bruno, R., and Toffano, G. (1991). *Cerebrovasc. Brain Metab. Rev.* **3**(2), 134–157.

Catania, M. V., Aronico, E., Sortino, M. A., Canonico, P. L., and Nicoletti, F. (1991). *J. Neurochem.* **56,** 1329–1335.

Ceccarelli, B., Aporti, F., and Finesso, M. (1976). *Adv. Exp. Biol. Med.* **71,** 275–293.

Chapman, A. G. (1992). *In* "Excitatory Amino Acids" (R. P. Simon, ed.), Fidia Res. Found. Symp. Ser., Vol. 9, pp. 265–271. Thieme, New York.

Chen, L., and Huang, L. Y. M. (1992). *Nature (London)* **356,** 521–523.

Choi, D. W. (1991). *In* "Excitatory Amino Acid Antagonists" (B. S. Meldrum, ed.), pp. 14–38. Blackwell, Oxford.

Comelli, M. C., Seren, M. S., Guidolin, D., Manev, R. M., Favaron, M., Rimland, J. M., Canella, R., Negro, A., and Manev, H. (1992). *NeuroReport* **3,** 473–476.

Conn, P. J., and Desai, M. A. (1991). *Drug Dev. Res.* **24**(3), 207–229.

Connor, J. A. (1991). Fidia Res. Found. Neurosci. Award Lec., Vol. 6, pp. 111–139. Thieme, New York.

Costa, E., Kharlamov, A., Guidotti, A., Hayes, R., and Armstrong, D. (1993). *Pathophysiol. Exp. Ther.* **4,** 17–23.

Domanska-Janik, K., and Zalewska, T. (1992). *J. Neurochem.* **58**(4), 1432–1439.

de Erausquin, G., Manev, H., Guidotti, A., Costa, E., and Brooker, G. (1990). *Proc. Natl. Acad. Sci. U.S.A.* **87,** 8017–8021.

Facci, L., Leon, A., and Skaper, S. D. (1990). *Neuroscience* **37,** 709–716.

Fagg, G. E., and Massieu, L. (1991). *In* "Excitatory Amino Acid Antagonist" (B. S. Meldrum, ed.), pp. 39–63. Blackwell, Oxford.

Favaron, M., Manev, H., Alho, H., Bertolino, M., Ferret, B., Guidotti, A., and Costa, E. (1988). *Proc. Natl. Acad. Sci. U.S.A.* **85,** 7351–7355.

Favaron, M., Manev, H., Siman, R., Bertolino, M., Szekely, A. M., de Erausquin, G., Guidotti, A., and Costa, E. (1990). *Proc. Natl. Acad. Sci. U.S.A.* **87,** 1983–1987.

Favaron, M., Rimland, J. M., and Manev, H. (1992). *Life Sci.* **50,** 189–194.

Fernandez, M. T., Zitko, V., Gascon, S., and Novelli, A. (1991). *Life Sci.* **49,** 157–162.

Fonnum, F. (1991). *In* "Excitatory Amino Acids" (B. S. Meldrum, F. Moroni, R. P. Simon, and J. H. Woods, eds.), Fidia Res. Found. Symp. Ser., Vol. 5, pp. 15–25. Raven Press, New York.

Gall, C., Sumikawa, K., and Lynch, G. (1990). *Proc. Natl. Acad. Sci. U.S.A.* **87,** 7643–7647.

Geisler, F. H., Dorsey, F. C., and Coleman, W. P. (1991). *N. Engl. J. Med.* **324**(26), 1829–1838.

Ghidoni, R., Riboni, L., and Tettamanti, G. (1989). *J. Neurochem.* **53**(5), 1567–1574.

Hadjiconstantinou, M., Rossetti, Z. L., Paxton, R. C., and Neff, N. H. (1986). *Neuropharmacology* **25,** 1075–1077.

Heafield, M. T., Fearn, S., Steventon, G. B., Waring, R. H., Williams, A. C., and Sturman, S. G. (1990). *Neurosci. Lett.* **110,** 216–220.

Higashi, H., and Yamagata, T. (1992). *J. Biol. Chem.* **267**(14), 9839–9843.

Higashi, H., Omori, A., and Yamagata, T. (1992). *J. Biol. Chem.* **267**(14), 9831–9838.

Hoermann, M. (1988). *In* "New Trends in Ganglioside Research: Neurochemical and Neurodegenerative Aspects" (R. W. Ledeen, E. L. Hogan, G. Tettamanti, A. J. Yates, and R. K. Yu, eds.), Fidia Res. Ser., Vol. 14, pp. 596–604. Liviana Press/Springer-Verlag, Padova/Berlin.

Joo, F., Tosaki, A., Olah, Z., and Koltai, M. (1989). *Brain Res.* **490,** 141–143.

Kemp, J. A., Foster, A. C., Leeson, P. D., Priestley, T., Tridgett, R., Iversen, L. L., and Woodruff, G. N. (1988). *Proc. Natl. Acad. Sci. U.S.A.* **85,** 8547–8550.

Kiedrowski, L., Manev, H., Costa, E., and Wroblewski, J. T. (1991). *Neuropharmacology* **30**(11), 1241–1243.

Koh, J. Y., Palmer, E., Lin, A., and Cotman, C. W. (1991a). *Brain Res.* **561,** 338–343.

Koh, J. Y., Palmer, E., and Cotman, C. W. (1991b). *Proc. Natl. Acad. Sci. U.S.A.* **88,** 9431–9435.

Kracun, I., Rosner, H., Drnovsek, V., Vukelič, Z., Cosovič, C., Trbojevič-Cepe, M., and Kubat, M. (1992). *Neurochem. Int.* **20**(3), 421–431.

Lipartiti, M., Lazzaro, A., Zanoni, R., Mazzari, S., Toffano, G., and Leon, A. (1991). *Exp. Neurol.* **113,** 301–305.

Manev, H., Favaron, M., Guidotti, A., and Costa, E. (1989). *Mol. Pharmacol.* **36,** 106–112.

Manev, H., Favaron, M., Vicini, S., and Guidotti, A. (1990a). *Acta Neurobiol. Exp.* **50,** 475–488.

Manev, H., Costa, E., Wroblewski, J. T., and Guidotti, A. (1990b). *FASEB J.* **4,** 2789–2797.

Manev, H., Favaron, M., Vicini, S., Guidotti, A., and Costa, E. (1990c). *J. Pharmacol. Exp. Ther.* **252**(1), 419–427.

Manev, H., Favaron, M., Siman, R., Guidotti, A., and Costa, E. (1991). *J. Neurochem.* **57**(4), 1288–1295.

Marin, P., Lafon-Cazal, M., and Bockaert, J. (1992). *Eur. J. Neurosci.* **4**(5), 425–432.

Masu, M., Tanabe, Y., Tsuchida, K., Shigemoto, R., and Nakanishi, S. (1991). *Nature (London)* **349,** 760–765.

Mattson, M. P. (1991). *Exp. Neurol.* **112,** 95–103.

McCulloch, J., Bullock, R., and Teasdale, G. M. (1991). *In* "Excitatory Amino Acids Antagonists" (B. S. Meldrum, ed.), pp. 287–326. Blackwell, Oxford.

McGeer, E. G., and Zhu, S. G. (1990). *Neurosci. Lett.* **112**(2–3), 348–351.

Meguro, H., Mori, H., Araki, K., Kushiya, E., Kutsuwada, T., Yamazaki, M., Kumanishi, T., Arakawa, M., Sakimura, K., and Mishina, M. (1992). *Nature (London)* **357,** 70–74.

Meldrum, B. S., Smith, S. E., Le Peillet, E., Moncada, C., and Arvin, B. (1992). *In* "Excitatory Amino Acids" (R. P. Simon, ed.), Fidia Res. Found. Symp. Ser., Vol. 9, pp. 235–239. Thieme, New York.

Milani, D., Guidolin, D., Facci, L., Pozzan, T., Buso, M., Leon, A., and Skaper, S. D. (1991). *J. Neurosci. Res.* **28,** 434–441.

Monyer, H., Sprengel, R., Schoepfer, R., Herb, A., Higuchi, M., Lomeli, H., Burnashev, N., Sakmann, B., Seeburg, P. H. (1992). *Science* **256,** 1217–1221.

Monyer, M. D., Sommer, B., Wisden, W., Veroorn, T. A., Burnashev, N., Sprengel, R., Sakmann, B., and Seeburg, P. H. (1992). *In* "Excitatory Amino Acids" (R. P. Simon, ed.), Fidia Res. Found. Symp. Ser., Vol. 9, pp. 29–33. Thieme, New York.

Moroni, F., and Pellicciari, R. (1992). *In* "Excitatory Amino Acids" (R. P. Simon, ed.), Fidia Res. Found. Symp. Ser., Vol. 9, pp. 89–93. Thieme, New York.

Nabeshima, T., Ogawa, S., Nishimura, H., Fuji, K., Kameyama, T., and Sasaki, Y. (1991). *Neurosci. Lett.* **122**(1), 13–16.

Nakanishi, S. (1992). *In* "Excitatory Amino Acids" (R. P. Simon, ed.), Fidia Res. Found. Symp. Ser., Vol. 9, pp. 21–22. Thieme, New York.

Nicoletti, F., Wroblewski, J. T., Fadda, E., Hynie, S., Alho, H., and Costa, E. (1989). *In* "Allosteric Modulation of Amino Acid Receptors: Therapeutic Implications" (E. A. Barnard and E. Costa, eds.), pp. 301–317. Raven Press, New York.

Olney, J. W. (1990). *Annu. Rev. Pharmacol. Toxicol.* **30,** 47–71.

Olney, J. W., Labruyere, J., and Price, M. T. (1989). *Science* **244,** 1360–1362.

Papo, I., Benedetti, A., Carteri, A., Merli, G. A., Mingrino, S., and Bruno, R. (1991). *Stroke* **22**(1), 22–26.

Pitto, M., Miglio, A., Kirschner, G., Leon, A., and Ghidoni, R. (1991). *Neurochem. Res.* **16**(11), 1187–1192.

Polo, A., Kirschner, G., Guidotti, A., and Costa, E. (1992). *Soc. Neurosci. Abstr.* (1), 1602.
Purpura, D. P. (1978). *Nature (London)* **276,** 520–521.
Qi, Y., and Xue, Q. M. (1991). *Mol. Chem. Neuropathol.* **14**(2), 87–97.
Randall, R. D., and Thayer, S. A. (1992). *J. Neurosci.* **12**(5), 1882–1895.
Rocca, W. A., Dorsey, F. C., Grigoletto, F., Gent, M., Roberts, R. S., Walker, M. D., Easton, J. D., Bruno, R., Carolei, A., Sancesario, G., and Fieschi, C. (1992). *Stroke* **23**(4), 519–526.
Rothstein, J. D., Martin, L. J., and Kuncl, R. W. (1992). *N. Engl. J. Med.* **326**(22), 1464–1468.
Sakimura, K., Morita, T., Kushiya, E., and Mishina, M. (1992). *Neuron* **8,** 267–274.
Scatton, B., Benavides, J., Schoemaker, H., and Carter, C. (1992). *In* "Excitatory Amino Acids" (R. P. Simon, ed.), Fidia Res. Found. Symp. Ser., Vol. 9, pp. 95–101. Thieme, New York.
Schneider, J. S., Pope, A., Simpson, K., Taggart, J., Smith, M. G., and DiStefano, L. (1992). *Science* **256,** 843–846.
Seren, M. S., Rubini, R., Lazzaro, A., Zanoni, R., Fiori, M. G., and Leon, A. (1990). *Stroke* **21**(11), 1607–1612.
Sheardown, M. J., Nielsen, E. O., Hansen, A. J., Jacobsen, P., and Honoré, T. (1990). *Science* **247,** 571–574.
Siliprandi, R., Canella, R., Carmignoto, G., Schiavo, N., Zanellato, A., Zanoni, R., and Vantini, G. (1992). *Visual Neurosci.* **8,** 567–573.
Simon, R. P., Chen, J., and Graham, S. H. (1992). *In* "Excitatory Amino Acids" (R. P. Simon, ed.), Fidia Res. Found. Symp. Ser., Vol. 9, pp. 241–246. Thieme, New York.
Skaper, S. D., Leon, A., and Toffano, G. (1989). *Mol. Neurobiol.* **3,** 173–199.
Somjen, G. G., Aitken, P. G., Balestrino, M., Herreras, O., and Kawasaki, K. (1990). *Stroke* **21,** Suppl. 11 (III), 179–183.
Sonnino, S., Cantú, L., Corti, M., Acquotti, D., Kirschner, G., and Tettamanti, G. (1990). *Chem. Phys. Lipids* **56**(1), 49–57.
Svennerholm, L. (1980). *In* "Structure and Function of Gangliosides" (L. Svennerholm, P. Mandel, H. Dreyfus, and P. F. Urban, eds.), pp. 533–544. Plenum Press, New York.
Szekely, A. M., Costa, E., and Grayson, D. R. (1990). *Mol. Pharmacol.* **38,** 624–633.
Tanabe, Y., Masu, M., Ishii, T., Shigemoto, R., and Nakanishi, S. (1992). *Neuron* **8,** 169–179.
Vaccarino, F., Guidotti, A., and Costa, E. (1987). *Proc. Natl. Acad. Sci. U.S.A.* **84,** 8707–8711.
Walker, M. D. (1991). *N. Engl. J. Med.* **324**(26), 1885–1887.
Wenthold, R. J., Yokotani, N., Doi, K., Wada, K., and Petralia, R. S. (1992). *In* "Excitatory Amino Acids" (R. P. Simon, ed.), Fidia Res. Found. Symp. Ser., Vol. 9, pp. 41–46. Thieme, New York.
Wroblewski, J. T., Raulli, R., Lazarewicz, J. W., Kiedrowski, L., Costa, E., and Wroblewska, B. (1991). *In* "Transmitter Amino Acid Receptors: Structures, Transduction and Models for Drug Development" (E. A. Barnard and E. Costa, eds.), Fidia Res. Found. Symp. Ser., Vol. 6, pp. 379–393. Thieme, New York.
Zukin, R. S., Pellegrini-Giampietro, D. E., McGurk, J. F., and Bennett, M. V. L. (1992). *In* "Excitatory Amino Acids" (R. P. Simon, ed.), Fidia Res. Found. Symp. Ser., Vol. 9, pp. 47–51. Thieme, New York.

ADVANCES IN LIPID RESEARCH, VOL. 25

The Role of Glycosphingolipids in Hypoxic Cell Injury

ADY KENDLER AND GLYN DAWSON

The Joseph P. Kennedy Jr. Mental Retardation Research Center
Departments of Pediatrics, Biochemistry, and Molecular Biology,
and the Committee on Neurobiology
University of Chicago, Chicago, Illinois 60637

I. Introduction

The majority of research on the biology of glycosphingolipids (GSL) to date has focused on exploring the functions of these lipids in normal physiology, i.e., the everyday life activities of the cell. Such research has suggested roles for GSL as receptors mediating interactions between cells and their environment (reviewed by Hakomori, 1990). Approaches to studying GSL in cell injury situations may prove useful in two ways: (1) the effects of injurious situations, such as hypoxia, can affect the synthesis and intracellular processing of GSL in unique and specific ways, which can in turn give clues as to the normal regulation of GSL metabolism; and (2) cell injury can provide insights about GSL biology through experiments in which the presence or absence of certain GSL appears to protect cells from injury. After a brief introduction to recent advances in the molecular mechanisms of cell injury, this article discusses evidence obtained in our laboratory concerning the effects of hypoxia on synthesis of the myelin GSL galactosylceramide (GalCer) in cultured oligodendrocytes and in the glioblastoma cell line G2620, and the insights this work provides about the regulation of GalCer metabolism. Finally, a discussion of the work of Costa and others is included, concerning the mechanisms by which GSL may serve to protect neural cells from injury, as may occur following brain ischemia (stroke).

II. Mechanisms of Cell Injury Can Be Very Specific

Recent advances in molecular pathology have shown that cell injury is a gradual process, one that involves, early in the course of injury, specific biochemical responses of cells to the particular type of injury. This topic has been discussed in Jennings (1981) and Farber (1979). Certain changes that occur early in injury are reversible, such as the depletion of ATP, mitochondrial swelling, and the activation of specific signal transduction pathways. Other, later events in injury are irreversible and lead to cell death, such as calcium phosphate crystal accumulation in mitochondria, or enzyme-mediated plasma membrane hydrolysis. The role of massive calcium influx in mediating injury has been extensively investigated (Jennings, 1981), as has the generation of free oxygen radicals and consequent peroxidation of membrane lipids (Gutteridge and Halliwell, 1990). Early in the course of injury, different cell types respond specifically to injury. Ischemic hippocampal neurons, for example, release large amounts of the transmitter glutamate, which is toxic at high concentrations and can bind to receptors on neuronal membranes and cause a deadly influx of calcium (Choi, 1988). Hepatocytes in the process of catabolizing a mildly toxic chemical, such as ethanol, can produce the much more severely damaging molecule acetaldehyde, which can lead to injury and cell death (Lieber, 1988). Since GSL are intimately involved in physiological processes such as signal transduction and regulation of cell growth and development (Hakomori, 1990), an understanding of the effects of injury on the biology of cell GSL may be valuable in understanding the normal metabolism and physiology of these lipids and the role they might play in the cellular response to injury.

III. Progressive Hypoxia Specifically Inhibits Synthesis of a Major Myelin Lipid

A. Introduction

Galactosylceramide (GalCer, galactosyl-*N*-acylsphingosine) is a GSL that is enriched in nervous system myelin (Morell, 1984). In the central nervous system, myelin is elaborated as a specialized extension of the oligodendrocyte (OLG) plasma membrane (Morell, 1984). The myelination of axons in the nervous system of the rat peaks at 15–25 days postnatally (Kishimoto *et al.*, 1979), and at this time OLG generate three times their weight each day in myelin lipids and proteins, many of which are unique to myelin (Lemke, 1988). GalCer comprises up to 30% of myelin lipid and, until recently, was considered to have a role only in the compaction of the myelin lamellae. GalCer, which is located on the external surface of myelin sheaths, was proposed to have a role in creating the rigidity and stability of the sheath through its hydrogen-bonding properties (Abrahammson *et al.,* 1976). Recently, studies utilizing GalCer-specific antibodies to probe the role of GalCer on the membrane sheets of cultured murine OLG have suggested that GalCer mediates a second messenger cascade in the cytosol of the OLG, including an influx of extracellular calcium and that these messages result in cytoskeletal changes within the membranes that may be involved in the wrapping and compaction of myelin (Dyer and Benjamins, 1990; Dyer, 1992). These authors propose that GalCer may be a receptor for an endogenous ligand, located on axons or on myelin membranes, and that binding of the ligand to myelin may be an important signal in initiating the myelination process (Dyer and Benjamins, 1991; Dyer, 1992).

B. Hypoxia Inhibits Synthesis of GalCer

We have studied the effects of progressive hypoxic injury on the synthesis of GalCer by cultured rat neonatal OLG (Kendler and Dawson, 1990,1992). These OLG express myelin-specific lipids and proteins in increasing amounts in a manner that parallels the temporal expression of myelin *in vivo* (Pfieffer, 1984). Thus, the cultured OLG system is considered to be a good model for myelinating OLG (Pfieffer, 1984; Wood and Bunge, 1984). Experiments in which cultured OLG were subjected to 6 hours of gradual, progressive, hypoxia in a GasPak (BBL, Beckton Dickinson) anaerobic chamber showed that this treatment did not injure the cells, with cellular ATP levels decreasing to 85% of controls, and no inhibition of the synthesis of membrane glycerophospholipids, the sphingolipid sphingomyelin, or glycoproteins detected. Similarly, OLG morphology was not affected at 6 hours of hypoxia. In contrast, at 12 hours of hypoxia, the OLG showed a 50% drop in ATP levels and swelling of the cell body with loss of processes. At 6 hours of hypoxia, synthesis of GalCer species (determined by [^{3}H]-palmitate incorporation, as described in Kendler and Dawson, 1990) was seen to be specifically inhibited, with the synthesis of the 2-hydroxylated fatty acid species of GalCer (HFA-GalCer, whose synthesis depends on molecular oxygen, see Fig. 1) inhibited by as much as 50% from controls. Synthesis of the nonhydroxylated fatty acid species of GalCer (NFA-GalCer) was inhibited by 25–40% (see Fig. 2). Concomitant with the decrease in synthesis of NFA-GalCer species was an *increase* in synthesis of NFA-ceramide, the ungalactosylated precursor of NFA-GalCer (see Fig. 2 for data, Fig. 1 for biosynthetic

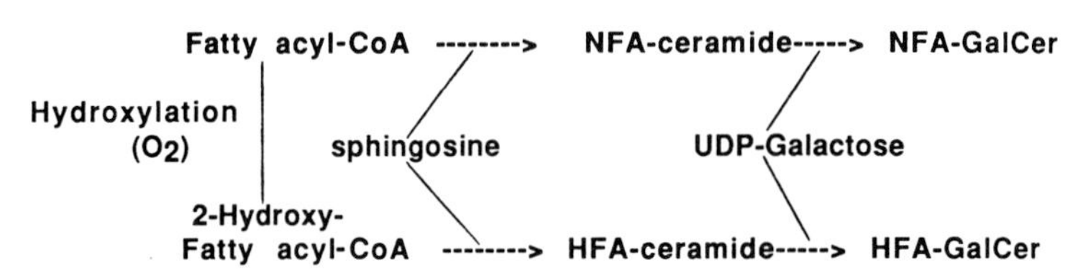

FIG. 1. Pathway of GalCer synthesis. Condensation of nonhydroxy-fatty acyl-CoA (NFA) or 2-hydroxy-fatty acyl-CoA (HFA) with sphingosine yields NFA-ceramide or HFA-ceramide, respectively. Galactosylation of ceramide via a UDP-Gal intermediate results in formation of GalCer (NFA or HFA species). Galactosylation is thought to be catalyzed by a UDP-galactose–ceramide galactosyltransferase enzyme (CGalT). Note that the rate-limiting step in HFA-GalCer synthesis is thought to be the hydroxylation of the free fatty acyl-CoA, while the galactosylation of NFA-ceramide is considered rate-limiting in NFA-GalCer synthesis (see text).

pathway). Since the galactosylation of NFA-ceramide by UDPgalactose–ceramide galactosyltransferase (CGalT, EC 2.4.1.62) is considered to be the rate-limiting step in NFA-GalCer synthesis (see Fig. 1; Radin, 1972), it appeared that hypoxia was inhibiting NFA-GalCer synthesis at its rate-limiting step. This inhibition of NFA-GalCer synthesis was reversible, and hypoxic OLG that were returned to fresh nonradioactive oxygenated media were seen to chase their accumulated [^{3}H]palmitate-labeled NFA-ceramide pool into NFA-GalCer species (Kendler and Dawson, 1990). We did not detect an increase in HFA-ceramide, the ungalactosylated precursor of HFA-GalCer, in hypoxia. Synthesis of HFA-GalCer is considered to be rate-limited *not* at the galactosylation of HFA-ceramide, but rather at the O_2-dependent 2-hydroxylation of fatty acids (Fig. 1; Kishimoto *et al.,* 1979; thus, it was not surprising to find in our system that hypoxia inhibited HFA-GalCer at its rate-limiting step (Kendler and Dawson, 1990). Assays of CGalT activity in homogenates of hypoxic OLG indicated that the galactosylation of NFA-ceramide was *not* due to a decreased CGalT activity in hypoxia; thus, we considered the possibility that NFA-ceramide was not being transported from its site of synthesis to its site of galactosylation.

C. Hypoxic Inhibition of the Transport of Ceramide from the ER to the Golgi

It is not clear where the site of galactosylation of ceramide resides within the subcellular membranes. A number of researchers have attempted to define the subcellular sites of the galactosylation of ceramide; the methods used focused on measuring *in vitro* CGalT activity in fractionated OLG membranes. While some researchers found CGalT activity to be enriched in the endoplasmic reticulum (ER; Carruthers and Carey, 1983), others found it enriched in the Golgi apparatus (Siegrist *et al.,* 1979) or in both organelles (Sato *et al.,* 1988). Inter-

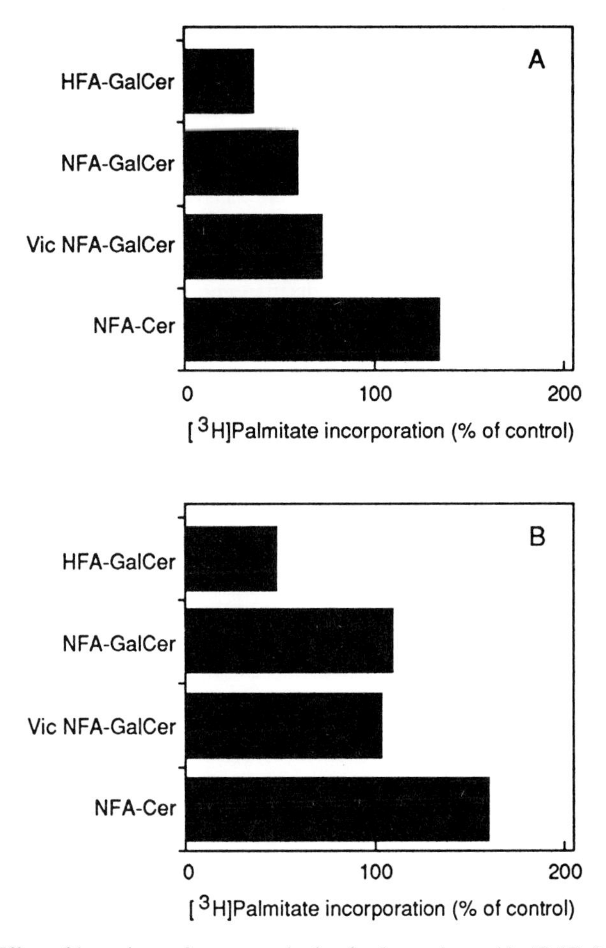

FIG. 2. Effect of hypoxia on *de novo* synthesis of galactosylceramide (GalCer) and ceramide (Cer). (A) Neonatal rat brain oligodendrocytes; (B) mouse oligodendroglioma cell line G2620. Cultures were labeled with [^{3}H]palmitate (2 μCi/ml) during 6 hours of increasing hypoxia and lipids were isolated by HPTLC as described previously (Kendler and Dawson, 1990). HFA, 2-Hydroxy fatty acid; Vlc, very long chain fatty acid. Very little radioactivity was detected in HFA-ceramide in either cell line.

pretation of results based on established CGalT methods is complicated by the fact that the enzyme assay is relatively unreactive toward the major species of ceramide (NFA-ceramide) in the brain (Brenkert and Radin, 1972; Sato *et al.*, 1988). We took advantage of the unique effects of hypoxia on the synthesis of GalCer in order to address the question of subcellular translocation of ceramide during GalCer synthesis. As described above (and shown in Fig. 2), progressive hypoxia caused an inhibition of GalCer synthesis and a simultaneous accumulation of NFA-ceramide; thus, we set out to determine which OLG subcellular fractions were accumulating ceramide during hypoxia. Subcellular fractions of control and hypoxic OLG were prepared as described in Kendler and Dawson (1992), and membranes in each fraction were assayed for specific ER and Golgi markers (Kendler and Dawson, 1992) as well as dpm of [^{3}H]palmitate-labeled GalCer and NFA-ceramide. Results (shown in Fig. 2) indicated that hypoxic cells contained 24% more NFA-ceramide than controls, and that the increase in NFA-ceramide correlated most closely with subcellular fractions containing the peak activity of the ER marker glucose-6-phosphatase ($r = 0.5$). GalCer synthesis in hypoxic OLG was inhibited by 50% throughout the subcellular gradients (Fig. 2). These results suggested that hypoxia was causing a reversible accumulation of NFA-ceramide in OLG, and that this accumulation was localized to the ER. Indeed, previous authors have shown evidence that ceramide synthesis is associated with the ER (Morell and Radin, 1969; Kishimoto, 1983), and investigations on the subcellular transport of a fluorescently labeled ceramide species, C_6-NBD-ceramide, have suggested that the ceramide accumulates in a "pre-Golgi" compartment in living cells, and that ceramide then is transported to the Golgi, where it is converted to more complex sphingolipids (van Meer *et al.*, 1987; Pagano, 1989).

D. Transport of Ceramide from the ER to the Golgi Is ATP-Dependent: Evidence in Semi-Intact OLG

We hypothesized that the small decrease in ATP (15% below controls) caused by hypoxia in OLG was the cause of the inhibition of transport of NFA-ceramide from the ER to the Golgi. Balch *et al.* (1986) have shown that the transport of the vesicular stomatitis protein "G" from the ER to the Golgi in cultured cells could be inhibited completely by a small (15%) drop in cell ATP levels. To test this possibility, we permeabilized OLG by the method of Beckers *et al.* (1987) and found that the semi-intact cells retained a fully functional CGalT enzyme, i.e., when incubated in buffer containing UDP-[^{3}H]Gal, the cells could incorporate the label into GalCer species (Kendler and Dawson, 1992). However, in order to obtain galactosylation of [^{3}H]palmitate-labeled NFA-ceramide (located in the ER) by permeabilized OLG, addition of ATP to the buffer was required. These data indicated that, in order for galactosylation of newly synthesized (ER-localized) ceramide to occur, the presence of ATP was necessary, supporting the hypothesis that transport of NFA-ceramide from its site of synthesis (ER) to its site of galactosylation (Golgi) was ATP-dependent. Thus, it was likely that inhibition of ceramide transport to the Golgi was inhibited by the small decrease in OLG ATP levels during reversible hypoxic injury.

IV. The Regulation of Glycosylceramide Synthesis in the Glioblastoma G2620

G2620 is a glioblastoma-derived cell line that is unique in that it synthesizes substantial amounts of both the OLG-specific GalCer and the predominant neuronal/astrocytic lipid glucosylceramide (GlcCer). Thus, we believe G2620 is a good system for dissecting the mechanisms by which the synthesis of these two GSL species is regulated.

A. Introduction

G26 is a glioma of central nervous system origin induced in C57B1/6 inbred mice by methylcolanthrine treatment (Zimmerman, 1955) and is classified as an immature glial cell with oligodendroglial and astrocytic properties. The solid tumor was subcloned, and a number of cell lines were developed which could be grown in cell culture (Sundaraj *et al.*, 1975). The subclone G2620 cells are rapidly dividing, bipolar cells that are triploid in chromosome number. G2620 cells express the glial-specific antigen NS-1, the OLG-specific enzyme CNPase, and a nervous system-specific soluble protein S-100 (Moore, 1965). Dawson (1979) has shown that G2620 synthesizes the myelin lipids GalCer and its sulfated derivative (sulfatide). G2620 is a cell of mixed phenotypes, synthesizing large amounts of glucosylceramide (GlcCer), a sphingolipid normally synthesized in large amounts by neurons, yet also synthesizing the OLG-specific GalCer. Dawson (1979) reported a ratio of GlcCer : GalCer of 3 : 1 synthesized by G2620 cells. GlcCer, being the precursor of a variety of a more complex GSL, such as the blood group antigens (hematosides) and gangliosides (Kishimoto, 1983), is synthesized by many tissues, such as fibroblasts, liver, and spleen. The glucosylation of ceramide to form GlcCer has been reported to be catalyzed by the enzyme activity UDPglucose–ceramide glucosyltransferase (CGlcT), an activity enriched in the microsomes of neurons in the central nervous system, but minimally expressed in OLG (Radin *et al.*, 1972). The specific activity of CGlcT has been seen to increase early in brain development, concomitant with maximal neuronal ganglioside synthesis, and preceding myelination. In contrast, the OLG-specific microsomal enzyme CGalT peaks in its activity in parallel with myelination (Brenkert and Radin, 1972; Kishimoto, 1983).

It is not known how the expression of GlcCer and GalCer is regulated in cells. Since neither glycosylating enzyme has been purified nor sequenced at the gene level, the two activities could conceivably be due to the same protein, with the specificity of action regulated by the availability of substrate (i.e., the levels of UDP-Glc and UDP-Gal). Alternatively, the two activities may reflect two unique gene products, whose expression is dependent on cell type (i.e., neuron versus OLG) and on developmental time.

B. Cultured G2620 Cells Synthesize Substantial Amounts of GalCer and GlcCer

In experiments described previously (Kendler and Dawson, 1992, 1993), we found that cultured G2620 cells, labeled with [^{3}H]palmitate in serum-free media (Cellgro, a biotin-supplemented F-12 media), incorporated the label primarily into two GlcCer species, a 24-carbon fatty acid species and an 18-carbon fatty acid species. Compared to the incorporation into GlcCer species (20,000 dpm/500 μg of cell protein), we found that approximately 8,000 dpm/500 μg of label was incorporated into GalCer species by G2620 cells. As mentioned in the above section, primary OLG cultures incorporated 200,000 dpm of [^{3}H]palmitate per 50 μg of protein into GalCer species, and 2,000 dpm/50 μg protein into GlcCer (1% of the incorporation into GalCer). Due to the small amount of detectable GlcCer synthesis in OLG, the G2620 cells presented a preferable system for the study of the coexpression of GalCer and GlcCer. Interestingly, if the G2620 cells were cultured in the presence of calf serum in the media, the labeling of the GalCer species was enhanced by 10-fold, while the GlcCer labeling increased by only 3-fold, suggesting that the serum was differentially enhancing the synthesis of GalCer over GlcCer.

C. Acute Hypoxia Does Not Injure G2620 Cells

G2620 cells incubated in serum-free media and subjected to 6 hours of hypoxia as described above did not show a decrease in cellular ATP, no morphological changes were seen, and there was no inhibition of cell division. This was also true of cells incubated in glucose-free media. G2620 cells treated with hypoxia and deprivation of glucose were seen to halt cell division, detach from the plate surface, and to contain reduced ATP levels, to 40% of controls. The G2620 cells therefore showed a resistance to injury by hypoxia or glucose deprivation alone, but were injured by the combination of the two. A resistance to hypoxic injury in tumors (the Warburg effect) has been reported previously (Warburg, 1930), and is due to the ability of solid tumors to shift entirely to anaerobic metabolism under hypoxic conditions, without incurring an oxygen debt.

D. Hypoxia Inhibits the Synthesis of HFA-GalCer and GlcCer Species in G2620 Cells

When incubated with [^{3}H]palmitate in serum-free Cellgro media for 6 hours of hypoxia, G2620 cells showed no inhibition of synthesis of GlcCer species nor of the NFA-GalCer species of GalCer. Only the HFA-GalCer species, which depends on molecular O_2 for its synthesis (see Fig. 1), was inhibited, to 67% of control values. This suggested that the same O_2-dependent fatty acyl-CoA

2-hydroxylase was involved in synthesizing HFA-GalCer in the G2620 cells as in the OLG. If G2620 cells were made hypoxic in serum-free EMS (Matalon's modified DMEM), however, an inhibition of both GlcCer species (to 58% of control for the C_{24}, and to 66% of control for the C_{18} species) was observed, as well as the inhibition of HFA-GalCer species, to 48% of controls. No effect on synthesis of the NFA-GalCer species was seen. Accompanying the decreased GlcCer synthesis was an increased labeling of NFA-ceramide, the immediate precursor of monoglycosylated GSL, by about 60%, suggesting that, under these conditions, hypoxia was causing an inhibition of the conversion of NFA-ceramide into GlcCer. Studies have suggested that the CGlcT enzyme activity is located in the Golgi (Coste *et al.*, 1986), and thus we hypothesized that, as seen in OLG (Kendler and Dawson, 1992), hypoxia was causing an inhibition of transport of ceramide from the ER to its site of glucosylation (the Golgi). Since no ATP depletion was detected in hypoxic G2620 cells, one cannot suggest that an ATP drop could cause the inhibition of the glucosylation of ceramide. It is possible, however, that a shift from reliance on oxidative phosphorylation to glycolysis as the primary ATP source could trigger a shutdown of a number of energy-dependent processes in the G2620 cells. One such process might be the use of ATP as a cofactor in the transport of ceramides (and other molecules) from the ER to the Golgi, resulting in ceramide accumulation and decreased GlcCer synthesis. Why no block in NFA-GalCer was seen is unclear; it is not known if the NFA-ceramide pool that serves as a precursor for GlcCer is distinct somehow from the NFA-ceramide pool destined for GalCer; perhaps transport of the GalCer precursor from ER to the Golgi is protected while transport of the GlcCer precursor is not. It does appear, however, that hypoxia specifically blocks HFA-GalCer synthesis in G2620 cells through O_2 depletion, and blocks GlcCer synthesis through the inhibition of precursor availability (i.e., NFA-ceramide). Although the G2620 cells were clearly able to synthesize significant amounts of both GalCer and GlcCer simultaneously, it is as yet unclear as to whether this was accomplished through coexpression of CGalT and CGlcT genes or through a metabolically regulated mechanism that utilizes a common glycosyltransferase enzyme.

E. The Availability of Cytosolic UDP-Hexose May Regulate Glycosylceramide Synthesis: Studies in Semi-Intact G2620 Cells

The next series of experiments were designed to explore whether synthesis of GlcCer and GalCer in G2620 cells could be controlled by the availability of the UDP-sugar substrate in the cell cytosol. It is well known that the interconversion of cytosolic UDP-Glc and UDP-Gal can be catalyzed by a UDP-hexose epimerase (Cohn and Segal, 1969). Neskovič *et al.* (1981) found that G2620 homogenates, when incubated with UDP-[^{3}H]Gal, incorporated radioactivity only into GlcCer, and not into GalCer. These authors reported that an active UDP-hexose epimerase activity was converting the UDP-Gal into UDP-Glc, which was then used to synthesize GlcCer. Such a mechanism in G2620 cells could be used to restrict glycosylceramide synthesis to primarily GlcCer, while limiting GalCer production to approximately one-third of GlcCer synthesis. In fact, when we permeabilized G2620 cells as described (Kendler and Dawson, 1992), and incubated them with UDP-[^{3}H]Gal, the cells synthesized labeled GalCer species, but not GlcCer. These results would be consistent with the fact that permeabilization causes cytosolic proteins to be washed out of the cells, including any UDP-hexose epimerase. Thus, it is possible that an active UDP-hexose epimerase activity may act to regulate the synthetic activity of a single nonspecific glycosyltransferase. Similarly, the cytosol of OLG may regulate epimerase activity so that UDP-Gal is the primary substrate available for glycosylation of ceramide, thus favoring the production of GalCer. The definitive answer to the question of whether there are two distinct enzymes will be solved completely only when the genes for these activities are finally sequenced. Mentioned above was our observation that, in the presence of serum, the proportion of GalCer to GlcCer labeling changed, with an enhancement of GalCer synthesis of 10-fold and an increase in GlcCer synthesis of 3-fold, changing the ratio of synthesis from 3 : 1 (GlcCer : GalCer) to 0.9 : 1. It is likely that factors present in serum may influence the balance of glycosylceramide synthesis in these cells. Dawson (1979) and Jungalwala *et al.* (1985) found that cortisol treatment of G26 clones stimulated the sulfation of GalCer to form sulfatide, a marker of OLG differentiation. The latter authors suggested that increased expression of sulfotransferase activity was responsible for the effect observed. It is of interest to explore in the future the mechanisms by which hormones may influence GalCer and GlcCer expression.

V. Gangliosides Influence the Response of Cells to Injury

The studies described above suggest that an understanding of the effects of hypoxia on the expression of important GSL in cultured cells may help to dissect the mechanisms of the synthesis and processing of these lipids, as well as to provide clues as to the responses of different cell types to injury. The studies discussed below (and elsewhere in this volume), however, were designed to determine more directly whether GSL (gangliosides, in particular) may play protective roles in the course of cell injury. The work of E. Costa *et al.* has provided evidence that both natural and semisynthetic GSL protect cultured cerebellar granule cells from glutamate-dependent cytotoxicity. The vulnerability of certain neuronal groups in the mammalian brain to ischemic–hypoxic injury has been linked to the observation that these neurons utilize glutamate or aspartate as their primary neurotransmitter (Choi, 1988; Rothman, 1985; Rothman and Olney, 1986). The findings that these amino acid transmitters were toxic in high doses to neurons *in vivo* (Rothman and Olney, 1986) led to the hypothesis that injury was causing a hyperrelease of these transmitters by cortical neurons, which then could stimulate receptors on the membrane surfaces of these neurons, and thus induce an unopposed and deadly influx of cations, such as calcium (the "excitotoxin hypothesis"; see Choi, 1988). Indeed, studies on neurons in culture have shown that high doses of glutamate can stimulate specific channels, resulting in a sustained increase in intracellular Ca^{2+} and delayed cell death (Choi, 1988; Abele *et al.*, 1990; Manev *et al.*, 1989; Favaron *et al.*, 1990). Pretreatment with glutamate antagonists could inhibit the cell injury process, and studies on their protective properties in brain injury in animals have been published (McDonald *et al.*, 1987; Stevens and Yaksh, 1990; Sheardown *et al.*, 1990). *In vivo* studies have suggested that nervous system injury could be ameliorated by the presence of sphingolipids (Mahadik and Karpiak, 1988). Costa *et al.* showed that pretreatment of cultured cerebellar neurons with ganglioside G_{M1} or its semisynthetic derivatives prevents the sustained increase in intracellular calcium induced by toxic doses of glutamate (Manev *et al.*, 1990); these authors propose that the mechanism of this protection lies in the ability of certain GSL to inhibit the translocation of protein kinase C from the cell cytosol to the membrane, a cell signaling event involved perhaps in the regulation of calcium fluxes in the cell. This work provides evidence that GSL, in their physiological roles as regulators of the cell's interaction with its environment, may play crucial roles in the cell's response to injury.

VI. Summary

Recent advances in the understanding of the molecular mechanism of cell injury have led to the realization that cell injury is a gradual process, which involves, in its early stages, specific biochemical responses of cells to the injury process. In this review, we have shown that the pursuit of the mechanisms by which inhibition of a specific GSL (GalCer) in hypoxic OLG occurs can lead to (1) an understanding of how the synthesis and transport of GalCer are regulated in the cell, and (2) evidence that the injury causes a potentially devastating (yet reversible) block in the production of this important lipid. This review has also discussed evidence that GSL, in their roles as regulators of the cell's interaction with its environment, can play important roles in determining the outcome of injurious processes.

References

Abele, A. E., Scholz, K. P., Scholz, W. K., and Miller, R. J. (1990). *Neuron* **2,** 413–419.

Abrahammson, S., Dahlén, B., Lofgren, H., Pascher, I., and Sundell, S. (1976). *Nobel Symp.* **34,** 1–21.

Balch, W. E., Elliot, M. M., and Keller, D. S. (1986). *J. Biol. Chem.* **261,** 14681–14689.

Beckers, C. J., Keller, D. S., and Balch, W. E. (1987). *Cell (Cambridge, Mass.)* **50,** 523–534.

Brenkert, A., and Radin, N. (1972). *Brain Res.* **36,** 183–193.

Carruthers, A., and Carey, E. M. (1983). *J. Neurochem.* **41,** 22–29.

Choi, D. W. (1988). *Neuron* **1,** 623–634.

Cohn, R., and Segal, S. (1969). *Biochim. Biophys. Acta* **171,** 333–341.

Coste, H., Martel, M.-B., Azzr, G., and Got, R. (1986). *Biochim. Biophys. Acta* **814,** 1–7.

Dawson, G. (1979). *J. Biol. Chem.* **254,** 155–162.

Dyer, C. A. (1993). *Mol. Neurobiol.* (in press).

Dyer, C. A., and Benjamins, J. (1990). *J. Cell Biol.* **111,** 625–633.

Dyer, C. A., and Benjamins, J. (1991). *J. Neurosci. Res.* **30,** 699–711.

Farber, J. (1979). *Science* **206,** 700–708.

Favaron, M., Manev, H., Siman, R., Bertolino, M., Szekely, A. M., de Erausquin, G. A., Guidotti, A., and Costa, E. (1990). *Proc. Natl. Acad. Sci. U.S.A.* **87,** 1983–1987.

Gutteridge, J. M. C., and Halliwell, B. (1990). *Trends Biochem. Sci.* **15,** 129–135.

Hakomori, S. (1990). *J. Biol. Chem.* **265,** 18713–18716.

Jennings, R. (1981). *Am. J. Pathol.* **102,** 239–291.

Jungalwala, F. B., Koul, O., Stoolmiller, A., and Sapirstein, V. S. (1985). *J. Neurochem.* **45,** 191–198.

Kendler, A., and Dawson, G. (1990). *J. Biol. Chem.* **265,** 12259–12266.

Kendler, A., and Dawson, G. (1992). *J. Neurosci. Res.* **31,** 205–211.

Kendler, A., and Dawson, G. (1993). In press.

Kishimoto, Y. (1983). *In* "The Enzymes" (P. D. Boyer, ed.), pp. 358–385. Academic Press, New York.

Kishimoto, Y., Akanuma, H., and Singh, I. (1979). *Mol. Cell. Biochem.* **28,** 93–105.

Lemke, G. (1988). *Neuron* **1,** 535–543.

Lieber, C. S. (1988). *N. Engl. J. Med.* **319,** 1639–1650.

Mahadik, S. B., and Karpiak, S. E. (1988). *Drug Dev. Res.* **15,** 337–360.

Manev, H., Favaron, M., Guidotti, A., and Costa, E. (1989). *Mol. Pharmacol.* **36,** 106–112.

Manev, H., Favaron, M., Vicini, S., Guidotti, A., and Costa, E. (1990). *J. Pharmacol. Exp. Ther.* **252,** 419–427.

McDonald, J. W., Silverstein, F. S., and Johnston, M. V. (1987). *Eur. J. Pharmacol.* **140,** 259–361.

Moore, B. W. (1965). *Biochem. Biophys. Res. Commun.* **19,** 739–744.

Morell, P. (1984). "Myelin." Plenum, New York.

Morell, P., and Radin, N. (1969). *Biochemistry* **8,** 506–512.

Neskovič, N. M., Rebel, G., Harth, S., and Mandel, P. (1981). *J. Neurochem.* **37,** 1363–1370.

Pagano, R. E. (1989). *Methods Cell Biol.* **29A,** 75–85.

Pfieffer, S. (1984). *Adv. Neurochem.* **5,** 233–298.

Radin, N. S. (1972). *In* "Sphingolipids, Sphingolipidoses, and Allied Disorders" (B. W. Volk and S. M. Aronson, eds.), pp. 475–486. Plenum, New York.

Radin, N. S., Brenkert, A., Arora, R. C., Sellinger, O. Z., and Flangas, A. L. (1972). *Brain Res.* **39,** 163–169.

Rothman, S. M. (1985). *J. Neurosci.* **5,** 1483-1489.

Rothman, S. M., and Olney, J. W. (1986). *Ann. Neurol.* **19,** 105–111.

Sato, C., Black, J. A., and Yu, R. K. (1988). *J. Neurochem.* **50,** 1887–1893.

Sheardown, M. J., Nielsen, E. O., Hansen, A. J., Jacobsen, P., and Honoré, T. (1990). *Science* **247,** 571–573.
Siegrist, H. P., Bukart, T., Wiesman, U. N., Herschkowitz, N. N., and Spycher, M. A. (1979). *J. Neurochem.* **33,** 497–504.
Stevens, M. K., and Yaksh, T. L. (1990). *J. Cereb. Blood Flow Metab.* **10,** 77–88.
Sundaraj, N., Schachner, N., and Pfeiffer, S. E. (1975). *Proc. Natl. Acad. Sci. U.S.A.* **72,** 1927–1931.
van Meer, G., Steltzer, E. H. K., Wijnaendts-van Resandt, R. W., and Simons, K. (1987). *J. Cell Biol.* **105,** 1623–1635.
Warburg, O. (1930). "The Metabolism of Tumors." Constable Press, London.
Wood, P., and Bunge, R. P. (1984). *In* "Oligodendroglia" (W. Norton, ed.), pp. 1–45.
Zimmerman, H. M. (1955). *Am. J. Pathol.* **31,** 1–29.

Bioactive Gangliosides: Differentiation Inducers for Hematopoietic Cells and Their Mechanism(s) of Actions

MASAKI SAITO

Division of Hemopoiesis
Institute of Hematology
Jichi Medical School
Tochigi-ken 329-04, Japan

I. Introduction

Glycosphingolipids (GSLs) comprise a family of complex lipids that are amphipathic molecules composed of both hydrophobic (ceramide) and hydrophilic (saccharides) moieties (Fig. 1). They are synthesized by a group of Golgi enzymes, glycosyltransferases, and are located almost exclusively in the outer leaflet of plasma membranes. In addition to conventional investigations on their chemical structures, dynamic studies are now taking place since they have been found to be involved in cell sociology, i.e., intercellular interactions and cell growth regulations, and characteristically change their composition and biosynthesis during cell development, differentiation, and oncogenic transformation even though they constitute only a small portion of the cell surface glycoconjugates. In addition, acidic GSLs (gangliosides) have attracted much interest

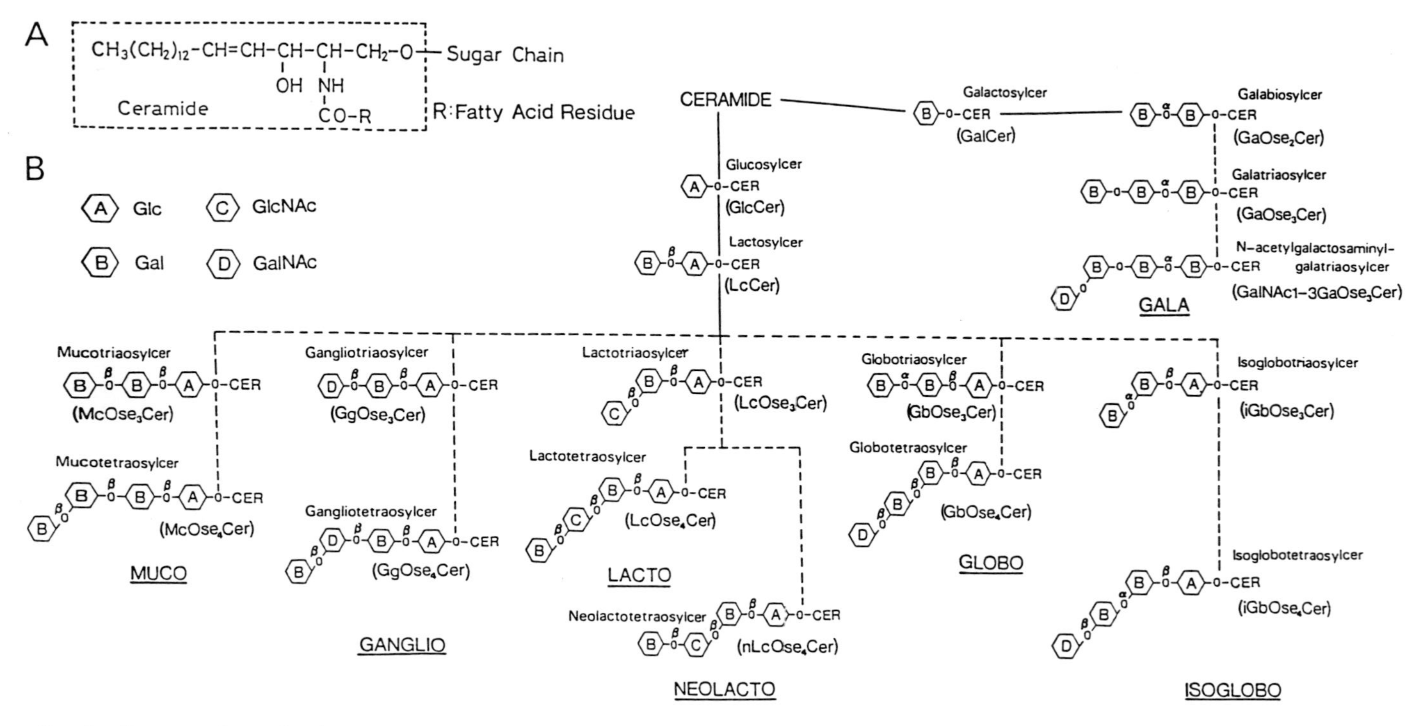

FIG. 1. Representative molecular structure (A) and molecular species of glycosphingolipids (GSLs) and their oligosaccharides (B). The terms capitalized and underlined in (B) represent the molecular series of glycosphingolipids, e.g., LACTO, GANGLIO, GLOBO.

since they have been shown to exhibit special receptor functions for exogenous bioactive factors such as bacterial toxins, hormones, and interferons.

GSLs are classified into three major molecular series, i.e., ganglio, globo, and lacto series, as well as other minor molecular types, according to their carbohydrate structures (Fig. 1). Various cells and tissues show characteristic compositions and specific structural features of their GSLs. For example, ganglioside profiles are characteristic for cell lineages and stages of differentiation of both normal and malignant hematopoietic cells and, therefore, serve as useful differentiation markers. These characteristics are also supported in terms of hydrolytic enzyme (glycosidase) activities, participating in the catabolism of gangliosides as well as synthesizing enzyme (glycosyltransferase) activities.

We have demonstrated that human myelogenous leukemia cells show distinct GSL, especially ganglioside, profiles, depending not only on differentiation stages but also on differentiation directions (Nojiri *et al.,* 1984). Further, we found that the acidic GSL molecules themselves are highly potent inducers for differentiation of human myeloid or monocytoid leukemia cell lines, such as HL-60, U937, and K562, and of fresh leukemia cells (Saito *et al.,* 1985; Nojiri *et al.,* 1986, 1988). In this article, the molecular structures and functions of bioactive gangliosides and the related synthetic compounds are discussed with special reference to the differentiation phenomena of human myelogenous leukemia cells, with some reference to some nonhematopoietic cells (Rodrig *et al.,* 1987; Tsuji *et al.,* 1983).

II. Sialosyl GSL (Ganglioside) Profiles as Differentiation Markers: Gangliosides Specifically Change during Differentiation of Myelogenous Leukemia Cells

The HL-60 cell line is of human acute myelogenous leukemia cell origin (Collins *et al.,* 1977) and undergoes morphological and functional differentiation in response to a wide range of chemicals, the differentiation proceeding along either the granulocytic or monocytic lineage, depending on the chemical inducers used (Nojiri *et al.,* 1982, 1984; Collins *et al.,* 1978; Kitagawa *et al.,* 1984; Rovera *et al.,* 1979). U937 cells are the monoblastoid cell line derived from human histiocytic lymphoma and exhibit a potential to differentiate morphologically and functionally into mature monocytic/macrophage-like cells (Sundström and Nilsson, 1976). K562 cells were established from the pleural effusion of a chronic myelogenous leukemia (CML) patient in blastic crisis, and found to be multipotent with respect to cell differentiation (Lozzio *et al.,* 1981). CML is a neoplastic disease that results from the development of an abnormal hematopoietic stem cell which gives rise to progenies that contain the Philadelphia chromosome. In CML, a large number of immature and mature granulocytic cells accumulate in the peripheral blood, and granulocytes are present at all the stages of development (Kitagawa *et al.,* 1984; Nojiri *et al.,* 1985). Most fresh leukemia cells (Ohta *et al.,* 1988) from acute myelogenous leukemia patients have been shown to differentiate into mature monocytic or granulocytic cells, responding to some of the chemical inducers.

Differentiation into monocytic or granulocytic mature cells was generally induced with 12-*O*-tetradecanoyl phorbol-13-acetate (TPA) or dimethylsulfoxide (DMSO) and retinoic acid (RA), respectively. The morphological assessment of differentiation was performed under a light and an electron microscope with the help of assays for naphthol-AS-D chloroacetate and α-naphthyl butyrate esterase activities (esterase double-staining method). Functional differentiation was assessed on the basis of phagocytic activity and superoxide-producing ability of the cells. Surface membrane antigens were assessed by cytofluorometry using the authorized monoclonal antibodies, including OKM1(CD11), OKB2(CD24), OKM5(CD36), and Mo2(CDW14).

When HL-60 cells were subjected to differentiation, they exhibited distinct GSL profiles, depending not only on the stage of differentiation but also on the direction of differentiation (Nojiri *et al.,* 1984) (Fig. 2). During the granulocytic differentiation induced by DMSO or RA, neolacto series gangliosides (NeuAc-nLc), including sialosylparagloboside (SPG), sialosylnorhexaosyl ceramide (SnHc), and other higher gangliosides having polylactosamine structures, characteristically increased with a concomitant decrease in the ganglio series ganglioside with a simple sugar chain, ganglioside G_{M3} (Fig. 2). This specific change in the ganglioside profile was also demonstrated both *in vitro* in granulocytic differentiation of the cells of another myeloid leukemia cell line, ML-1, and *in vivo* in differentiation of CML cells in chronic phase (Nojiri *et al.,* 1985). The significant decrease in G_{M3} with concomitant increase in neolacto gangliosides having longer sugar moieties was seen with immature and mature granulocytic cells at all the stages of differentiation accumulated in the peripheral blood of CML patients (Nojiri *et al.,* 1985), which gave a typical *in vivo* model for human granulocytic differentiation.

In marked contrast to the result with granulocytic differentiation, a remarkable increase in G_{M3} was demonstrated during monocytic/macrophage-like differentiation of HL-60 cells induced by TPA, and there was a concurrent decrease in the neolacto gangliosides having longer sugar moieties, such as SPG, SnHc, and NeuAcnLcOse$_n$Cer (n = 8 or more) (Nojiri *et al.,* 1984) (Fig. 2). Simultaneously, ceramide dihexoside (CDH), the precursor of G_{M3}, decreased markedly during this differentiation (Nojiri *et al.,* 1984). Therefore, a significant increase in the biosynthesis of G_{M3} was suggested to be part of the process of monocytic differentiation of HL-60 cells. The result was supported by the recent finding that a sialosyltransferase [CDH:CMP-NeuAc sialosyltransferase (EC 2.4.99.1)] synthesizing G_{M3} was significantly enhanced during the differentiation induced

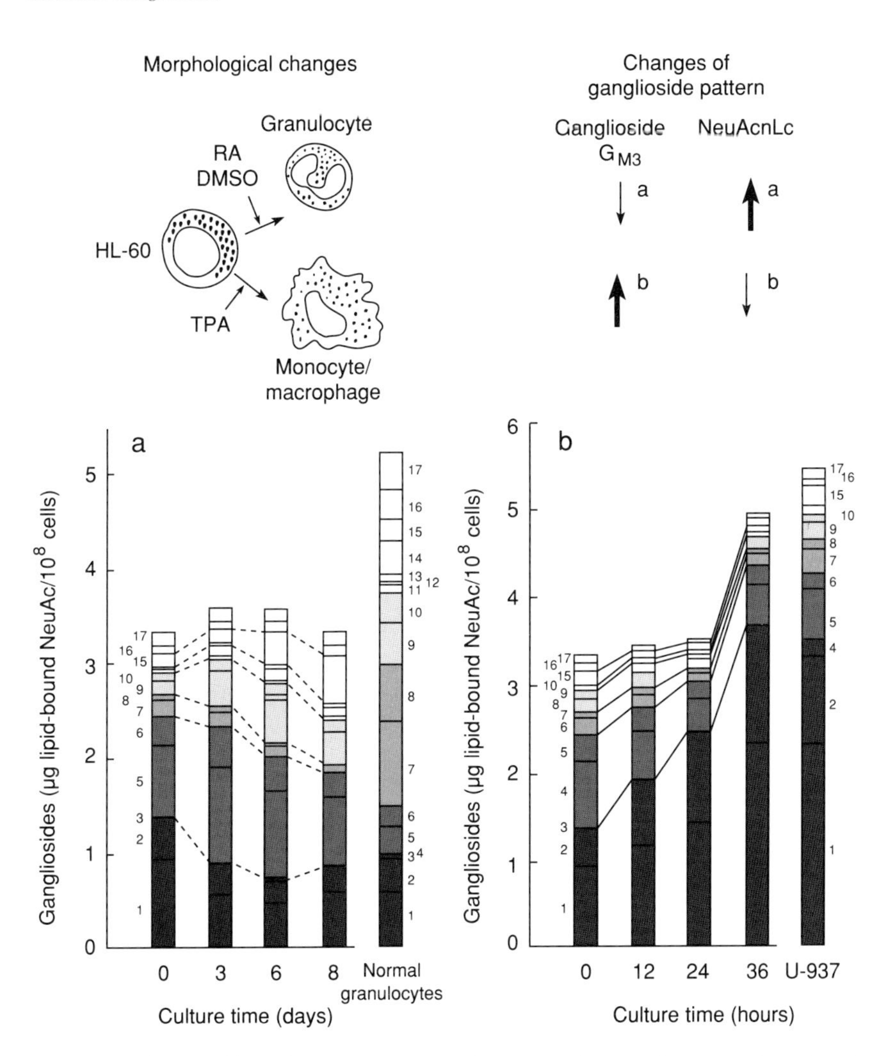

FIG. 2. Bipotent differentiation of human myelogenous leukemia cell line HL-60 cells and changes of ganglioside pattern dependent on differentiation directions. (a) Ganglioside patterns specific for granulocytic differentiation with DMSO; (b) ganglioside patterns specific for monocytic/macrophage-like cell differentiation with TPA. The numbering of columns indicates the molecular species of gangliosides as follows: 1 and 2, ganglioside G_{M3}; 3 to 17, neolacto series gangliosides (NeuAcnLc). RA, Retinoic acid; DMSO, dimethylsulfoxide; TPA, 12-*O*-tetradecanoyl phorbol 13-acetate. These data were published in Nojiri *et al.* (1984).

by TPA (Nakamura *et al.*, 1992) (Fig. 3). Similar results with respect to the specific change in the ganglioside profile were obtained during monocytic/macrophage-like differentiation of other myeloid and monocytoid leukemia cell lines such as ML-1, K562, KG-1, and THP-1.

III. Bioactive Gangliosides toward Myelogenous Leukemia Cells

A. Monosialosyl G_{M3} Is Highly Potent in Inducing Differentiation along a Monocyte/Macrophage Lineage

Our major interest initially was focused on a remarkable increase in G_{M3} found in the process of differentiation of myelogenous leukemia cells into mature monocytic/macrophage-like cells. In search of a biological significance of such a profound change of level of a sialo-GSL molecule in cell differentiation processes, we performed the following experiments. The exogenous addition of the purified G_{M3}, which was prepared from human or canine erythrocyte ghosts and was found to carry *N*-acetylneuraminic acid as a sialic acid residue (NeuAc-type G_{M3}: *N*-acetyl-α-D-neuraminosylgalactosylglucosyl ceramide), to HL-60 and U937 cell cultures resulted in strong monocytic differentiation of both HL-

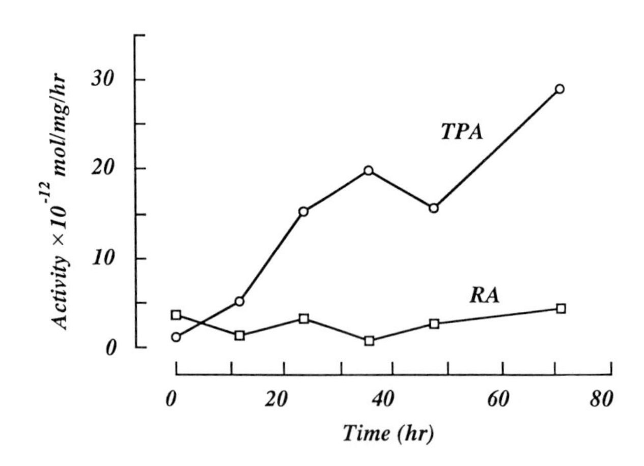

FIG. 3. Changes in G_{M3} synthase activity during differentiation induced by TPA and RA. HL-60 cells were cultured in the presence of 8×10^{-9} *M* TPA (○) or 10^{-6} *M* RA (□) for up to 3 days. Enzyme assays were performed under optimal conditions (Nakamura *et al.*, 1992). Each point is the mean of two determinations.

60 and U937 cells (Saito *et al.*, 1985; Nojiri *et al.*, 1986) (Fig. 4). To facilitate clearer and unambiguous analyses of the effects of gangliosides, it is necessary to cultivate the cells under serum-free culture conditions because it was shown that serum proteins (albumin, etc.) (Schwarzman *et al.*, 1983) bind and incorporate sialoglycosphingolipids into mammalian cells. The monocytic differentiation of HL-60 cells was well demonstrated in terms of striking changes in morphological characteristics, such as loss of cytoplasmic basophilia, lobulation of nuclei, and the presence of somewhat vacuolated cytoplasms with ruffled surface membranes, whereas the differentiation of U937 cells was characterized by an increase in cell size, a decrease in the nuclear–cytoplasmic ratio, paler cytoplasms, more prominent granules, and a greater degree of vacuolization in the cytoplasm (Fig. 4). Simultaneously with monocytic/macrophage-like differentiation, the growth of both cell lines was dose-dependently inhibited with the NeuAc-type ganglioside G_{M3}, and its inhibition was almost complete at 50 μM of this ganglioside molecule (Nojiri *et al.*, 1986). Such biological activities were demonstrated in a very specific manner relative to the chemical structure of G_{M3}, especially its sugar moiety, since other ganglio gangliosides, such as G_{M1}, G_{D1a}, G_{D1b}, G_{T1b}, and the brain ganglioside mixture, instead showed a rather stimulatory action on cell growth. Furthermore, free NeuAc had no effect on it at the same concentration. Synthetic ganglioside G_{M3}, which was successfully synthesized by Sugimoto and Ogawa (1985), was found to be almost equivalent to natural G_{M3} with respect to biological activities. Interestingly, a marked preference for α-linkage of the sialic acid residue was exhibited with respect to the growth-inhibitory and differentiation-inducing activities of G_{M3}; many fewer effects were observed with the β-anomer. NeuGc-type G_{M3} (*N*-glycolyl-α-D-neuraminosylgalactosylglucosyl ceramide) and Tay-Sachs ganglioside (G_{M2}) were also found to be much less active, compared with natural NeuAc-type G_{M3}.

The monocytic/macrophage-like differentiation with exogenous G_{M3} was histochemically demonstrated in terms of a remarkable increase in the number of α-naphthylbutyrate esterase (NBE)-positive cells (a maximum of 84.8% at 50 μM) and in the intensity of NBE staining (Fig. 4). This esterase activity was completely inhibited by sodium fluoride (NaF), indicating the specificity for the monocytic lineage. In addition, differentiation with exogenously added G_{M3} was also evaluated functionally, with a significant increase in the capacity to phagocytose latex or yeast particles, and immunochemically by cytofluorography in terms of expression of the cell surface antigens specific for mature monocytes/macrophages. The treatment of HL-60 cells with 50 μM G_{M3} for 6 days resulted in a marked increase in mature monocyte-specific antigens that were detectable with monoclonal antibodies (MAbs) such as OKM1 (CD11) and OKM5 (CD36), while a concomitant decrease in the granulocyte-specific epitope was recognized by the MAb OKB2 (CD24) (Nojiri *et al.*, 1986).

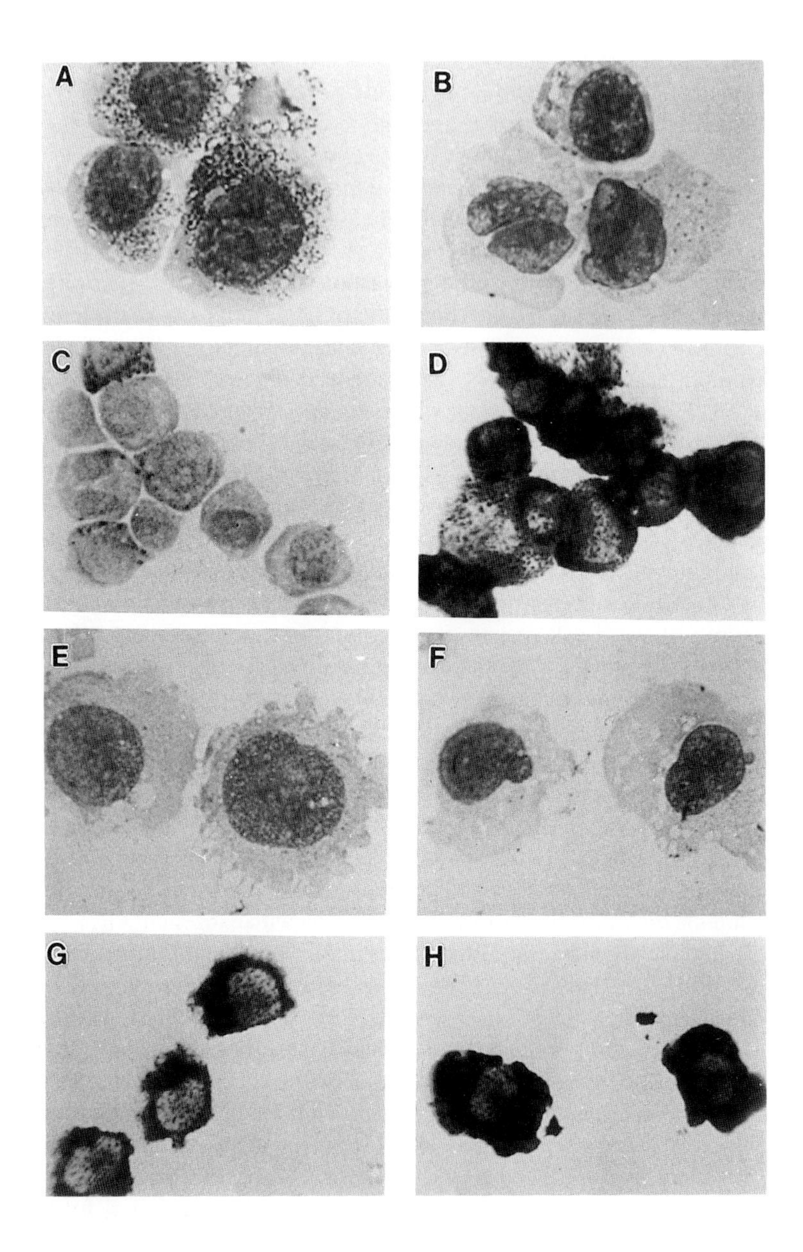
A
B
C
D
E
F
G
H

B. Neolacto Gangliosides Induce Granulocytic Differentiation: Sugar Moieties Might Be Crucial for Determination of the Differentiation Directions

Neolacto series gangliosides (NeuAc-nLc) with a linear poly-*N*-acetyllactosaminyl oligosaccharide backbone structure, especially those with sialic acid joined to this backbone by an α2–3 linkage, have been demonstrated to characteristically increase during the granulocytic differentiation of HL-60 cells induced by DMSO or RA (Nojiri *et al.*, 1984, 1985) (Fig. 2). When the cells were cultured in the presence of neolacto gangliosides prepared from mature granulocytes, they were found to differentiate into mature granulocytes on the basis of changes in morphology, surface membrane antigens, nonspecific esterase activity, and the activities of phagocytosis and respiratory burst (Nojiri *et al.*, 1988) (Fig. 5). The differentiation of cells was dependent on the concentration of the gangliosides added, and was accompanied by an inhibition of cell growth. With the results of biological activities of G_{M3} taken together, these findings suggest that the particular ganglioside molecules play an important role in the regulation of cell differentiation and that the appearance or accumulation of particular ganglioside molecules on the cell surface membrane not only triggers differentiation but also determines the direction of differentiation in HL-60 cells (Fig. 6).

On the basis of the following findings, it is now suggested that gangliosides having more than two repeats of the *N*-acetyllactosaminl unit with a sialic acid residue linked by an α2–3 linkage at the nonreducing end, such as VI^3NeuAc-$nLcOse_6$Cer, VIII^3NeuAcn$LcOse_8$Cer, and X^3NeuAcn$LcOse_{10}$Cer, may be essential for induction of granulocytic differentiation: (1) sialylparagloboside (IV^3NeuAcn$LcOse_4$Cer; 1.5 μM) showed apparently monocytic or myelomonocytic/hybrid (but not granulocytic) differentiation-inducing activity: (2) a series of neolacto-type gangliosides with α2–6 sialic acid linkage did not increase remarkably in concentration during the granulocytic differentiation of HL-60 cells induced by DMSO or RA. If fact, IV^6NeuAcn$LcOse_4$Cer, which is one of the major ganglioside constituents of normal granulocytes, decreased slightly

---

FIG. 4. Differentiation of HL-60 and U937 cells along the monocyte–macrophage lineage induced with ganglioside G_{M3}. Shown are the morphological (A,B,E, and F) and cytochemical (C,D,G, and H) changes of HL-60 (A–D) and U937 (E–H) cells observed during differentiation with G_{M3}. The cells were cultured for 6 days in the presence (B,D,F, and H) or absence (A,C,E, and G) of G_{M3} and stained with Wright–Giemsa staining solution (A,B,E, and F). α-Naphthylbutyrate esterase activity was also detected according to the esterase double staining method after the culture (C,D,G, and H). These data were published in Nojiri *et al.* (1986).

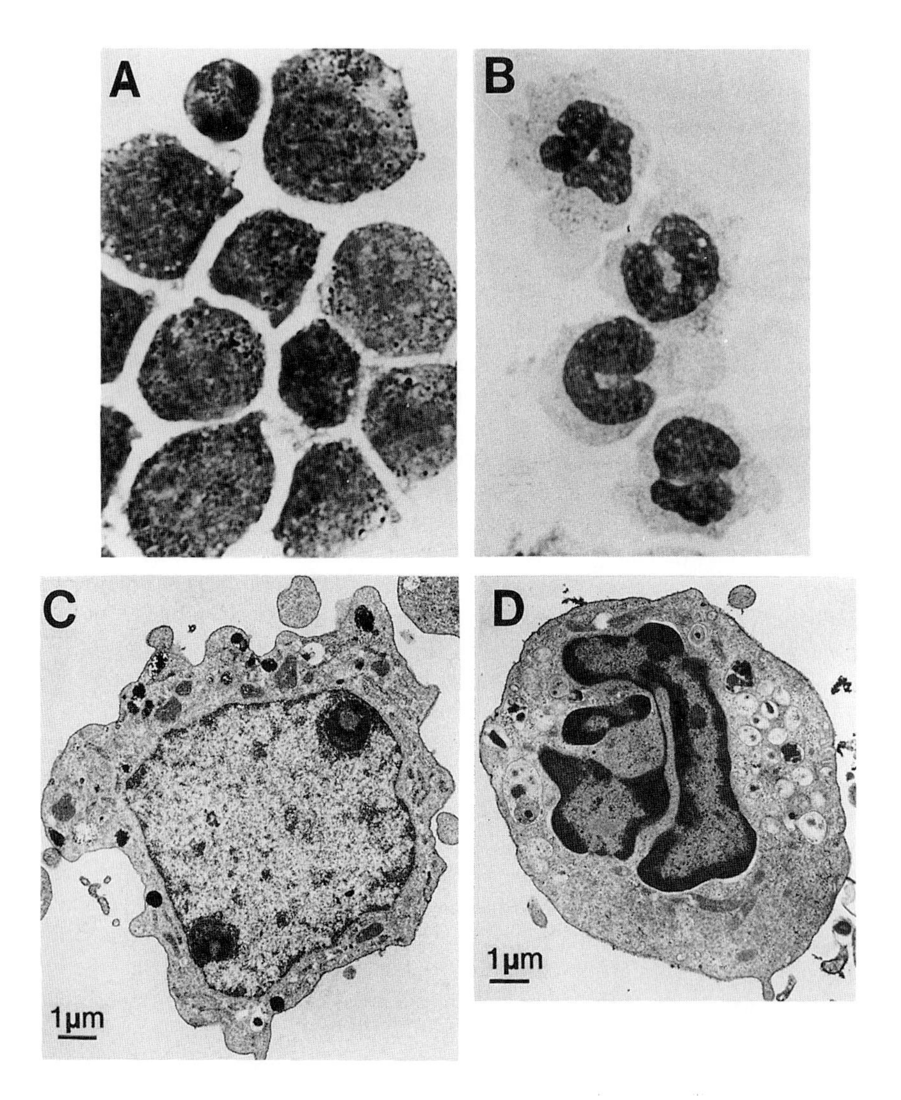

FIG. 5. Differentiation of HL-60 cells along a granulocytic lineage induced with neolacto series gangliosides (NeuAcnLc). Shown are light (A,B) and electron (C,D) microscopic pictures: (A) Control cells (×500), (B) HL-60 cells differentiated by NeuAcnLc (×500), (C) control cells (×9000), (D) HL-60 cells differentiated by NeuAcnLc (×9000). In C, the nuclear chromatin is primarily euchromatic, and prominent nucleoli are situated adjacent to the nuclear membrane. Peroxidase-positive primary granules can be detected in the cytoplasm. The cell in D displays evidence of nuclear and cytoplasmic differentiation characteristic of mature granulocytes. Considerable nuclear segmentation and increased definition of heterochromatic areas are observed. Although secondary granules are lacking, a great many glycogen particles are present in the cytoplasm. The data were published in Nojiri *et al.* (1988).

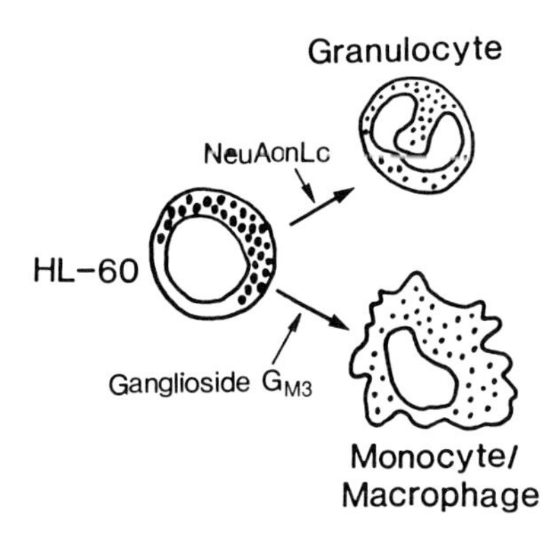

FIG. 6. Bipotent induction of differentiation of HL-60 cells with bioactive gangliosides such as ganglio series G_{M3} and neolacto series gangliosides (NeuAcnLc).

(Nojiri *et al.,* 1984, 1985). All these findings suggest that the differentiation-inducing activity of gangliosides is not due to a simple detergent-like effect or to nonspecific stimulation but to a specific action which is highly regulated by specific carbohydrate moieties (Figs. 6 and 7).

C. Retinoic Acid-Resistant Mutant HL-60 Cells: Hyposialylation of GSLs and Granulocytic Differentiation Induction with Neolacto Gangliosides

In order to clarify roles of ganglioside in the induction of differentiation further, neutral and acidic GSLs in a mutant HL-60 subline, which was selected by cultivating the original wild-type HL-60 cells with RA sequentially increased up to a concentration of 1 μ*M* (Gallagher *et al.,* 1985) and was absolutely resistant to differentiation induction by RA, were analyzed in comparison with those in the parental cells. In contrast to wild-type HL-60 cells, which show a significant increase in neolacto-type gangliosides when induced to differentiate with DMSO or RA (Nojiri *et al.,* 1984, 1985) and, in turn, differentiate into mature granulocytes when the cells are cultured in the presence of neolacto gangliosides (Nojiri *et al.,* 1988), the mutant cells showed a marked decrease in the ganglioside content (approximately one-sixth of that of the parental HL-60 cells), especially in that of neolacto gangliosides, and did not show any increase in the content of gangliosides when cultured with RA (Fig. 8). However, when the mutant HL-60 cells were cultivated in the presence of neolacto gangliosides, they were evidently found to differentiate into mature granulocytes on the basis of

(A) Ganglio Series Ganglioside G_{M3} and Its Derivatives

G_{M3} (Differentiation induction, Growth inhibition)

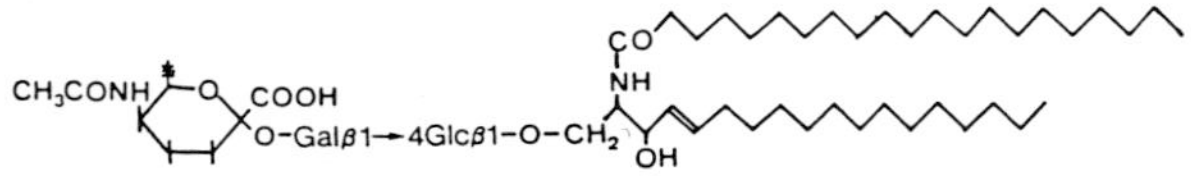

CH_3CONH COOH O-Galβ1→4Glcβ1-O-CH_2 CO NH OH

Lyso-GM3 (Growth inhibition)

CH_3CONH COOH O-Galβ1→4Glcβ1-O-CH_2 NH_2 OH

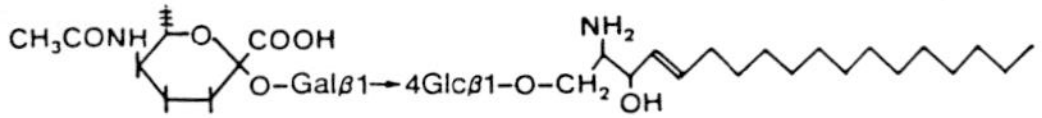

De-N-acetyl GM3 (Growth promotion)

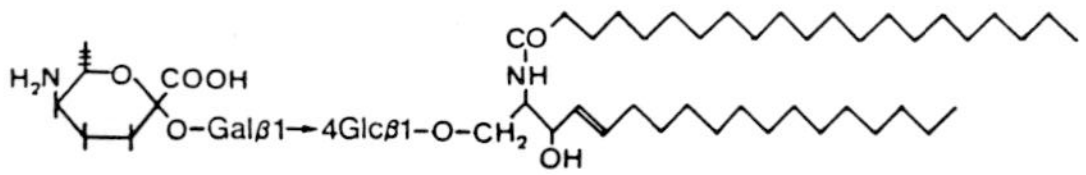

H_2N COOH O-Galβ1→4Glcβ1-O-CH_2 CO NH OH

De-N-acetyl Lyso-GM3 (Growth inhibition)

H_2N COOH O-Galβ1→4Glcβ1-O-CH_2 NH_2 OH

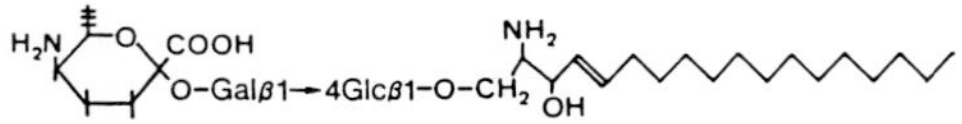

(B) Other Ganglio Series Gangliosides

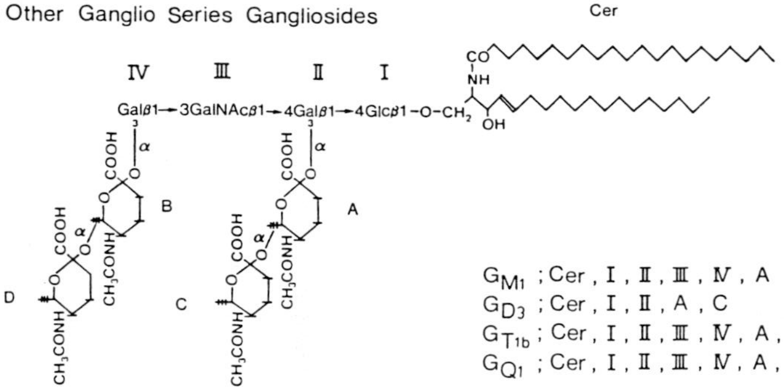

Cer

IV III II I

Galβ1→3GalNAcβ1→4Galβ1→4Glcβ1-O-CH_2 CO NH OH

A, B, C, D: COOH, CH_3CONH, α

G_{M1} ; Cer, I, II, III, IV, A (Growth promotion / inhibition)

G_{D3} ; Cer, I, II, A, C (Differentiation induction)

G_{T1b} ; Cer, I, II, III, IV, A, B, C (Differentiation induction, Growth promotion)

G_{Q1} ; Cer, I, II, III, IV, A, B, C, D (Differentiation induction, Growth promotion)

(C) Neolacto Series Gangliosides

(Differentiation induction, Growth inhibition)

3/6(Galβ1→4GlcNAcβ1→3)$_n$ Galβ1→4Glcβ1-O-CH_2 CO NH OH

COOH α CH_3CONH

(n=1-4)

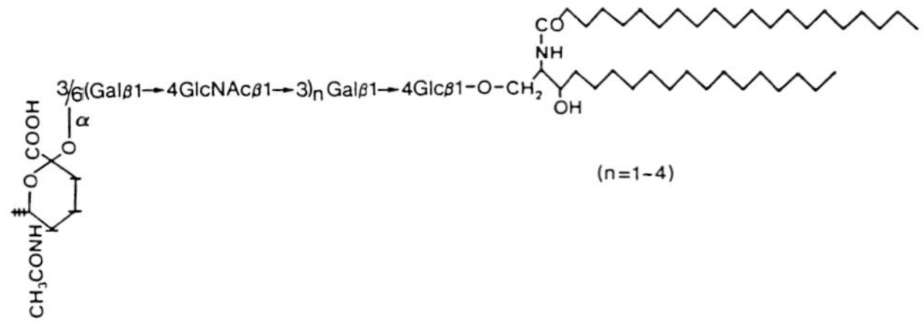

(D) Synthetic Amphipathic Compounds (Neosialoglycolipids)

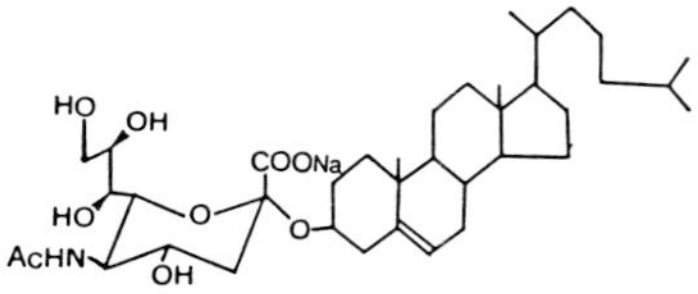

HO OH COONa HO AcHN OH

α-Sialocholesterol (Differentiation induction, Growth inhibition)

HO OH COONa C-O-CO-R_1 HC-O-CO-R_2 CH_2 HO AcHN OH

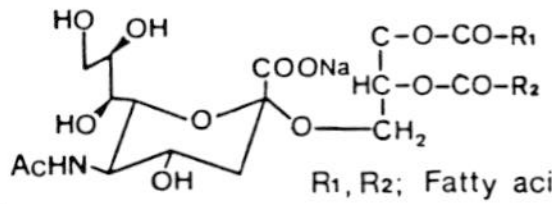

R_1, R_2; Fatty acid residues

α-Sialoglyceride (Differentiation induction)

changes of morphology, surface membrane antigens, and the activity of respiratory burst (Kitagawa *et al.,* 1989). Differentiation of the mutant cells was dependent on the concentration of gangliosides and was accompanied by an inhibition of cell growth. These findings suggest that the synthesis of endogenous neolacto gangliosides is essential for RA-induced granulocytic differentiation, that this step could be bypassed or replaced by exogenously added neolacto gangliosides, and that the defective synthesis of neolacto gangliosides is responsible for the failure of differentiation induction in RA-resistant HL-60 cells by RA (Kitagawa *et al.,* 1989).

The decreased content of gangliosides in RA-resistant HL-60 cells may result from a general decrease of GSL synthesis but not from a specific defect in the terminal sialylation because neutral GSLs, the precursors of gangliosides, did not accumulate in the mutant cells and the total content was slightly less than that of the wild-type parental cells (Kitagawa *et al.,* 1989).

IV. Sialosyl Neoglycolipids: Synthetic Amphipathic Sialo Compounds Are Potent Differentiation Inducers

Recent advances in the synthetic chemistry of oligosaccharides have enabled us to synthesize GSLs and their related amphipathic compounds. Sugimoto and Ogawa (1985) were the first to succeed in synthesizing G_{M3} at much higher efficiency compared to previous methods. We found that the activity of this synthetic NeuAc-type G_{M3} in HL-60 cells with respect to both differentiation induction and growth inhibition was almost comparable to natural G_{M3}, and that the α-anomeric compound, which has a sialic acid residue linked to the terminal galactose moiety with α-linkage at the nonreducing end of the oligosaccharide chain, was much more potent in inducing differentiation in HL-60 cells than the β-anomer. This indicates that the differentiation-inducing activity as well as the growth-inhibitory action of G_{M3} on the leukemic cells was stereospecific. Such is also the case with certain types of amphipathic, synthetic compounds containing sialic acid residues (Fig. 7), which have been deliberately synthesized (Sato *et al.,* 1987) on the basis of observations that naturally occurring sialyl amphipathic compounds (gangliosides) play important roles in the regulation of growth and differentiation of some mammalian cells of neuronal, renal, and hematopoietic origins (Nojiri *et al.*, 1986, 1988; Rodrig *et al.,* 1987; Tsuji *et al.,*

FIG. 7. Molecular structures of bioactive sialoglycosphingolipids (gangliosides) and synthetic amphipathic sialo compounds. Major biological functions are represented in the parentheses. (A) Ganglio series G_{M3} and its derivatives; (B) other ganglio series gangliosides; (C) neolacto series gangliosides. (D) synthetic amphipathic sialo compounds (neosialoglycolipids).

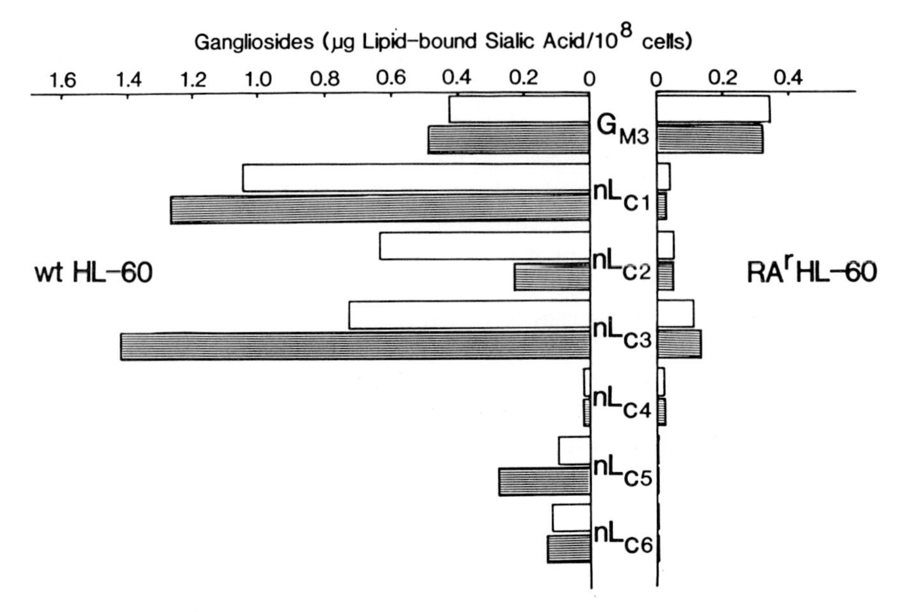

FIG. 8. Ganglioside composition of wild-type parental (wt HL-60) and retinoic acid-resistant (RA^rHL-60) HL-60 cells. Values represent the means of two determinations in three separate experiments and are expressed as micrograms of lipid-bound sialic acid per 10^8 cells. Standard deviations were always less than 10%. Cells were cultivated with (▤) or without (□) retinoic acid (1 μM) for 4 days. nL_{c1}, $IV^3NeuAcnLcOse_4Cer$; nL_{c2}, $IV^6NeuAcnLcOse_4Cer$; nL_{c3}, IV^3NeuAc-$nLcOse_6Cer$; nL_{c4}, $IV^6NeuAcnLcOse_6Cer$; nL_{c5}, $IV^3NeuAcnLcOse_8Cer$; nL_{c6}, IV^6NeuAc-$nLcOse_8Cer$ [The nomenclature for glycosphingolipids follows the recommendations of the Nomenclature Committee of the International Union of Pure and Applied Chemistry (1977).]

1983). Accordingly, we have found that the α-anomer of sialocholesterols, which are absolutely synthetic and are amphipathic compounds composed of a hydrophobic moiety (cholesterol) and a hydrophilic sugar residue (sialic acid) with either an α or a β isomeric linkage (Sato *et al.*, 1987), was both a potent differentiation inducer and a growth inhibitor for human myelogenous leukemia cells, and that the β-anomer was much less active (Saito *et al.*, 1990). Neither free sialic acid nor cholesterol by themselves induced the differentiation. We also found that another synthetic sialocompound, α-sialodiglyceride, was potent in the induction of HL-60 cell differentiation on the basis of changes in surface membrane antigens, nonspecific esterase activity, and the activity of respiratory burst as well (Saito *et al.*, 1990). The differentiation of cells, accompanied by an inhibition of cell growth, was dependent on the concentration of the compound. The β-anomer was also shown to be much less active.

These findings, then, indicate that the stereospecific structure or physical properties may be important for the differentiation-inducing activity of sialo compounds on human myelogenous leukemia cells (Saito *et al.*, 1990). However, such a preference for α-anomeric linkage was never observed in the induction of neuritogenesis in a neuroblastoma cell line with sialocholesterols (Tsuji *et al.*, 1988).

V. Bioactive Gangliosides as Differentiation Inducers for Cultured Cells of Neuroblastoma and Renal Origins

Gangliosides are currently of great interest since they are present at high levels in brain tissues and they form complex structures in synaptic membranes of nerve cells (Wiegandt, 1982). Historically, the function of gangliosides in regulation of nerve cell growth has been investigated by adding them to the serum-free culture medium for nerve cells and measuring the resulting changes in cell number and neurite outgrowth. Tsuji *et al.* (1983) found that tetrasialoganglioside G_{Q1b}, when exogenously added, could promote cell proliferation and neurite outgrowth in two human neuroblastoma cell lines, GOTO and NB-1. This biological activity turned out to be strictly attributable to the molecular structure of the oligosaccharide moiety of G_{Q1b}, the ceramide moiety being necessary for full expression of the activity (Nakajima *et al.*, 1986). It is suggested that a receptorlike mechanism would be involved in the action G_{Q1b} at the cell membrane since the G_{Q1b} oligosaccharide was demonstrated to inhibit the activity of G_{Q1b} at a concentration a few times greater than that of G_{Q1b}. Further, G_{D1a} was shown to be necessary for prolonged G_{Q1b}-driven nerve growth promotion in neuroblastoma cell lines, which indicates the synergistic effect of G_{Q1b} and G_{D1a} on proliferation of the cells and neurite outgrowth (Arita *et al.*, 1984).

The epithelial cell line MDCK cells derived from proximal or distal renal tubules of the canine kidney, which provide an excellent model for the study of epithelial transport phenomena (Misfeldt *et al.*, 1976), were found to be differentiated to increase the dome formation with G_{M3} and G_{D3} (Rodrig *et al.*, 1987). It is suggested that the tissue-specific gangliosides function as inducers or mediators of the dome formation, the mechanism of which might involve adenylate cyclase or another transmembrane biosignal-transducing system (Rodrig *et al.*, 1987).

It is of great importance that a novel ectotype protein kinase, which was stimulated by gangliosides such as G_{Q1b}, G_{T1a}, and G_{D1a}, was discovered in the plasma membrane of human neuroblastoma cells (Tsuji *et al.*, 1985), and that a similar ganglioside-stimulated protein kinase was also partially purified from guinea pig brains (Chan, 1987). Characterization of such ganglioside-modulated protein phosphorylations might be crucial for elucidating and understanding the mechanism of function of gangliosides at the molecular level.

VI. Mechanism(s) of Differentiation Induction by Bioactive Sialoglycolipids

A. Metabolic Flow of Bioactive Ganglioside Biosynthesis

We have demonstrated lineage-specific GSL expression during bidirectional differentiation of human myelogenous leukemia cell line HL-60 cells (Nojiri *et al.,* 1984). This shift of GSL expression is also of great interest because HL-60 can be differentiated by the enriched ganglioside fractions. There have been reports on GSL core structure switching from one series to another not only in human myelogenous cells but also in various other cells. Changes from globo to lacto series that were associated with the viral transformation of hamster NIL cells were investigated (Gahmberg and Hakomori, 1975). The shift from lacto to ganglio and then to globo series structure was also reported during differentiation of murine myelogenous leukemia M1 cells along a macrophage lineage (Kannagi *et al.,* 1983). Another example was the switch from globo to lacto and ganglio series core structures in NTERA-2 clone D1 embryonal carcinoma cells during differentiation (Chen *et al.,* 1989). The latter example was further investigated and the enzymatic basis of the marked shift was revealed (Chen *et al.,* 1989). However, the NTERA-2 clone D1 cells are induced to differentiate into a variety of somatic cell types by RA, and the changes observed represent an average change of enzymatic activity for a heterogeneous cell population. On the other hand, the shift of glycolipid expression in HL-60 cells during differentiation is lineage specific. In this context, the elucidation of the enzymatic basis of this lineage-specific glycolipid expression is of great interest.

We have investigated two glycosyltransferase systems. The first system CMP-NeuAc:LacCer α2–3sialyltransferase (G_{M3} synthase) is responsible for the first step in the synthesis of ganglio GSLs. This enzyme activity increased strikingly during monocytic differentiation, while the activity level remained unchanged during granulocytic differentiation (Nakamura *et al.,* 1992). The second enzyme system, for synthesis of neolacto glycosphingolipids, includes the upstream β1–3GlcNAcT (Lc_3Cer synthase) and the downstream glycosyltransferases: β1–4GalT (for synthesis of Galβ1–4GlcNAcβ1–3Galβ1-structure), elongation GlcNAcT (for synthesis of GlcNAcβ1–3Galβ1–4GlcNAcβ1-structure), nLc_4Cer α2–3sialyltransferase (for synthesis of NeuAcα2–3Galβ1–4GlcNAcβ1-structure), and nLc_4Cer α2–6sialyltransferase (for synthesis of NeuAcα2–6Galβ1–4GlcNAcβ1-structure). The upstream Lc_3Cer synthase was downregulated during monocytic differentiation, while the activity was upregulated during granulocytic differentiation (Nakamura *et al.,* 1992). The downstream glycosyltransferases were upregulated or remained unchanged during differentiation into both lineages. These changes in glycosyltransferase activities, and the proposed regulatory mechanism of the total metabolic flow of glycolipid biosynthesis, are summarized in Fig. 9. Preferential activation of elongation β1–3GlcNAcT and Lc_3Cer synthase during granulocytic differentiation was observed in our present study (Nakamura *et al.*, 1992). However, two β1–3GlcNAc transferase activities behaved in a different manner: (1) The activity of elongation β1–3GlcNAcT was elevated in the late stage (72 hours) of granulocytic differentiation, while Lc_3Cer synthase was upregulated in the early stage (36 hours) as well as in the late stage of differentiation. (2) During monocytic differentiation, elongation β1–3GlcNAcT was activated slightly, while Lc_3Cer synthase was downregulated (Nakamura *et al.*, 1992). It has not been reported that elongation β1–3GlcNAcT is enzymatically different from Lc_3Cer synthase (Chen *et al.*, 1989; Basu and Basu, 1984), and the difference has not been studied carefully. However, our recent study suggested that β1–3GlcNAcT, which recognizes the terminal

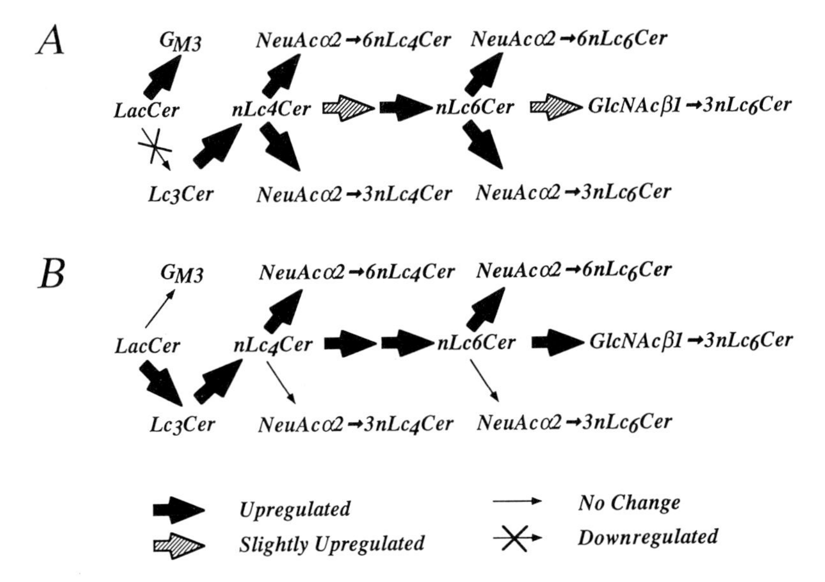

FIG. 9. Proposed mechanism of the regulation of glycosphingolipid biosynthesis during differentiation induced by TPA and RA in HL-60 cells. (A) The regulation in monocytic differentiation of HL-60 cells. (B) The regulation in granulocytic differentiation of HL-60 cells. Each arrow represents the change in glycosyltransferase activity during differentiation of HL-60 cells. During monocytic differentiation, G_{M3} synthase is upregulated, resulting in the dramatic G_{M3} increase, and the upstream Lc_3Cer synthase is downregulated, resulting in the decrease of the neolacto series glycosphingolipids, although the downstream glycosyltransferases are ready to catalyze their reactions. During granulocytic differentiation, Lc_3Cer synthase is upregulated together with the activation of the downstream glycosyltransferases, resulting in the considerable increase of NeuAcnLc, while G_{M3} synthase is unchanged, resulting in relative G_{M3} decrease.

disaccharide unit Galβ1–4GlcNAc, does not show the catalytic activity on the terminal Galβ1–4Glc unit (Gu *et al.*, 1992). Although this hypothesis is in good agreement with our present results, further studies are required.

In the undifferentiated cells, a low level of nLc_4Cer α2–3sialyltransferase or almost the same level of this enzyme activity as nLc_4Cer α2–6-sialyltransferase conflicts with our previous finding that gangliosides with NeuAcα2–3Galβ1–structure are more abundant than that with NeuAcα2–6Galβ1–sequence (Nojiri *et al.*, 1984) (Table I). However, preferential production of NeuAcα2–3Galβ1-structures could be accomplished if the K_m value for nLc_4Cer is significantly lower for α2–3-sialyltransferase than for α2–6-sialyltransferase. Another possibility is that the K_m of sialidase for NeuAcα2–6Galβ1-sequence is significantly lower than that for NeuAcα2–3Galβ1-structure. Comparative kinetic values have not yet been obtained for these enzymes from HL-60 cells. Moreover, during granulocytic differentiation, nLc_4Cer α2–6sialyltransferase but not nLc_4Cer α2–3sialyltransferase was activated in spite of a considerable increase in NeuAcα2–3Galβ1-structures and decrease in NeuAcα2–6nLc_4Cer. However, overall synthesis of NeuAcα2–3Galβ1-structures can be explained and should be upregulated by the activation of the key Lc_3Cer synthase, regardless of whether one of the downstream enzymes, nLc_4Cer α2–3sialyltransferase, remains unchanged (Fig. 9B). The relationship between nLc_4Cer α2–6sialyltransferase and its enzymatic product has not yet been revealed. The specific expression mechanisms of these sialylated neolacto series glycolipids remain uncertain and require further study.

Our current study indicates that the switching of two key glycosyltransferase activities, Lc_3Cer synthase and G_{M3} synthase, is playing an important role in regulating the total metabolic flow of GSL biosynthesis (Nakamura *et al.*, 1992). According to the K_m values which were determined in the present study, there was no significant difference between the values of both glycosyltransferases competing for the same acceptor, LacCer (Nakamura *et al.*, 1992). Therefore, the upregulation of Lc_3Cer synthase and the unchanged G_{M3} synthase activities in granulocytic differentiation were not simply explained by an increased LacCer concentration, although LacCer content in granulocytoid cells was much higher than in monocytoid cells (Table I). Thus, Lc_3Cer synthase was activated by a certain mechanism in granulocytic differentiation, and the reaction catalyzed by Lc_3Cer synthase can be regarded as a rate-limiting step in the synthesis of neolacto glycolipids. The idea of regulation of GSL biosynthesis at the branching step has been suggested for the determination of which series a certain ganglioside molecule is directed toward (asialo a and b series) (Pohlentz *et al.*, 1988). However, the alteration of a single glycosyltransferase activity can be responsible not only for the synthesis of ganglio gangliosides but also for the synthesis of diverse neolacto structures. Similar results were described previously using human colonic adenocarcinoma by Holmes *et al.* (1987). It is of in-

Table I
GLYCOSPHINGOLIPIDS OF HL-60 CELLS DURING MONOCYTIC AND GRANULOCYTIC DIFFERENTIATION[a]

	Expression[b]		
Glycosphingolipid	Undifferentiated	Monocytic (36 hr)	Granulocytic (day 3)
	μg lipid-bound hexose/10^8 cells		
GalCer	1.02	3.11	0.53
LacCer	8.14	4.29	10.75
Lc_3Cer	0.31	0.30	0.31
nLc_4Cer	1.06	0.27	1.64
nLc_6Cer	0.01	0.01	0.01
	μg lipid-bound NeuAc/10^8 cells		
G_{M3}	1.38	3.67	0.92
NeuAcα2→3nLc_4Cer	1.05	0.69	1.43 (0.98 day 8)[c]
NeuAcα2→6nLc_4Cer	0.20	0.19	0.21
NeuAcα2→3nLc_6Cer	0.21	0.18	0.50
NeuAcα2→6nLc_6Cer	0.01	0.01	0.02
Longer gangliosides[d]	0.43	0.19	0.53

[a]Data are summarized from our earlier investigation (Nojiri *et al.*, 1984).
[b]Each datum is the abundance of each glycosphingolipid in HL-60 cells calculated from chromatoscanning results of HPTLC analysis.
[c]Since the abundance of NeuAcα2→3nLc_4Cer is unique, datum on day 8 is also shown in the same table.
[d]These are unidentified gangliosides, which have longer sugar chains, including NeuAcα2→3nLc_8Cer and NeuAcα2→6nLc_8Cer.

terest that synthesis of neolacto glycolipids controlled by activation of a single enzyme is observed during cellular differentiation as well as during oncogenesis.

The molecular mechanism of the activation of the two key glycosyltransferases remains totally unknown. The effect of RA and TPA on these enzymes might be an indirect one, since the enzymatic activity from crude homogenate of undifferentiated HL-60 cells was not altered by RA and TPA. RA interacts with three nuclear receptors, RA receptors α, β, and γ (Petkovich *et al.*, 1987; Brand *et al.*, 1988; Krust *et al.*, 1989), and RA receptors are thought to activate gene expression and regulate cell differentiation (Umesono *et al.*, 1988). It has been reported that the recognition site of a transcriptional factor, AP-1, can act as a TPA-inducible enhancer (Lee *et al.*, 1987). Actually, organization of the β-galactoside α2–6sialyltransferase gene has been revealed, and the promoter region contains the consensus binding site of AP-1 (Svensson *et al.*, 1990). RA and TPA treatment may result in specific changes in the transcription of glycosyltransferases. Alternatively, the glycosyltransferase activities may be controlled posttranslationally by glycosylation (Ivatt, 1981) or by phosphorylation (Burczak *et al.,* 1984). Acquisition of primary gene structure information and the pure enzyme proteins of glycosyltransferase would make it possible to ascertain these mechanisms.

B. Incorporation and Action of Exogenous Bioactive Gangliosides

Our recent data indicate that tritium-labeled G_{M3} was more rapidly, compared with the inactive G_{M1}, incorporated into the cells and that its major parts were instantly transferred to lysosomal compartments and metabolized rapidly, resulting in its high turnover rate, although a minor part of it remained in the membrane for longer (Nakamura *et al.,* 1989) (Fig. 10). Furthermore, it was demonstrated that *de novo* synthesis of G_{M3} was significantly stimulated by its exogenously added molecule. These results suggest that rapid turnovers and newly synthesized molecules of G_{M3} might play crucial roles in triggering the induction of differentiation (Nakamura *et al.,* 1989).

On the other hand, it has been demonstrated (Nojiri *et al.,* 1991) that a kind of ganglioside, 2→3-sialosylparagloboside (2→3SPG), which significantly increased during myelomonocytic (hybrid-type) differentiation of HL-60 cells that was induced by the bioactive vitamin D_3 [1α,25-$(OH)_2D_3$], could induce myelomonocytic differentiation as well as growth inhibition in HL-60 cells; subsequently, it was then found that 2→3SPG could strongly inhibit the tyrosine kinase activity of the β-subunit (95 kDa) of insulin receptors, without affecting the insulin binding to its receptor. 2→6SPG was found to be much less active. This autophosphorylation of the β-subunit was also inhibited significantly in the presence of 50 μM G_{M3}, which induces the monocytic differentiation as well as the complete growth inhibition of HL-60 and U937 cells. With these data, it is suggested that bioactive gangliosides might be interacting with the β-subunit of insulin receptors, and consequently, that growth inhibition and differentiation induction might be brought about (Fig. 11).

VII. Discussion

Glycosphingolipids exogenously added to the culture medium are reported to be incorporated into plasma membrane, the lipophilic ceramide moiety being inserted into the lipid layer (Kanda *et al.,* 1982; Laine and Hakomori, 1973). Taken together, all our current observations strongly suggest that the expression of particular ganglioside molecules having specific oligosaccharide chains on the cell surface membrane not only constitutes the differentiation-associated phenotype but also acts as a potent trigger for induction of differentiation of human

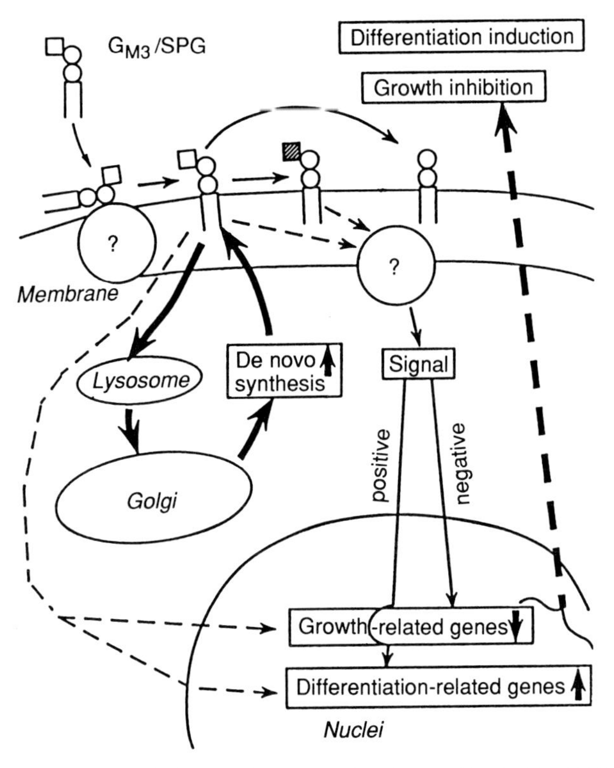

FIG. 10. Hypothetical mechanism(s) of action of bioactive ganglioside G_{M3}/SPG in differentiation–induction. The double bond of the sphingosine base of the bioactive gangliosides was labeled with NaB^3H_4 and, using the tritium-labeled gangliosides, the turnover (binding, incorporation, and metabolism) in HL-60 cells was investigated (Nakamura *et al.*, 1989). ◇○○⊏, G_{M3}; ○○⊏, ceramide dihexoside; ◈○○⊏, unidentified derivatives.

myeloid leukemic cells along a particular cell lineage. It is also possible that the direct interaction between ganglioside micelles and cell surface membrane triggers the differentiation. In either case, the specific oligosaccharide structure may be intimately involved in the differentiation-inducing activity of gangliosides. We found that the synthetic ganglioside G_{M3} exhibited differentiation-inducing activity toward HL-60 cells that was almost comparable to that of native G_{M3}. We also tested various synthetic sialocompounds that are physicochemically related to gangliosides, and found that some of these compounds, including α-sialocholesterol and α-sialodiglyceride, exhibited the differentiation-inducing activity.

Exogenous gangliosides are also reported (1) to inhibit the action of several growth factors, as well as the tyrosine kinase activity associated with the growth

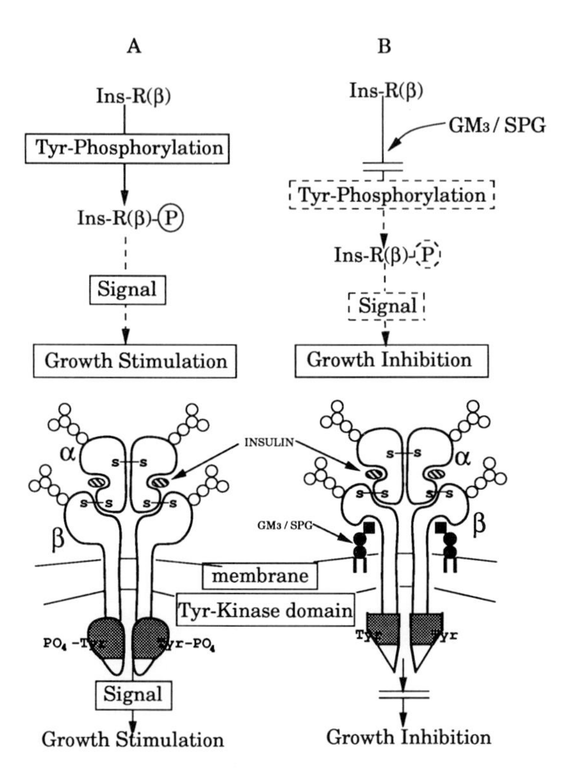

FIG. 11. Interaction between bioactive ganglioside G_{M3}/SPG and insulin-receptor (Ins-R). Autophosphorylation of the β-subunit of the insulin receptor [Ins-R(β)] was specifically inhibited by G_{M3} or α2→3SPG, and, simultaneously, both growth inhibition and differentiation induction into monocytic or myelomonocytic (hybrid-type) mature cells were demonstrated in HL-60 cells (Nojiri *et al.*, 1991).

factor receptors (Bremer *et al.*, 1986); (2) to inhibit the proliferation of lymphocytes stimulated by lectins, antigens, or interleukin 2 (Robb, 1986); (3) to sensitize tumor cells to growth inhibitors (Kinders *et al.*, 1982); and (4) to stimulate the proliferation of astroglial (Katoh-Semba *et al.*, 1986) and neuroblastoma cells (Tsuji *et al.*, 1983). The β subunit of cholera toxin, which binds specifically to G_{M1} on the cell surface membrane, has been shown to stimulate or inhibit the proliferation of thymocytes (Spiegel *et al.*, 1985) or mouse 3T3 cells (Spiegel and Fishman, 1987). In addition to the growth inhibitory activities, G_{M3} and the neolacto gangliosides have been found to induce the differentiation of HL-60 cells (Nojiri *et al.*, 1986, 1988). Furthermore, it has been reported that various synthetic sialo compounds, including sialocholesterols, are potent inducers for neuritogenesis in a neuroblastoma cell line (Neuro2a) (Tsuji *et al.*, 1988). These observations strongly suggest that gangliosides play an important role in the regulation of cell differentiation as well as cell growth, and that certain types of synthetic sialo compounds, such as α-sialocholesterol and α-sialodiglyceride, could act as naturally occurring gangliosides do. These synthetic sialo compounds provide useful tools for elucidating roles of gangliosides and mechanisms of cell differentiation.

The major gangliosides found in mature granulocytes are greatly reduced in myeloid leukemia cells (Saito *et al.*, 1982). Thus, an attractive hypothesis is that the incomplete expression of ganglioside molecules, which are essential for promotion of cell differentiation, might be at least partly involved in the arrest of differentiation of acute leukemic cells. Our current results show that some types of leukemic cells could undergo differentiation into functionally mature cells when appropriate ganglioside molecules are added exogenously. We also found that neolacto gangliosides could induce the granulocytic differentiation of mutant HL-60 cells which were resistant to differentiation induction by RA (Gallagher *et al.*, 1985), and that the ganglioside content of these cells was much lower than that of parental HL-60 cells (Nojiri *et al.*, 1988).

The differentiation and growth inhibition mechanisms of leukemia cells induced by gangliosides have not been fully analyzed. Only the following have been clarified: (1) G_{M3} was characteristically bound and incorporated into HL-60 cells at a concentration which caused growth inhibition and cell differentiation (Nakamura *et al.*, 1989); (2) protein kinase C is regulated not only by TPA (Castagna *et al.*, 1982) but also by gangliosides (Kreutter *et al.*, 1987). Nonetheless, it was suggested that protein kinase C does not play a major role in monocytoid differentiation of HL-60 cells by G_{M3}, since the differentiation was not inhibited by sphinganine (Stevens *et al.*, 1989), an inhibitor of protein kinase C (Merrill and Stevens, 1989). It is obvious that gangliosides cannot explain all of features of cellular differentiation of human hematopoietic cells. However, our findings strongly suggest that gangliosides play a crucial role at least in determination of differentiation direction (Saito *et al.*, 1985; Nojiri *et al.*, 1986, 1988; Kitagawa *et al.*, 1989; Nakamura *et al.*, 1991). Together with these observations, the switching of the two key glycosyltransferase activities demonstrated recently may be one of the most important parts of the determining system of differentiation direction into monocytoid or granulocytoid lineages in human myeloid cells.

Our recent experiments also showed that α-sialocholesterol, like neolacto gangliosides, induced the granulocytic differentiation of HL-60 cells, suggesting that α-sialocholesterol and neolacto gangliosides share a common pathway, or, more likely, use the same pathway for inducing the differentiation of HL-60 cells. Further studies with various types of synthetic sialo compounds may elucidate the structure–activity relationship in the differentiation induction as well as the determination of differentiation direction in HL-60 cells. More detailed future investigations into the mechanism of ganglioside-induced differentiation of HL-60 cells may shed light on the pathophysiology of differentiation and proliferation of leukemic cells.

References

Arita, M., Tsuji, S., Omatsu, M., and Nagai, Y. (1984). *J. Neurosci. Res.* **12,** 289–297.

Basu, M., and Basu, S. (1984). *J. Biol. Chem.* **259,** 12557–12562.

Brand, N., Petkovich, M., Krust, A., Chambon, P., de Thé, H., Marchio, A., Tiollais, P., and Dejean, A. (1988). *Nature (London)* **332,** 850–853.

Bremer, E. G., Schlessinger, J., and Hakomori, S. (1986). *J. Biol. Chem.* **261,** 2434–2440.

Burczak, J. D., Soltysiak, R. M., and Sweeley, C. C. (1984). *J. Lipid Res.* **25,** 1541–1547.

Castagna, M., Takai, Y., Kaibuchi, K., Sano, K., Kikkawa, U., and Nishizuka, Y. (1982). *J. Biol. Chem.* **257,** 7847–7851.

Chan, K.-F. J. (1987). *J. Biol. Chem.* **262,** 5248–5255.

Chen, C., Fenderson, B. A., Andrew, P. W., and Hakomori, S. (1989). *Biochemistry* **28,** 2229–2238.

Collins, S. J., Gallo, R. C., and Gallagher, R. E. (1977). *Nature (London)* **270,** 347–349.

Collins, S. J., Ruscetti, F. W., Gallagher, R. E., and Gallo, R. C. (1978). *Proc. Natl. Acad. Sci. U.S.A.* **75,** 2458–2462.

Gahmberg, C. G., and Hakomori, S. (1975). *J. Biol. Chem.* **250,** 2438–2446.

Gallagher, R. E., Bilello, P. A., Ferrari, A. C., Chang, C.-S., Yen, R.-W. C., Nickols, W. A., and Muly, E. C., III (1985). *Leuk. Res.* **9,** 967–986.

Gu, J., Nishikawa, A., Fujii, S., Gasa, S., and Taniguchi, N. (1992). *J. Biol. Chem.* **267,** 2994–2999.

Holmes, E. H., Hakomori, S., and Ostrander, G. K. (1987). *J. Biol. Chem.* **262,** 15649–15658.

Ivatt, R. J. (1981). *Proc. Natl. Acad. Sci. U.S.A.* **78,** 4021–4025.

Kanda, S., Inoue, K., Nojima, S., Utsumi, H., and Wiegandt, H. (1982). *J. Biochem. (Tokyo)* **91,** 1707–1718.

Kannagi, R., Levery, S. B., and Hakomori, S. (1983). *Proc. Natl. Acad. Sci. U.S.A.* **80,** 2844–2848.

Katoh-Semba, R., Facci, L., Skaper, S., and Varon, S. (1986). *J. Cell. Physiol.* **126,** 147–153.

Kinders, R. J., Rintoul, D. A., and Johnson, T. C. (1982). *Biochem. Biophys. Res. Commun.* **107,** 663–669.

Kitagawa, S., Ohta, M., Nojiri, H., Kakinuma, K., Saito, M., Takaku, F., and Miura, Y. (1984). *J. Clin. Invest.* **73,** 1062–1071.

Kitagawa, S., Nojiri, H., Nakamura, M., Gallagher, R. E., and Saito, M. (1989). *J. Biol. Chem.* **264,** 16149–16154.

Kreutter, D., Kim, J. Y. H., Goldenring, J. R., Rasmussen, H., Ukomadu, C., DeLorenzo, R. J., and Yu, R. K. (1987). *J. Biol. Chem.* **262,** 1633–1637.

Krust, A., Kastner, P., Petkovich, M., Zelent, A., and Chambon, P. (1989). *Proc. Natl. Acad. Sci. U.S.A.* **86,** 5310–5314.

Laine, R. A., and Hakomori, S. (1973). *Biochem. Biophys. Res. Commun.* **54,** 1039–1045.

Lee, W., Mitchell, P., and Tjian, R. (1987). *Cell (Cambridge, Mass.)* **49,** 741–752.

Lozzio, B. B., Lozzio, C. B., Bamberger, E. G., and Feliu, A. S. (1981). *Proc. Soc. Exp. Biol. Med.* **166,** 546–550.

Merrill, A. H., Jr., and Stevens, V. L. (1989). *Biochim. Biophys. Acta* **1010,** 131–139.

Misfeldt, D. S., Hamamoto, S. T., and Pitelka, D. R. (1976). *Proc. Natl. Acad. Sci. U.S.A.* **73,** 1212–1216.

Nakajima, J., Tsuji, S., and Nagai, Y. (1986). *Biochim. Biophys. Acta* **876,** 65–71.

Nakamura, M., Ogino, H., Nojiri, H., Kitagawa, S., and Saito, M. (1989). *Biochem. Biophys. Res. Commun.* **161,** 782–789.

Nakamura, M., Kirito, K., Yamanoi, J., Wainai, T., Nojiri, H., and Saito, M. (1991). *Cancer Res.* **51,** 1940–1945.

Nakamura, M., Tsunoda, A., Sakoe, K., Gu, J., Nishikawa, A., Taniguchi, N., and Saito, M. (1992) *J. Biol. Chem.* **267,** 23507–23514.

Nojiri, H., Takaku, F., Tetsuka, T., and Saito, M. (1982). *Biochem. Biophys. Res. Commun.* **104,** 1239–1246.

Nojiri, H., Takaku, F., Tetsuka, T., and Saito, M. (1984). *Blood* **64,** 534–541.

Nojiri, H., Takaku, F., Ohta, M., Miura, Y., and Saito, M. (1985). *Cancer Res.* **45,** 6100–6106.

Nojiri, H., Takaku, F., Miura, Y., and Saito, M. (1986). *Proc. Natl. Acad. Sci. U.S.A.* **83,** 782–786.

Nojiri, H., Kitagawa, S., Nakamura, M., Kirito, K., Enomoto, Y., and Saito, M. (1988). *J. Biol. Chem.* **263,** 7443–7446.

Nojiri, H., Stroud, M., and Hakomori, S. (1991). *J. Biol. Chem.* **266,** 4531–4537.

Nomenclature Committee of the International Union of Pure and Applied Chemistry (1977). *Lipids* **12,** 455–468.

Ohta, M., Takaku, F., Miura, Y., Kitagawa, S., and Saito, M. (1988). *Jpn. J. Cancer Res.* **79,** 350–358.

Petkovich, M., Brand, N. J., Krust, A., and Chambon, P. (1987). *Nature (London)* **330,** 444–450.

Pohlentz, G., Klein, D., Schwarzmann, G., Schmitz, D., and Sandhoff, K. (1988). *Proc. Natl. Acad. Sci. U.S.A.* **85,** 7044–7048.

Robb, R. (1986). *J. Immunol.* **136,** 971–976.

Rodrig, N., Osanai, T., Iwamori, M., and Nagai, Y. (1987). *FEBS Lett.* **221,** 315–319.

Rovera, G., O'Brien, T. G., and Diamond, L. (1979). *Science* **204,** 868–870.

Saito, M., Nojiri, H., Takaku, F., and Minowada, J. (1982). *Adv. Exp. Med. Biol.* **152,** 369–384.

Saito, M., Terui, Y., and Nojiri, H. (1985). *Biochem. Biophys. Res. Commun.* **132,** 223–231.

Saito, M., Nojiri, H., Ogino, H., Yuo, A., Ogura, H., Itoh, M., Tomita, K., Ogawa, T., Nagai, Y., and Kitagawa, S. (1990). *FEBS Lett.* **271,** 85–88.

Sato, S., Fujita, S., Furuhata, K., Ogura, H., Yoshimura, S., Itoh, M., and Shitori, Y. (1987). *Chem. Pharm. Bull.* **35,** 4043–4048.

Schwarzman, G., Hoffmann-Bleihauer, P., Schubert, J., Sandhoff, K., and Marsh, D. (1983). *Biochemistry* **22,** 5041–5048.

Spiegel, S., and Fishman, P. H. (1987). *Proc. Natl. Acad. Sci. U.S.A.* **84,** 141–145.

Spiegel, S., Fishman, P. H., and Weber, R. J. (1985). *Science* **230,** 1285–1287.

Stevens, V. L., Winton, E. F., Smith, E. E., Owens, N. E., Kinkade, J. M., Jr., and Merrill, A. H., Jr. (1989). *Cancer Res.* **49,** 3229–3234.

Sugimoto, M., and Ogawa, T. (1985). *Glycoconjugate J.* **2,** 5–9.

Sundström, C., and Nilsson, K. (1976). *Int. J. Cancer* **17,** 565–577.

Svensson, E. C., Soreghan, B., and Paulson, J. C. (1990). *J. Biol. Chem.* **265,** 20863–20868.

Tsuji, S., Arita, M., and Nagai, Y. (1983). *J. Biochem. (Tokyo)* **94,** 303–306.

Tsuji, S., Nakajima, J., Sasaki, T., and Nagai, Y. (1985). *J. Biochem. (Tokyo)* **97,** 969–972.

Tsuji, S., Yamashita, T., Tanaka, M., and Nagai, Y. (1988). *J. Neurochem.* **50,** 414–423.

Umesono, K., Giguere, V., Glass, C. K., Rosenfeld, M. G., and Evans, R. M. (1988). *Nature (London)* **336,** 262–265.

Wiegandt, H. (1982). *Adv. Neurochem.* **4,** 149–223.

INDEX

D

E

T

V

CONTENTS OF PREVIOUS VOLUMES